A Handbook for Directors of Quantitative and Mathematics Support Centers

August 25, 2016

The current maintainer of this work is Michael Schuckers (`schuckers at gmail dot com`).

First edition: August 2016

Coulombe, Grace, O'Neill, Mary B., and Schuckers, Michael
A Handbook for Directors of Quantitative and Mathematics Support Centers
Includes illustrations, bibliographical references and index.
ISBN 978-0-9776744-4-2
`scholarcommons.usf.edu/qmasc_handbook`

Neck Quill Press, 2016

For QMaSC Directors.

- G.L.C.

For my grandchildren: Always in my heart.

- M.B.O'N.

For Daniel, Christopher and Gabriella $\mathcal{FWMWH}$.

- M.E.S.

This material is based upon work supported, in part, by the National Science Foundation under Grant DUE-1255945. Any opinions, findings, and conclusions or recommendations expressed in this material are those of the author(s) and do not necessarily reflect the views of the National Science Foundation

Acknowledgments

This book is the result of the work and the efforts of so many people. As we sometime joke, 'it takes a village to run a QMaSC.' We could not have done this work without the help of so many people.

We want to thank Lee Zia and Ron Buckmire, Program Managers at the National Science Foundation, and the National Science Foundation (NSF) for funding this project. Lee and Ron gave helpful suggestions and nudged our vision along so that it came to fruition.

We want to thank all of the authors whose chapters appear in this Handbook. It has been a pleasure working with them. We have been consistently very impressed by the enthusiasm, collegiality, patience, and timeliness of our collaborators. They have made this project a joy. As always, when we spend time with other people who direct QMaSCs, we are enriched by the conversations and the interactions. Thanks also go to Cat McCune and Patrice Conrath for the work that they did in helping to lead parts of the workshop that launched this handbook. They made valuable contributions.

Cat McCune and Mary's husband, John O'Neill, did more than their share of help with the editing process and we are very grateful for that assistance. Mega-kudos to them.

The workshop that led to this handbook was greatly aided by the organizational efforts of McKayla Nuffer. McKayla, we thank you so much.

Matt Salomone deserves special thanks for designing the cover art for this book and for allowing us to use it.

Thanks also to the team at the University of South Florida who are publishing this handbook through their Scholar Commons (`http://scholarcommons.usf.edu/qmasc_handbook/`) especially Jason Boczar.

Michael wants to thank Tim Morse, Alli Morrow, McKayla Nuffer, Kerrin Ehrensbeck, Patti Lock, Ed Harcourt, Devon Stein and Carol Smith, all from St. Lawrence University, for their support, advice and guidance since this journey began. He thanks Martie and Gregg Peterson for their generosity to the Peterson Quantitative Resource Center at St. Lawrence University. He also wants to thank his family, especially Stephanie, for the patience they have shown during this process.

Mary thanks Grace Coulombe and Michael Schuckers, whose humor and intelligence made it a pleasure to work on this project. Thanks to Hamilton College and to the Dean of the Faculty for their generous support, and to the QSR Advisory Committee, especially Professor Tim Kelly. Finally, Mary would like to thank her husband and Professor of English John H. ONeill, who was a member of the original Quantitative Literacy Committee at Hamilton College, and whose support and encouragement have been invaluable.

Grace thanks Hillory Oakes and Misty Beck, of the writing program at Bates College, as well as, John O'Neill and Catherine McCune for their support and advice on writing and editing. For their expertise and assistance in helping to secure the grant to fund this project, Grace thanks Phil Walsh and Joseph Tomaras, from Bates, as well as, Carol Smith and Devon Stein, from St. Lawrence. Grace also thanks Michael Schuckers and Mary O'Neill, without their energy and passion this handbook would never have existed.

Finally, we want to thank the mentors and the tutors that have been part of our QMaSCs at Bates College, at Hamilton College, and at St. Lawrence University. Our experiences with them provide the context and the capacity for so much of what we do.

Contents

Part I

Introduction

Preface

© Michael Schuckers,
St. Lawrence University

This handbook is the culmination of something that started in the Spring of 2011 at a Northeast Consortium for Quantitative Literacy (NECQL) meeting at the University of Massachusetts Boston. For several years, Mary, Grace, and I, the editors of this Handbook, had had informal discussions about the lack of resources available for managing tutoring centers like the ones that we directed. This was especially of note to me since, in 2007, I had started the Quantitative Resource Center at St. Lawrence University from scratch. We wanted to create a source of information for the people who lead or direct or administer math-related centers like ours. In particular, we wanted to create something that would be very useful to someone just starting as a new QMaSC director.

Mary O'Neill, Grace Coulombe and I put together two panels on how to lead what we came to call a QMaSC or Quantitative and Mathematical Support Center: the aforementioned one at UMass - Boston and another the following January at the Joint Mathematics Meeting. We got some good comments and suggestions at those sessions and we joked about doing more. At the end of the next NECQL meeting, Mary stood up and said we were writing a book. Grace and I were a *little* surprised but we rolled with it. A couple weeks later, Mary and Grace came to St. Lawrence University for a couple of days and we put together an outline of a grant proposal. We pitched the idea of a national workshop to the National Science Foundation (NSF), they liked the idea and had some good suggestions. Specifically, they were interested in a diverse range of institutions being involved and challenged us to get commitments from center directors at research institutions and community colleges and at schools all over the country. This was in large part due to the fact that Mary, Grace and I all ran centers at liberal arts colleges in the northeastern United States.

Suggested Citation: Michael Schuckers (2016), "Preface", in G. Coulombe, M. O'Neill, M. Schuckers (Eds.) *A Handbook for Directors of Quantitative and Mathematical Support Centers*, Neck Quill Press, `http://scholarcommons.usf.edu/qmasc_handbook`.

This material is based upon work supported, in part, by the National Science Foundation under Grant DUE-1255945. Any opinions, findings, and conclusions or recommendations expressed in this material are those of the author(s) and do not necessarily reflect the views of the National Science Foundation

We reached out to many, many QMaSC directors around the country and we were able to build a diverse set of people willing to come to Hartford, CT for a day on July 31, 2013. The directors that participated can be seen in the photo from the workshop in Figure 1.

Figure 1: QMaSC Workshop Participant, July 31, 2013

Having built a list of committed QMaSC directors, we were fortunate to have NSF fund our proposal in late 2012. We spent the next year recruiting more QMaSC directors to be part of the workshop and organizing the details of the workshop. Our chapter authors all wrote a draft of the chapters to which they were assigned prior to coming to Hartford. (Hartford was chosen because MathFest started there the next day.) Before everyone arrived in Hartford, we forwarded these drafts to other directors, they read the drafts and at the workshop we discussed feedback on individual chapters. After the workshop, these critiques were collected and given to the chapter authors who then incorporated them and revised their chapters. What follows in this handbook is the result of that workshop. In addition, we had several QMaSC directors write case studies of their centers. Since the chapters focus on an individual aspect of directing a QMaSC, we thought that having these case studies would give readers a sense of how these various aspects of managing a center are accomplished within a single QMaSC.

At the workshop, the last item we discussed (before adjourning for some social time) was the possibility of a network of QMaSC directors. That network would be a way to continue the conversation but also provide recognition for the professional work that QMaSC directors do. Such a network is still a work in progress. We have yet to find even an imperfect way to make it happen, but we are hopeful that it will eventually happen.

We are excited to present this handbook to you and hope you will find it helpful.

-Canton, NY, June 28, 2016

Introduction

© Mary B. O'Neill,
Hamilton College

1 Introduction

This Handbook, incorporating the contributions of many colleagues, is intended to be a resource for Quantitative and Mathematics Support Center (QMaSC) directors and staff, whether they are heading an existing program, starting a program, or updating a program. Directors and administrators who wish to establish support centers that represent the mission of their institutions will find examples of successful practices, including planning an inviting space, hiring staff and tutors, managing budgets, working with the campus community, incorporating technology, developing assessment practices, and coordinating with stakeholders.

QMaSCs exist in some form at many U.S. colleges and universities, and are pivotal for student retention for majors within the Science, Technology, Engineering and Mathematics (STEM) pipeline, such as physics, mathematics and engineering. But QMaSCs can also support courses with quantitative content within the core curriculum outside of the STEM field. Many colleges and universities administer Quantitative Literacy and Quantitative Reasoning programs aimed at assessing and building the quantitative skills of all of their students. Traditional support uses tutoring, review sessions, and workshops. Advances in technology have added to the types of support a QMaSC can offer.

Each of the sections below provides a short summary of each chapter appearing in that part of this Handbook.

Suggested Citation: Mary B. O'Neill (2016), "Introduction", `http://dx.doi.org/10.5038/9780977674435.ch1` in G. Coulombe, M. O'Neill, M. Schuckers (Eds.) *A Handbook for Directors of Quantitative and Mathematical Support Centers*, Neck Quill Press, `http://scholarcommons.usf.edu/qmasc_handbook`.

This material is based upon work supported, in part, by the National Science Foundation under Grant DUE-1255945. Any opinions, findings, and conclusions or recommendations expressed in this material are those of the author(s) and do not necessarily reflect the views of the National Science Foundation

2 Leadership and Management

Because a flourishing QMaSC depends on successful leadership and management, our Handbook devotes its first section to that topic. Gail Hilyard explains why proper strategic planning and management are needed to establish the goals and objectives of center services. Ja'Wanda Grant points out that the placement of a QMaSC in the institution's hierarchy clarifies the importance of that support. Ibrahim Rosic reveals how an institution can address diversity to help various groups of students achieve their academic goals. Maria Belk gives a concise and thorough explanation of how important it is to keep up-to-date with computers and technology for administrative tasks and to enhance student learning. In her chapter on effective management, Rosalie Tepper suggests that finding a balance between setting and enforcing expectations while supporting center staff is crucial for a successful program. Recognizing that many students have math anxiety, Gail Hilyard details how institutions can help students overcome this barrier to success.

3 Community Interactions

Community interactions are of vital importance to the success of a QMaSC, from coordinating with administration to establishing a virtual presence on campus. Frederick Van Swearingen begins this part of the Handbook by discussing the value of coordinating services with other academic support groups on campus, keeping in mind relationships and structures within the institution. Laurie Wern explains how to develop a learning community that is welcoming and threat-free, by having center staff constantly update and review the center's promotional and informational material. Matthew Salomone and Kathryn Bjorge emphasize the importance of establishing dialogue with faculty that will lead to more effective teaching in quantitative and math courses. Eric Gaze continues in this vein by discussing methods for collaborating with faculty about the ways services are provided to varied campus populations. McKayla Nuffer concludes the Community Interactions part by stressing the importance of a virtual presence using technology in the form of websites and social media to publicize a QMaSC's services.

4 Staffing, Hiring, and Training

Using each institution's established resources for staffing, hiring, and training, a QMaSC must look to its mission and needs in order to find and hire administrative staff and to hire qualified tutors and mentors. Christine Tinsley gives an overview of the hiring of administrative staff and tutors in a university setting, which can include graduate students as tutors. Jonathan Nelson discusses tutoring practices and methods that a director of a QMaSC should address when hiring undergraduate tutors. Cymra Haskell outlines specific examples of the art and practice of tutoring, showing how tutors learn from their experiences.

5 Center Assessment

Internal and external assessment are similar in that data must be collected to determine how well a QMaSC is meeting the needs of its constituencies. In this section, Tiana Bosley describes tools used in obtaining data to make management decisions, to improve a center's program, and to highlight best practices. Dennis Lewandowski continues the discussion of assessment by explaining how having a system in place to gather information from center staff helps to assess how goals are accomplished and where improvement may be needed. Rachel Dunwell, Grace Coulombe, and Michael Schuckers stress the importance of developing the appropriate assessment tools to develop materials for the creation of an internal review, a necessary component, for outside review panels and for institutional accreditation.

6 Starting a Center

When an institution establishes a new QMaSC or updates an existing center, many issues come into play, including the design of center space, the definitions of the roles of center staff, and the hiring of the center director. George Recck emphasizes the importance of an appropriate job description in hiring the right person to lead the center. Darry Andrews and Carolyn Johns discuss how the size and mission of the institution affect staff roles and duties. Eileen Phillips explores ways of planning or updating a QMaSC space. She shows that success of a center depends on how it is set up, and where it is located, demonstrating a commitment by the institution to this kind of academic support.

7 Case Studies

From private four-year colleges and universities to public universities and community colleges, the methods of supporting students and faculty in STEM fields and beyond can be quite different. In this part of the handbook, authors from across this spectrum of institutions describe their respective QMaSCs. The goal for each QMaSC is always the same — to give students the best experience possible. The ten case study authors describe how their QMaSCs have addressed the topics discussed in the rest of this Handbook. A director or QMaSC staff member will see how these particular institutions have handled issues of leadership and management; community interactions; staffing, hiring, and training; assessment; and starting or updating a center. The individual authors write about their own institutions, adding commentary on special features unique to their own colleges or universities.

8 The Future

The editors of this Handbook seek to bring to light the many and varied aspects of running a QMaSC. This document is meant to start a conversation about QMaSCs. We hope it has answered some questions about leading a QMaSc and that it may prompt new questions leading to a continued discussion of what makes a functional and successful QMaSC.

Part II

Center Leadership and Management

Effective Management

© Rosalie Tepper,
Shoreline Community College

1 Introduction

As in any workplace, effective management is needed to ensure smooth operations. One aspect of this is staff management: rules and expectations need to be set, communicated, and enforced, while maintaining an environment in which staff feel valued and supported. Another aspect is determining what services a QMaSC will provide, then implementing and maintaining them.

2 Managing Staff

The employees of a Quantitative and Mathematics Support Center (QMaSC) are often managed by the director of the center, or occasionally a director's assistant. For directors who are new to managing staff, it is worth taking some time to develop a management style that will guide the director's interactions with employees. One aspect of this is finding the right balance between friendliness and professionalism, which can vary due to a director's personality and the level of the staff. To establish a clear boundary, at a center in which staff members include community college or undergraduate students who may be without much prior work experience, the director may benefit from refraining from conversations that are not work related, and from socializing outside of work with the staff. Even so, a friendly, approachable demeanor can still be maintained so that staff members are comfortable seeking assistance from the director when needed. Some

Suggested Citation: Rosalie Tepper (2016), "Effective Management", http://dx.doi.org/10.5038/9780977674435.ch2 in G. Coulombe, M. O'Neill, M. Schuckers (Eds.) *A Handbook for Directors of Quantitative and Mathematical Support Centers*, Neck Quill Press, http://scholarcommons.usf.edu/qmasc_handbook.

This material is based upon work supported, in part, by the National Science Foundation under Grant DUE-1255945. Any opinions, findings, and conclusions or recommendations expressed in this material are those of the author(s) and do not necessarily reflect the views of the National Science Foundation

directors with non-student or graduate student employees have an occasional staff social activity such as a happy hour. In any case, the end result should be the same — to create a positive work environment in which employees respect the director's authority, see the director as accessible, feel supported but not micromanaged, and understand that good job performance is necessary and failing to do so will have consequences.

Center directors typically communicate the rules and practices expected of the staff in multiple ways, usually including both a written document and a training meeting (if the training meeting is required, it must be paid time) [1]. The written document is emailed to staff members, distributed in paper form, posted in the center, posted online, or some combination of these possibilities. In some centers, employees are required to initial or sign the document to acknowledge that they have read and understood it. Possible topics include:

1) General conduct expectations, such as being on time, refraining from inappropriate language, refraining from cell phone use or other distracting behaviors, dressing appropriately, avoiding strong perfumes or scents, etc. Also, appropriate vs. inappropriate interactions between tutor and tutee (for example, refraining from physical contact, and avoiding language, intonation, or gestures that could be misunderstood as offensive or as sexual harassment).

2) What breaks are allowed during work (if any). Directors should be aware of school policies and state laws regarding breaks; for example, Washington State requires 10 minutes (consecutive or broken up) paid break time for every 4 consecutive hours worked, and employees cannot be scheduled for more than 5 consecutive hours without a 30 minute meal break [2]. An institution's human resource office should be able to provide this information.

3) How employees report their hours for pay (i.e. instructions for filling out and submitting time sheets).

4) Instructions on what an employee should do if he or she needs to miss a shift due to a conflict known in advance or due to an emergency situation.

5) Beginning and end-of-shift duties.

6) Reminders regarding tutoring practices, such as what the employee should do when stuck on a problem, distinguishing appropriate and inappropriate levels of help to provide, or what an employee working in a drop-in tutoring environment should do when not actively helping a student. For more information about training employees regarding tutoring methods, refer to section IV of this handbook.

7) Maintaining student confidentiality, potentially including restrictions on sharing information such as a student's grade on an assessment, in accordance with the Family Educational Rights and Privacy Act [3] .

8) How to handle on-site emergencies when higher-level supervision isn't present, if applicable. This might include distinguishing when to call campus security vs. 911, and how to call either from a campus phone.

9) Appropriate and inappropriate uses of the center's resources such as phones and computers. For state schools, this may be limited by state law in addition to center's or school's policies. For example, Washington State's legal code allows for brief, infrequent personal use of state-owned resources if, "there is little or no cost to the state," and the use is not for personal financial gain [4].

Employees with little or no prior work experience might need a bit more guidance and patience than more experienced employees. If employees are carefully selected and expectations are clearly communicated, problems with employees will be infrequent. However, some centers do experience problems such as employees who repeatedly arrive late to work, submit timesheets late or inaccurately, etc. In these situations, protocols vary among centers for addressing these issues, in particular whether initial interventions are written and/or verbal and the level of formality of the process. If the reprimand is due to an incident that was reported by a student or another staff member, it is possible that the report might be somewhat biased or based on a misunderstanding. A conversation with the employee allows him or her to share the other side of the story. In any situation, it is important to be clear about the nature of the issue, the expectations for improvement, and the potential consequences for failure to improve. If it seems that the employee might not understand why the behavior was problematic, explaining the consequences of the issue helps to maintain a good working relationship rather than risk being perceived as authoritarian. As an example, an employee who is routinely late to work in a drop-in tutoring center can be told that his or her tardiness reduces the availability of help available before he or she arrives, causing students to have to wait longer to get their questions answered. This in turn creates a stressful situation for other staff members who are trying to keep up with demand.

It is advisable to keep written records on situations and resulting written or in-person disciplinary interactions that might lead to an employee's dismissal. Doing so will help the director to remember what was discussed, and it might become useful at some point to have evidence of the situation. A warning to a staff member whose employment will be terminated if he or she repeats the problematic behavior should be issued in writing, and the employee should sign a copy acknowledging receipt of the warning.

If staff members are hourly employees, formal procedures such as written warnings and keeping records of misconduct can be useful, but are not necessary for terminating an employee in many instances. Employees of this type are often "at-will" employees, meaning that they do not have a contract stating their length of employment and they can be terminated for almost any reason (or even for no reason) aside from discrimination. Due to this, employees can legally be dismissed with little or no explanation, but documentation could be useful if an employee claimed his or her termination was discriminatory. Exceptions to at-will employment aside from discrimination vary in different states, so directors should be aware of local laws. If a director is considering terminating an employee and wants guidance on how to proceed, his or her institution's human resource office should be a good place to turn.

While setting and enforcing expectations takes some care, the positive aspects of the center will outweigh the negatives. Ideally, the center will become a resource which is beloved by students due to the great work of the staff. There are several ways to maintain a positive working environment to encourage and sustain a high level of work. One is having leadership that is responsive to the employees' needs. For example, a director who also works with students (as opposed to having primarily administrative duties) can pitch in "on the floor" if the center gets busy and tutors cannot keep up with students' questions. In addition to reducing the students' wait times, it diffuses a situation in which tutors feel pressured to work faster and as a result become stressed and less able to offer high-quality tutoring. It also gives the director valuable perspectives on what the tutors experience in day-to-day situations. A responsive director should also let employees know that he or she is happy and willing to help them if they are stuck on a problem or have other questions. Reminding them of this helps encourage employees who might otherwise be shy about approaching the director, especially if he or she seems busy, to seek out the help they need.

To make sure employees understand their importance to the center, the director should provide positive reinforcement. One part of this is by letting the employees know how much their work is appreciated, when speaking to them both in groups and individually. Also, the director should collect feedback from students using the center and anonymously share comments with tutors to pass on positive feedback and identify areas for staff development. The occasional gesture such as bringing in cookies is sure to be taken as a welcome sign of appreciation. If employees have ideas they perceive as improvements for the center (whether regarding tutoring methods or something more operational), the director should take the time to listen to and consider the ideas to show that he or she values their input, even if the suggestion is not ultimately implemented. In group meetings, allowing time for employees to talk about successes they have had in explaining concepts or how they have handled certain student situations helps to further develop their sense of value to the center while also spreading good ideas to other staff members.

3 Program Management

A Quantitative and Mathematics Support Center might consider offering any of a variety of activities to support students. The most common type of support is a drop-in center in which tutoring staff are available to answer questions. Some centers also offer one-on-one tutoring appointments, workshops, study groups, and/or Supplemental Instruction. Informing faculty which of these services the center provides allows them to promote the activities in their classes.

Drop-in tutoring is the easiest to implement, as it does not require content preparation (unlike workshops, for example). Also, once a staff schedule is established, it can remain fairly constant during the school term, though some centers track their usage throughout the day and alter the schedule to better align staffing levels with student demand. A drop-in environment typically serves as a study space in which students can work individually or with classmates. To allow students to distinguish tutors from other students when they have questions, tutors typically wear nametags. The most common way for students to access help when they have a question is simply to raise their hands. It should be recommended that when tutors are not actively working with students, they look around the center for students with confused or frustrated expressions; these students might benefit from getting help, but may be shy about asking or unaware that a tutor is available. Help in the drop-in environment is often provided on a problem-by-problem basis; tutors assist students on just a problem or two and then encourage them to try working independently using what they have learned, especially when there are other students waiting for help. Some drop-in centers are divided by level (e.g. pre-college level and college level, where applicable) or more specifically by course. This sort of division can increase collaboration among students in the same class and increase efficiency in serving students especially in centers where different staff members are hired for different strengths.

Some students may want more contact time with a tutor than the drop-in environment typically provides. A center may offer longer one-on-one or small group tutoring sessions (often twenty minutes to an hour) for these students, though at some schools another campus program provides this service. Appointments are often arranged and managed using programs such as TutorTrac. Readers considering implementing this type of support can refer to section II of this handbook. In a center which offers both drop-in tutoring and one-on-one appointments, tutors in the drop-in environment should be advised to keep an eye out for students who seem like they could use more intensive assistance and recommend one-on-one appointments to those students.

Workshops are an efficient way to help multiple students at a time. In some places, they are delivered by the same employees who also provide general tutoring, though elsewhere, they are delivered by staff members with additional training, faculty members, or the director of the center. Workshops may be given on a wide variety of topics; when deciding what workshops to offer, the director can ask faculty members teaching the courses supported by the center what they perceive their students need. Some workshops are not course-specific, covering topics such as study skills or test-taking strategies. At schools where a specific calculator (e.g. a TI-84) is recommended or required for one or more courses, a workshop at the beginning of the term on using the calculator is another option. Some centers also deliver course-specific workshops, either regular supplements to lectures focusing on topics known to be challenging or less frequent exam reviews. Involving faculty in the planning of course-specific workshops is useful for finding out what topics to emphasize and ensuring that methods used are consistent with lectures.

Some centers organize study groups, an opportunity for students taking the same course to meet and work together. Study groups can be on a drop-in basis, or students may sign up at the beginning of the term with the expectation that they will attend all sessions. In the latter case, students might sign up via a sheet sent around the classroom, in the center, or via the center's website. Study groups vary in format; some involve reviews of material given by the staff member leading the group, similar to a workshop but allowing more time for students to work together on problems, while others are more discussion-based, led by a facilitator who is trained to foster conversations in which students assist and learn from each other. Topics beyond the basic material, such as study strategies and problem solving skills, are sometimes discussed. The facilitator may occasionally attend classes or have contact with faculty teaching the course to keep the content aligned with the course.

Supplemental Instruction (SI) is an academic support system targeting "historically difficult courses" developed at the University of Missouri-Kansas City (UMKC) in 1973 and adopted by hundreds of schools internationally [5]. SI involves regular group sessions, "in which students compare notes, discuss readings, develop organizational tools, and predict test items," led by trained SI leaders [5]. SI leaders are typically students who previously took the course themselves and finished with a high grade, often recommended by faculty teaching the course. They attend all class meetings, and are trained to facilitate discussions during SI sessions in which they guide students to work together and share strategies, though leaders do not provide additional instruction on the material or help with homework[6]. The International Center for Supplemental Instruction at UMKC offers regular Supervisor trainings for schools implementing the program [7].

4 Conclusion

Taking time both to manage staff as described in this chapter and to thoughtfully select and implement services provided by the center is well worth a QMaSC director's efforts and has multiple benefits in the long run. A center with staff that understand what is expected of them, along with well-implemented programs, will allow students to efficiently get the quality assistance that they need while minimizing the number of problems that arise.

5 Bibliography

[1] 29 C.F.R. § 785.29 (2011).

[2] Wash. Admin. Code § 296-126-092 (1976).

[3] 20 U.S.C. § 1232g (1974).

[4] Wash. Admin. Code § 292-110-010 (2009).

[5] University of Missouri-Kansas City, "UMKC supplemental instruction program." http://www.umkc.edu/asm/umkcsi/index.cfm, 2016.

[6] The International Center for Supplemental Instruction, "Frequently asked questions." http://info.umkc.edu/si/faq/, 2016.

[7] University of Missouri-Kansas City, "UMKC si supervisor trainings in Kansas City." http://info.umkc.edu/si/training-in-kansas-city/, 2016.

Responding To Institutional Needs

© Ja'Wanda S. Grant,
Vassar College

1 Introduction

The needs of each institution of higher education can be extracted from the core mission of the college. This mission can be expressed in a variety of ways and is usually published online for the public, and also included in student and faculty handbooks or governance documents. In a random internet search, institutional missions were found in the form of statements of principles and practices [1], core beliefs [2], and mission and strategic plans [3]. Each of these statements is an expression of the institutional values, and the mission of the QMaSC should mirror these values.

Some QMaSC directors have been recruited from within the institution because of their passion for quantitative literacy and education. Those internal candidates are usually already familiar with the curriculum, the students, and the administration on their campus. Others, who are recruited externally, may only know what they have researched about the college and what they learn during the interview process. This chapter will help external hires as they learn the institutional culture, and internal recruits as they transition into the administrative capacity of a QMaSC director.

2 QMaSC Placement and Mission

A QMaSC might be housed in one specific department, such as Mathematics or Statistics. A QMaSC could also be housed in a larger academic support center and report to the director of that

Suggested Citation: Ja'Wanda S. Grant (2016), "Responding To Institutional Needs", http://dx.doi.org/10.5038/9780977674435.ch3 in G. Coulombe, M. O'Neill, M. Schuckers (Eds.) *A Handbook for Directors of Quantitative and Mathematical Support Centers*, Neck Quill Press, http://scholarcommons.usf.edu/qmasc_handbook.

This material is based upon work supported, in part, by the National Science Foundation under Grant DUE-1255945. Any opinions, findings, and conclusions or recommendations expressed in this material are those of the author(s) and do not necessarily reflect the views of the National Science Foundation

center. Others are completely stand-alone entities that report directly to deans or other academic administrators. As the QMaSC director learns the organizational structure in which the QMaSC resides, it will become easier to focus on very specific goals and a clearly defined mission.

The QMaSC at Vassar College (Quantitative Reasoning Center), established in 2011, is a part of the Learning, Teaching and Research Center with the following mission:

> The Learning, Teaching, and Research Center (LTRC) is dedicated to addressing the needs of Vassar's diverse student body. Our goal is to enable students to maximize their unique educational experiences at Vassar College. We therefore provide an *extensive range of academic resources* that foster the fundamental aim of a liberal arts education: *to facilitate the intellectual and professional growth of ethical, informed, and reflective students* who can engage creatively with important social issues. The LTRC houses a peer-staffed Writing Center and Quantitative Reasoning Center, including a Supplemental Instruction Program for select quantitative analysis (QA) courses. We also offer expert learning support with a focus on developing individual academic skills. [4]

Here are a few programs or initiatives offered by the QMaSC at Vassar that align with the overarching LTRC mission:

- <u>An extensive range of academic resources</u>
  - Drop-in tutoring offered Sundays, 1:00 pm–11:00pm, and Monday–Thursday, 3:00 pm – 11:00 pm
  - Supplemental Instruction for selected quantitative courses
  - Designated tutors for selected quantitative course
- <u>Facilitate intellectual growth of students</u>
  - Opportunities for students to review subjects and experience deep learning by teaching peers
- <u>Facilitate professional growth of students</u>
  - Annual training for in-house staff (tutors and supplemental instruction leaders)
  - Training for departmental academic interns who tutor for select course
  - Collaborations with regional tutoring programs for an annual tutor training conference in which tutors present workshops, preside over sessions, and receive credits towards certification hours where eligible (see Appendix A for the 2014 conference agenda)

If a QMaSC is a stand-alone center or housed in a department that may not have such a clear "mission statement," the director should refer to the faculty governance/handbook, student handbook, or other similar documents to learn the college's values and mission.

3 Inherited Mission(s)

In some cases, the QMaSC is expected to fulfill certain needs that are already identified, which may include the merging or transfer of pre-existing services or programming. For example, one of the purposes of the Vassar Quantitative Reasoning Center (QRC) is to support the quantitative analysis (QA) courses designed to equip all students with quantitative reasoning skills. Vassar defines quantitative reasoning as "the ability to understand and evaluate arguments framed in quantitative or numerical terms; to analyze subject matter using quantitative techniques; to construct and evaluate quantitative arguments of one's own; and to make reasoned judgments about the kinds of questions that can be effectively addressed through quantitative methods" [5]. Various courses across the curriculum are tagged "QA" and are often encountered by students pursuing STEM degrees (**S**cience, **T**echnology, **E**ngineering, and **M**ath). In an effort to increase retention of students in the STEM pipeline, Vassar implemented the Supplemental Instruction (SI) Program. As the QRC is commissioned to support the QA curriculum, SI is an inherited QRC program.

SI is a nationally renowned program originally established at the University of Missouri-Kansas City (UMKC) that targets historically difficult courses and provides students the opportunity to engage with their course materials outside the classroom [6]. At Vassar, students voluntarily attend weekly review sessions facilitated by SI Leaders, who are undergraduate peers who have taken the courses and done well.

The QMaSC director should learn about programs or initiatives that should be implemented or sustained upon arrival. Some of these programs, such as SI, offer training that will help the QMaSC director learn more about the programs' mission and structure. The SI Supervisor training is offered annually at UMKC and is an invaluable resource for those QMaSC directors who supervise SI, or for QMaSC directors who plan to implement SI to support the established mission of their center [7].

4 Identifying And Responding to Institutional Needs

The QRC at Vassar came about as a result of a series of faculty conversations regarding ways to support and improve the QA program. An article published prior to the opening of the QMaSC highlights the foundational mission of the QRC (see Appendix B). Some QMaSC directors may not have such a template to follow, but as evident in the push for the Vassar QRC, there is usually a cohort of faculty and administrators who could offer guidance.

The following template for Vassar identifies the main categories that embody the institutional needs of any college/institution–students, staff, faculty, and the administration:

Student Needs

Responding to student needs takes many forms. At Vassar, the QRC is charged with supporting the courses that have been tagged as QA. In order to graduate, a student must take at least one QA designated course, which include many introductory courses required for STEM majors and medical school. This requirement can be taken at any time, but the majority of students enrolled in these courses are freshman who have to adjust to the new amount of schedule freedom, the culture of the institution, and expectations of different faculty, while balancing social and extracurricular interests. The SI program connected to many of these courses is a great way to support freshman as they transition into traditionally difficult STEM courses (including General Chemistry, Organic Chemistry, Calculus, and Physics) and to college life. SI offers a way for freshman (and other students) to meet their classmates in an informal environment with emphasis on collaborative learning and study skills.

Student Populations Served

A QMaSC director should determine if there are certain student populations that will require specific types of support. Based on the mission of the institution, these populations will vary. The spectrum of students a QMaSC might support include:

- High-achieving/honor students
- Underprepared Students
- Minority/Underrepresented Students
- Students with Disabilities
- Veterans
- Adult/non-traditional students
- First generation students
- Students with math anxieties

The above list is not all-inclusive, but does cover the spectrum of students that might be supported by a QMaSC. Depending on the student population at the institution, the QMaSC could support students as early as pre-admission or even post-graduation. For instance, a community college QMaSC might offer pre-admission support for college placement tests, such as ACT® Compass or Accuplacer. The Math Center at Holyoke Community College offers math prep classes that prepare students for the Accuplacer test [8].

Staffing Needs

In defining ways to respond to the needs of each and all students groups, the director will need to determine what support will be provided for the QMaSC to implement and sustain its initiatives.

At Vassar, peer educators staff the main QRC programs with financial support from the LTRC and one of the dean's offices. Research and discussions about additional support mechanisms are currently underway (including alumni donations/gifts and grant-funding).

A QMaSC could be staffed with peer educators and/or professional peer tutors or instructors. The staffing will determine what types of programs or support the QMaSC will offer. It is common for professional QMaSC staff to teach courses designed to target students needing more review of foundational math topics (from arithmetic to precalculus). Community colleges, which often serve underprepared students or adult students who have been out of school for extended periods, offer a variety of these developmental courses in math or statistics. Small highly selective liberal arts colleges also offer courses to groups of students who might not have the skills they need to transition smoothly into the standard quantitative courses. The Quantitative Reasoning program at Wellesley College, for example, administers a quantitative skills assessment at orientation, and if students do not pass, they enroll in the QR Basic Skills Course (QR 140), which is taught by a professional staff member [9].

Staff Recruitment

The QMaSC director, when responsible for hiring staff, is responsible for ensuring that they are trained and have support to do the job(s) for which they are hired. At Vassar, the SI leaders and tutors participate in a pre-semester training program each fall. Returning staff members are called upon to lead and mentor the new/junior staff members, and other campus offices are invited to inform staff where to direct students when they are not able to meet their needs. Appendix C includes a copy of the most recent two-day training agenda for the fall of 2013. Because the QRC at Vassar supports quantitative courses across the curriculum, the training is open to students within departments who also serve as peer tutors for quantitative courses not supported by QRC staff.

The QMaSC director should rely on faculty and peer educators to help recruit students who have done well in their quantitative and math courses and who show potential to serve as model students and peer educators.

In addition to recruiting staff who have the content mastery and communication skills required, it is important to have a diverse staff. Diversity in a QMaSC is not always easy to achieve or sustain, since the qualifications are restricted to a select pool of candidates. The campus offices of affirmative action and human resources may help with strategies to ensure that the hiring process is accessible and equitable for all candidates.

A QMaSC that houses a service where students are offered assistance with homework or in preparation for graded work must take care to ensure that the institutional guidelines for academic integrity are upheld [10]. Time should be spent in QMaSC staff training discussing academic honesty and how to uphold it when supporting students with course work. The published statement of the college's honor code could be included in a staff handbook.

Academic Use of Technology

Institutions of higher education use many forms of technology to recruit, market, advertise, and support their mission. Students are often required to use these various types of technology in quantitative and math courses, some of which they are savvy with and others where they might have some challenges [11]. A QMaSC with a mission to support specific courses should be aware of programs students are expected to utilize. QMaSC staff working with these students should be proficient in these programs, and the QMaSC should have the software available when appropriate.

Technology support could present itself as more of an academic support need especially for students in distance education programs; therefore the QMaSC director should make sure all staff are given the appropriate access and training [12].

Numerous course management programs are online (e.g., Moodle or Blackboard), and students should learn those systems to find lecture notes and assignments, and sometimes to interact with the instructor and or classmates. The technical support teams on each campus usually offer basic support for such programs, however, some quantitative and math courses use specialized programs or software for online coursework (i.e. Mastering Physics) or for data analysis (i.e. SPSS or R). Select Q-Tutors at Vassar can help with content in various quantitative courses and are given some teaching assistant privileges for respective online course programs.

Microsoft Excel is also a common tool used in quantitative and math courses. The Vassar Q-Center usually employs a few peer tutors who can help students with their basic Excel questions in addition to specific subject areas. Many of the students at Vassar have their own laptops and the software needed for their courses, so the Vassar QRC does not offer a computer lab with these programs. At other institutions, the QMaSC might provide a computer lab with these programs for students who do not have access on their own.

Technology Used in Outreach

A QMaSC can be a hidden treasure on a college campus. Students are bombarded with lots of information when they arrive. And even when students have been on a campus for several semesters, they may be unaware of all of the resources that are available to them.

A QMaSC should take advantage of the technology that students are socially drawn to, such as Facebook and Twitter. These social media outlets can connect students to the QMaSC in a less intimidating way and allow students to communicate freely with each other. The Vassar QRC uses Facebook to advertise center hours, schedule changes, closures, and special events. Students who have a Facebook account can see the most up-to-date feeds, managed by a senior peer tutor selected by the QMaSC director (see Appendix F for Vassar Q-Center Facebook Sample).

In addition to the Facebook page, the Vassar QRC has an institutionally supported website (`ltrc.vassar.edu/qrc`) for prospective students, other departments, and other institutions or community members who may want to learn more about the Vassar QRC. The QRC staff is also added to a "Vassar QRC" course Moodle Page where they share resources, particularly session review materials from previous SI leaders.

Faculty Needs

Faculty needs are often intertwined with the student needs. With growing course enrollments, departmental staffing constraints, and varying preparation of enrolled students, faculty may rely heavily on support from the QMaSC. Some Math or Statistics departments house courses taught by QMaSC staff, connecting the QMaSC more intimately to the curriculum.

Alternatively, a QMaSC may host workshops that offer review of materials that courses do not dedicate class time to covering. Faculty may require some students to attend, whereas in some cases they may also be strictly voluntary. SI or other peer-led review/study sessions housed in a QMaSC can be built in to a course and bear weight on the final course grade.

The QMaSC director should meet with appropriate department chairs or course coordinators frequently to stay in tune with their needs. A great way to strengthen the relationship between the QMaSC and the faculty is to share the QMaSC vision at faculty departmental or divisional meetings so they know which needs the QMaSC can address.

Communication of the QMaSC vision is particularly important for new faculty as they are transitioning into new responsibilities and often have other pressing demands that keep their focus on teaching and/or research. A great place to connect with new faculty is during their orientation because they will get to meet the QMaSC director and find ways to collaborate with the QMaSC to in turn help meet student needs.

Returning faculty may be interested in QMaSC-sponsored sessions or workshops that address their needs regarding teaching pedagogy for quantitative and math courses, or even ways to incorporate technology. This type of faculty development may not be in the mission of the QMaSC. If another center or office hosts such faculty development programming, a QMaSC director could collaborate or participate as an opportunity to learn more about the faculty needs, challenges, and current best practices.

Administration Needs

A QMaSC director may report to a dean or other administrative director in student affairs, or the chair of a Mathematics or Statistics department. Many of these reporting structures require a type of report, in the form of progress report, quarterly reviews, or annual reports. These reports help administrators stay in tune with what is happening in the departments/programs they oversee, evaluate budgetary needs, measure program effectiveness, and monitor if programming is meeting expectations as stated in the mission or vision statement. There are various ways in which QMaSC programs and initiatives can be assessed. An office of institutional research is also a good reference point for QMaSC directors seeking to implement the appropriate assessment tools.

At Vassar, the QRC reports directly to the director of the Learning, Teaching and Research Center (LTRC). Reports to the LTRC director are also shared with the dean in the Office of the Dean of Studies, which is a part of the larger Dean of the College division and the Office of the Dean of the Faculty. Since these offices currently fund the majority of the QRC programs, the reports usually highlight the use of services and the growth and/or development of programs that support the mission of the institution.

Assessment data at Vassar, which would include more feedback from faculty, the staff, and a peer tutor self-review, serves as a professional development tool. This will allow the QMaSC director to better support staff as they grow in their roles as peer educators, and to respond appropriately to the needs of the institution.

The Vassar SI program provides an end-of-term survey that is administered in class along with the course evaluations (which provides a high response rate). In the fall semester of 2012, the QRC invested in the Remark Software by Gravic, Inc [14]. Paper forms are created so that selected fields can be scanned, then transferred to a spreadsheet and analyzed by Remark (sample SI evaluation form in Appendix E). In addition to SI program evaluations, the SI leaders are observed in the form of a teaching observation each semester by the QMaSC director and separately by another peer SI leader. SI leaders also take attendance at each session and during their office hours in the Q-Center. Attendance data is then compared to final grades, but to date, no significant correlation has been found between session attendance and grades. The Vassar QRC is collaborating with a faculty statistician to do more extensive analysis of the collected data, possibly correlating it with admissions data for high school GPA and standardized test scores.

5 Responding to Changing Institutional Needs

It is impossible to meet every institutional need, especially because they change so often. Departments add, modify, and discontinue quantitative and math courses based on departmental and larger curricular needs. Faculty who are hired may have short-term or adjunct appointments, making faculty development initiatives unproductive. Since students are only at the institution for about four years (two years at community colleges), staffing a QMaSC requires constant recruitment and training. The pre-college skill sets of each admitted class of students vary, requiring more attention to these ever-changing needs. Staffing and resource needs can be stifled by budgetary constraints. Most importantly, if the mission of an institution evolves, the mission of QMaSC should follow suit.

QMaSC directors should maintain regular communications with their reporting directors and administrative or faculty governing bodies to learn about the evolving institutional changes. Professional organizations are also a great resource for strategies to address institutional changes, for in higher education they seem to follow current trends. And finally, a QMaSC director should utilize the appropriate assessment tools that will best inform the development of current strategic plans.

As the QMaSC director begins to launch or further develop programs, further training or certification may be required. The QMaSC director will also be assessed in the form of an administrative review of some sort; therefore, an additional mission of the QMaSC should be to keep all staff equipped to do their jobs effectively. Peer tutoring staff can be trained in-house or sent to other trainings or conferences (such as the MATE conference for tutors, see Appendix A). The QMaSC director should stay connected with professional organizations, such as Mathematics Association of America, Northeast Consortium on Quantitative Literacy, and the National Numeracy Network, to name a few. Membership and participation in these organizations will keep the QMaSC director connected to the larger QMaSC community, network with peers, and often learn of best practices that will help achieve the QMaSC mission.

6 QMaSC Intersections

Given the breadth of needs in a given student population, a QMaSC director should stay informed about resources available on campus that intersect with the daily work of the QMaSC. Below is a sample list of offices that can offer support to a variety of students who may also use QMaSC resources:

- First Year Programs
- Career Development
- Disability Services
- Other tutoring/academic support spaces (i.e. departmental)
- Psychological Counseling Services
- Dean of Studies
- Programs designed for first generation, low-income, or underrepresented students
- Veterans Support Services

The QMaSC director might network with these offices by nature of the organizational structure, or perhaps by just meeting with representatives individually to share the work of the QMaSC and learn about the work of the respective offices. The challenges that students face on each campus are often quite similar, but can vary within populations. Open communication with various offices is a great way to learn of ways to reach students who may otherwise not reach out independently for support from the QMaSC.

7 Bibliography

[1] Earlham College, "Statement of principles and practices," February 2014. Accessed from http://www.earlham.edu/about/mission-beliefs/principles-practices/ February 2014.

[2] The College of New Jersey, "Core beliefs." Accessed from http://tcnj.pages.tcnj.edu/about/mission/core-beliefs/ February 2014.

[3] Shoreline Community College, "Mission and strategic plan." Accessed from http://www.shoreline.edu/about-shoreline/strategic-plan.aspx February 2014.

[4] Vassar College, "Learning, teaching, and research center mission statement." Accessed from http://ltrc.vassar.edu/ February 2014.

[5] Vassar College 2013-2014 Catalogue, "Quantitative course requirement." Accessed from http://catalogue.vassar.edu/academic-information/degrees-and-courses-of-study/requirements-for-the-degree-of-bachelor-of-arts/quantitative-courses.html February 2014.

[6] The International Center for Supplemental Instruction, "Overview of supplemental instruction." http://info.umkc.edu/si/, 2016.

[7] The International Center for Supplemental Instruction, "Overview." http://info.umkc.edu/si/training-in-kansas-city/, 2016.

[8] Holyoke Community College, "Math center resources." Accessed from http://www.hcc.edu/courses-and-programs/academic-resources/academic-support/math-center February 2014.

[9] Wellesley College, "Quantitative reasoning program." Accessed from http://www.wellesley.edu/qr February 2014.

[10] G. Gilbert, B. Day, A. Murillo, J. Patton, A. Sibley-Smith, and B. Smith, "Promoting and sustaining an institutional climate of academic integrity," 2007. Accessed from http://www.asccc.org/sites/default/files/publications/academic-integrity-2007_0.pdf February 2014.

[11] G. Salas and J. Alexander, "Technology for institutional enrollment, communication, and student success," *New Directions for Student Services*, vol. 124, pp. 103–116, 2008.

[12] S. L. Restauri, "Creating an effective online distance education program using targeted support factors," *TechTrends*, vol. 48, no. 6, pp. 32–39, 2004.

[13] "How to create forms using Google Docs," 2014. Accessed from https://www.google.com/forms/about/ June 2016.

[14] Gravic, Inc., "Remark Software." Accessed from http://www.gravic.com/remark/Office-OMR-Scanning-Software/.

8 Appendix A: Mid-Hudson Tutor Educators (MATE) 2014

Figure 1: Tutor Training Conference Agenda (1)

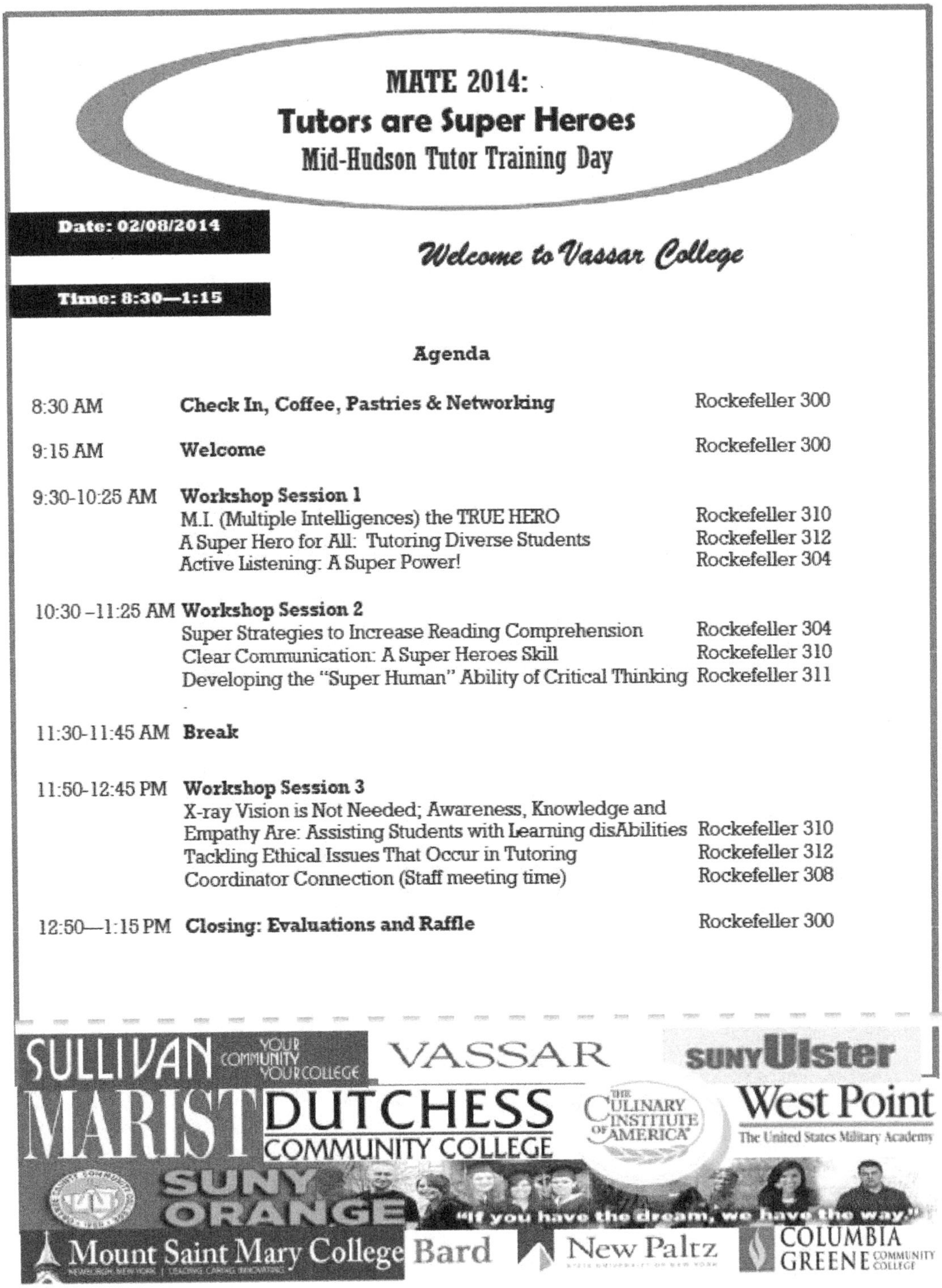

MATE 2014:
Tutors are Super Heroes
Mid-Hudson Tutor Training Day

Date: 02/08/2014

Time: 8:30—1:15

Welcome to Vassar College

Agenda

8:30 AM	**Check In, Coffee, Pastries & Networking**	Rockefeller 300
9:15 AM	**Welcome**	Rockefeller 300
9:30-10:25 AM	**Workshop Session 1**	
	M.I. (Multiple Intelligences) the TRUE HERO	Rockefeller 310
	A Super Hero for All: Tutoring Diverse Students	Rockefeller 312
	Active Listening: A Super Power!	Rockefeller 304
10:30 -11:25 AM	**Workshop Session 2**	
	Super Strategies to Increase Reading Comprehension	Rockefeller 304
	Clear Communication: A Super Heroes Skill	Rockefeller 310
	Developing the "Super Human" Ability of Critical Thinking	Rockefeller 311
11:30-11:45 AM	**Break**	
11:50-12:45 PM	**Workshop Session 3**	
	X-ray Vision is Not Needed; Awareness, Knowledge and Empathy Are: Assisting Students with Learning disAbilities	Rockefeller 310
	Tackling Ethical Issues That Occur in Tutoring	Rockefeller 312
	Coordinator Connection (Staff meeting time)	Rockefeller 308
12:50—1:15 PM	**Closing: Evaluations and Raffle**	Rockefeller 300

SULLIVAN YOUR COMMUNITY YOUR COLLEGE
VASSAR
SUNY Ulster
MARIST
DUTCHESS COMMUNITY COLLEGE
THE CULINARY INSTITUTE OF AMERICA
West Point The United States Military Academy
SUNY ORANGE "If you have the dream, we have the way."
Mount Saint Mary College
Bard
New Paltz
COLUMBIA GREENE COMMUNITY COLLEGE

Figure 2: Tutor Training Conference Agenda (2)

Workshop Descriptions

*Required for CRLA Level 1 certification

M.I. (Multiple Intelligences) the TRUE HERO— Are you a HERO of the Multiple Intelligences? Well if you need more assistance this workshop will help you fight for the civilians (our tutees) to surpass these Villains (classes). This workshop will provide a broader understanding discussing different learning styles among tutees. We will discuss how each style "likes" to learn and which style preference and modes in which particular tutees learn best to maximize their learning potential. So our tutees can successfully "fight" through their exams. *Heather Breittholz, SUNY Orange*

A Super Hero for All: Tutoring Diverse Students—This presentation will help student tutors develop new strategies and techniques to deal with the challenges of tutoring students from diverse cultural, academic and socioeconomic backgrounds. Although many student tutors themselves come from diverse backgrounds; there are common tutoring and proofreading issues that student tutors will encounter and should be prepared to face in order to successfully become a Super Hero to their tutee. *Karen Tomkins-Tinch, Richard Cusano, Julius Cox, Elizabeth Hamberger, Marist College*

Active Listening: A Super Power!*– How well do you really listen? Do you show the tutee that you care about being there with them? This workshop is designed to develop necessary skills in active listening. Techniques discussed and practiced will include paraphrasing, minimal encouragements, use of "I" statements, emotional labeling, open ended questions, effective pauses, and mirroring. *Cynthia Worrad, Dutchess Community College*

Super Strategies to Increase Comprehension When Reading Texts—Just opening a textbook and reading without proper preparation does not increase retention and recall in the classroom or during tests. Be a hero: Learn how to help students increase comprehension when reading texts. Pre-reading preparation includes organization, activating schema, "diagramming the question," and reading actively. Workshop will focus on "diagramming the question," which applies best to humanities courses such as history, psychology, or sociology. Learn by doing with opportunities for hands-on application. *Kathleen Eagan, West Point*

Clear Communication: A Super Hero's Skill— Ever try to teach someone how to tie your shoe? At first it may seem to be an easy task, but when actually explaining the steps as someone is following your verbal instructions, you may feel you need super human powers! This interactive workshop highlights the importance of carefully choosing how you communicate when explaining content or giving instructions. Participants will explore different ways to communicate as teams try to construct a tool needed to survive on an island. *Jennifer Wrage, Dutchess Community College*

Developing the "Super Human" Ability of Critical Thinking*— Good thinking skills are a must in college as well as in life. To conquer life's hurdles, students need to be able to reason, apply, evaluate and problem solve. Learn how to help your tutees develop their critical thinking skills. *Joseph Dunstan, Dutchess Community College*

Tackling Ethical Issues That Occur in Tutoring—Situations arise in tutoring that require tutors to call upon ethical standards of conduct in order to support "truth and justice". In this workshop tutors discuss situations they have encountered or expect to occur. Frequent ethical issues may include responsibility, respect, fairness, professionalism, and confidentiality. Identifying and dealing with appropriate boundaries is a common issue. *Bianca Keesler, Dutchess Community College*

X-ray Vision is Not Needed; Awareness, Knowledge and Empathy Are: Assisting Students with Learning disAbilities—Many students "battle" learning disabilities. Super tutors "see" the signs of these "invisible disabilities" and are equipped to help their tutees "win" the learning battle. This presentation will introduce the tutor to common indicators of learning disabilities and provide strategies for assisting tutees. Clips from Richard Lavoie's video "How Difficult Can This Be?" will provide an awareness as to what it is like to be temporarily learning disabled. *Sally Weglinski, Dutchess Community College*

Coordinator Connection—A time for coordinators to meet and discuss issues and best practices

9 Appendix B: Articles highlighting the need for a QMaSC at Vassar College

The Miscellany News

Since 1866 | miscellanynews.com

April 15, 2010

Volume CXLIII | Issue 20

Students push to reinstate old SAVP position

"So much surrounds the venue, and we have gotten so much positive feedback from local businesses who are excited to have so many Vassar students in the area."

—Peter Denny '10, Director of ViCE

Civic Center to host legendary Flaming Lips

Local school in danger of closing

J. Cole, Big Daddy Kane to emcee

VC to form center for QA courses

Inside this issue

5 Features — Vassar's Educational Outreach programs grow in community

6 Features — Students turn dorm room into Turkish tea room

18 Sports — An update on the crew program from its rowers

April 15, 2010 NEWS Page 3

Vassar to create Quantitative Reasoning Center

CENTER continued from page 1

Vassar's New Quantitative Reasoning Center

As with the current Writing Center, the Quantitative Reasoning Center will aim to

- enhance students' quantitative reasoning abilities
- provide peer-to-peer tutoring for quantitative classes
- contribute to faculty development initiatives for teaching quantitative reasoning across the curriculum

The director of the Center will

- train and oversee student peer tutors
- meet with individual students
- consult with faculty teaching courses that fulfill the College's Quantitative Reasoning requirement
- teach one or two courses each year

News Briefs

Teachers focus on new media

Vassar plans carbon analysis

MISCELLANY NEWS | VASSAR COLLEGE

10 Appendix C: Fall 2013 Peer Educator Training Agenda, Vassar College

QRC PEER EDUCATOR TRAINING 2013

TRAINING SCHEDULE

Friday, August 30

(WITH NOTES FOR TRAINER/DIRECTOR)

9:00 am	Welcome
	Introductions
9:30 am	Opening Activity
9:45 am	Peer Educators
	About the QRC
10:00 am	Know Your Value
10:10 am	Role Overview
10:20 am	Powerful Learning Experiences - (Group Activity)
11:00 am	Differentiated Instruction: One Size Does Not Fit All (Neva Lozada)
12:00 pm	Lunch Break (one wrap/cookie per person)
12:45 pm	Video Clips (Key Components To Effective Tutor)
1:00 pm	Video Reflection
1:10	Staff Handbook
	Responsibilities
	General Guidelines
1:20	Opening and Closing a Session
1:40	Video – Remain Professional
	(http://www.youtube.com/watch?v=6oT5WJ0hM5A)
	Video Reflection
1:50	Tutor Session Guidelines Continued
2:00	Scenario Discussions (Led by Returning Tutors)
2:10	Exercise Summary – Visit Overall Goals of Tutoring/Handbook
2:20	Staff Handbook
	Policies/Procedures – End with Scheduling
2:40	PAPERWORK
	Student Employment Forms
	Confidentiality Agreement
	Employment Agreement
2:50 pm	Scheduling!
3:20 pm	Closing Remarks
	Class Visits (template email for Faculty from Paige)
	POSSE VETERANS
	WEBSITES (ltrc and FB)

QRC PEER EDUCATOR TRAINING 2013

DAY 2: SI LEADERS

Saturday, August 31

Agenda Outline

Meet Your Peers

Personal Goals

The SI Program (p.10-11)

Class Introductions

Tasks of the SI Leader (p.12-13)

SI leader and Professor (p.16-17)

SI Leader and Student (p.27-28)

Opening the SI Session (p.29)

Conducting the Session (p.30)

Helpful Resources

Nuts & Bolts

Policies & Procedures

11 Appendix D. Vassar College Peer Educator Application

VASSAR COLLEGE

Quantitative Reasoning Center
Q-Tutor Job Application

Please return the completed application form to Susan Ward, Library 122 (her desk is in the front of the Writing Center/Q-Center space in the Library). She is usually at her desk between 9am-12pm and 2-5pm, Monday-Friday. Applications are due by Friday, January 31 at 4:30 pm.

General Information

Last ______________________ First ______________________

Student ID Number (999): ______________

Cellular phone number: ______________

Current Classification (circle one): Freshman Sophomore Junior Senior

Which course(s) are you applying to be a Q-Tutor for? (*i.e. Econ 100, Math 220, Math 221…*)
***Preference will be given to students who can work full time (8hrs/week)**

____

____

Academic Information

Major (s): ______________________ Correlate(s): ______________________

Cumulative GPA: ______________

Please list grades received in the course(s) for which you are applying to tutor. If you did not take the course at Vassar, please list grades for courses in the subject area(s) you are interested in

	Course	When	Grade
Example	*Chem 108*	*SP-2011*	*A*
1.	________	________	________
2.	________	________	________
3.	________	________	________
4.	________	________	________
5.	________	________	________
6.	________	________	________

VASSAR COLLEGE

Quantitative Reasoning Center
Q-Tutor Job Application

Experience
Please provide brief answers to the following questions.

1. Why do you want to be a Q-Tutor?

2. What academic, work, or personal experiences do you see as potentially helpful in preparing you to be an effective peer tutor?

3. Did you visit the Q-Center when taking the course for which you are applying to be a tutor?

4. If you answered yes to #3, please describe your experience. If you answered no, please share your reasons for not attending.

5. What would you like to gain from the experience of being a Q-Tutor?

VASSAR COLLEGE

Quantitative Reasoning Center
Q-Tutor Job Application

References

List as references two members of the Vassar faculty; at least one should be for the subject in which you would like to be an SI Leader. If both are in the same department, that is fine.

Name ______________________________ Dept. ______________

Name ______________________________ Dept. ______________

Scheduling In the space below please provide your SPRING 2014 CLASS SCHEDULE

Course (*Chem 244*) Time (*12:00-12:50pm*) Days (*MWF*)

1.

2.

3.

4.

5.

******In the space below note any commitments I should be aware of (i.e soccer, EMT) and the days/times you are committed. I will compare all of the selected Q-Tutors' schedules to set our regular group meeting time for the semester.**

******Special notes below (anything else I need to know??)**

12 Appendix E. Vassar Supplemental Instruction Evaluation for Remark

Fall 2013 Supplemental Instruction Evaluation

Please Provide Your Name and Session Information (Omit Any Non-Letter Characters in Your Name)

FILL IN THE OVALS (DO NOT CIRCLE THEM!!!):

CORRECT ● INCORRECT (O)

Your Name							
First Name (first 4 letters)				Last Name (first 4 letters)			
O A	O A	O A	O A	O A	O A	O A	O A
O B	O B	O B	O B	O B	O B	O B	O B
O C	O C	O C	O C	O C	O C	O C	O C
O D	O D	O D	O D	O D	O D	O D	O D
O E	O E	O E	O E	O E	O E	O E	O E
O F	O F	O F	O F	O F	O F	O F	O F
O G	O G	O G	O G	O G	O G	O G	O G
O H	O H	O H	O H	O H	O H	O H	O H
O I	O I	O I	O I	O I	O I	O I	O I
O J	O J	O J	O J	O J	O J	O J	O J
O K	O K	O K	O K	O K	O K	O K	O K
O L	O L	O L	O L	O L	O L	O L	O L
O M	O M	O M	O M	O M	O M	O M	O M
O N	O N	O N	O N	O N	O N	O N	O N
O O	O O	O O	O O	O O	O O	O O	O O
O P	O P	O P	O P	O P	O P	O P	O P
O Q	O Q	O Q	O Q	O Q	O Q	O Q	O Q
O R	O R	O R	O R	O R	O R	O R	O R
O S	O S	O S	O S	O S	O S	O S	O S
O T	O T	O T	O T	O T	O T	O T	O T
O U	O U	O U	O U	O U	O U	O U	O U
O V	O V	O V	O V	O V	O V	O V	O V
O W	O W	O W	O W	O W	O W	O W	O W
O X	O X	O X	O X	O X	O X	O X	O X
O Y	O Y	O Y	O Y	O Y	O Y	O Y	O Y
O Z	O Z	O Z	O Z	O Z	O Z	O Z	O Z

Session Information	
CHEM-108-31 MWF 0930AM-1020AM Haley Merritt	O
CHEM-108-33 MWF 1200PM-1250PM Natalie DiCenzo	O
CHEM-125-31 TR 0900AM-1015AM Mohammed Abdelaziz	O
CHEM-125-32 TR 1030AM-1145AM Mohammed Abdelaziz	O
CHEM-244-31 MWF 1030AM-1120AM Colin White-Dzuro	O
CHEM-244-32 MWF 0930AM-1020AM Marcos Vargas	O
CHEM-244-33 MWF 0830AM-0920AM Anthony Khoudary	O
CHEM-244-34 MWF 1200PM-1250PM Colin White-Dzuro	O
MATH-121-01 MWF 0930AM-1020AM Noelle Sawyer*	O
MATH-121-02 MWF 1030AM-1120AM Noelle Sawyer*	O
MATH-125-01 MWF 0930AM-1020AM Laura Schachter	O
MATH-125-02 MWF 1030AM-1120AM Christopher Brown	O

PLEASE FILL IN THE CIRCLE THAT **BEST** REPRESENTS YOUR RESPONSE.

Your class year:		Are you pre-med?	
Freshman	O		
Sophomore	O	Yes	O
Junior	O	No	O
Senior	O		

Please evaluate your SI leader and the SI sessions you attended.						
1 – Strongly Disagree; 5 – Strongly Agree	1	2	3	4	5	N/A
The class instructor encouraged students to attend SI:	0	0	0	0	0	0
SI sessions helped me understand this subject better:	0	0	0	0	0	0
SI helped improve my problem-solving skills:	0	0	0	0	0	0
My grades improved because I attended SI sessions:	0	0	0	0	0	0
Attending SI encouraged me to study with s:	0	0	0	0	0	0
SI sessions have made me confident in my own abilities:	0	0	0	0	0	0
The SI leader was organized and prepared for each session:	0	0	0	0	0	0
The SI leader used the time well:	0	0	0	0	0	0
The SI leader made me feel welcome and at ease:	0	0	0	0	0	0
The SI leader was confident and knowledgeable of the subject:	0	0	0	0	0	0

Please provide reasons for not attending some or all of the SI sessions offered for your class.			
	Yes	No	N/A
I did not need the extra sessions	0	0	0
I needed more one-on-one help	0	0	0
I prefer individual study time only	0	0	0
The times/days were inconvenient	0	0	0
I felt uncomfortable asking for help	0	0	0
Sessions frequently canceled/rescheduled	0	0	0

Did you visit your professor during office hours?		Did you visit your SI Leader during office hours?		Did you visit department interns during office hours?		Did you use the Q-Tutors in the Q-Center?	
Yes	0	Yes	0	Yes	0	Yes	0
No	0	No	0	No	0	No	0

Any Comments or Feedback?

13 Appendix F. Vassar College Q-Center on Facebook

Addressing Diversity

© Ibrahim Rosic,
Mohawk Valley Community College

1 Introduction

As American society is increasingly becoming more diverse, postsecondary education institutions are making every effort to admit students with a variety of ages, experiences, talents, viewpoints, and cultural and educational backgrounds. Research shows that bringing diversity to higher education enhances personal growth opportunities for the students. For this reason, many colleges and universities strive to build and keep a diverse student population.

The majority of these diverse student populations enter the college campuses underprepared and unaware of expectations for college mathematics or quantitative coursework. As a result, student persistence and completion rates in higher education are low, particularly among underprepared students, returning adult students, culturally diverse students, and students with disabilities. These students are usually at risk of leaving college before they complete their education [1].

Quantitative and Math Support Centers (QMaSCs) play a significant role in helping these different groups of students achieve their academic goals. By staffing centers with diverse and well-trained tutors, providing a relaxed and attentive environment, collaborating with faculty members and other student support offices on the campus, and effectively promoting tutoring services, the success of centers working with a diverse student population could significantly improve.

Suggested Citation: Ibrahim Rosic (2016), "Addressing Diversity", http://dx.doi.org/10.5038/9780977674435.ch4 in G. Coulombe, M. O'Neill, M. Schuckers (Eds.) *A Handbook for Directors of Quantitative and Mathematical Support Centers*, Neck Quill Press, http://scholarcommons.usf.edu/qmasc_handbook.

This material is based upon work supported, in part, by the National Science Foundation under Grant DUE-1255945. Any opinions, findings, and conclusions or recommendations expressed in this material are those of the author(s) and do not necessarily reflect the views of the National Science Foundation

2 Traditional Underprepared Students

A traditional, underprepared college student could be defined as a high school graduate entering postsecondary education who is required to complete remedial or developmental work before taking college-level courses. The majority of underprepared students enter college lacking the background and motivation to succeed in, specifically, college-level math courses. The lack of basic academic skills puts these students at higher risk of leaving college before receiving their degrees [1]. "The success or failure of these students is directly related to the effectiveness and availability of tutorial service" [2].

Meeting the needs of underprepared students is one of the main challenges facing QMaSCs. "These students usually do not seek out assistance on their own and getting them to the academic support centers is a great accomplishment" [2]. Thus, center directors need to impose proactive efforts to reach out to these students and encourage them to utilize available support. First, the director should have tutors, other administrative staff members, themselves, or some combination of the three visit quantitative and mathematic courses and present the tutoring services to students. Awareness of possible additional support in learning mathematics helps create a sense of safety for underprepared students. Second, the director needs to establish a working relationship with faculty members teaching First-Year Experience Seminars, or some program for first-semester students, and encourage them to inform new students about math tutoring available on the campus. Third, tutor availability should be mentioned in the college catalogue and other college information brochures, the college website, student newsletters, and orientation sessions for new students. The QMaSC director could also use social networking sites, such as Facebook and Twitter, to further promote academic support among traditional students. Lastly, the director of the QMaSC should maintain a close relationship with other campus services such as counseling, advising, and housing to promote academic support through these offices.

At Mohawk Valley Community College (MVCC), the First-Year Experience Seminar (College Seminar) class is designed to connect new students with academic-support professionals and other resources available on the campus. During the first and second week of each semester, instructors of First-Year seminars bring their students to the Learning Center for a tour. By coming to the Learning Center, students have a chance to meet with tutors and become acclimated to the center for the future. In addition, students taking the class must complete an assignment that requires interaction with academic support professionals outside the classroom. After implementing these efforts in college seminar classes, the number of underprepared students utilizing Math Lab services had significantly increased.

Once the underprepared students actually come to the support centers, they usually have difficulty focusing on the problem at hand, and may not understand the role of the tutor [2]. To eliminate any confusion or insecurities, at the beginning of the first session, tutors should spend some time to get to know their students, their background, and how they feel about math. In addition, the student should sign a tutoring contract that clearly indicates the role of a tutor, and student expectations. During a tutoring session, students should feel comfortable asking questions, and expressing their frustration without being criticized. Tutoring must be focused on guiding and empowering the students, not completing problems for them. When working with the underprepared student, the tutor should pay attention to the student's study skills, note taking skills, and how much work is done by the student [2]. If necessary, the tutor should spend some time helping the student to acquire these basic academic skills.

Research shows that peer tutoring could be very effective for underprepared traditional college students. "The best answer to the question of what is the most effective method of teaching is that it depends on the goal, the student, the content and the teachers ... but the next best answer is students teaching other students" [3]. Research also shows that traditional college students are more comfortable learning with their peers, such that, "students can ask peers questions and not worry about 'looking stupid' or having their lack of understanding impact their grade" [4]. As a result, the director should staff the QMaSC with qualified peer tutors. Tutors should be selected based on their quantitative background and empathetic ability, and represent student diversity on the campus whenever possible. Peer tutors are students and may lack basic tutoring skills. Hence, it is essential to train peer tutors to work with underprepared students. The training should include topics such as establishing a comfortable tutoring environment, tutoring techniques, assisting students with math anxiety, assisting students with English as a Second Language, and understanding policies and procedures.

In an effort to adequately assist underprepared students, centers should provide walk-in tutoring and appointment-based tutoring. In walk-in tutoring programs, tutors will assist individual students over a short period of time, offer feedback, and answer specific questions. Tutors may work with multiple students simultaneously or in small groups. Students do not need to schedule an appointment for walk-in tutoring. As many students seek assistance shortly before their assignments are due, walk-in tutoring provides flexibility. In a survey conducted at the MVCC Math Lab, 78% of traditional students stated that they prefer walk-in tutoring over appointment-based tutoring [5]. In appointment-based tutoring programs, a tutor and a student work in a one-on-one setting. Appointment sessions are usually 30 minutes long and are arranged for the student and tutor at a mutually convenient time. In connection with the type of tutoring session, another important factor influencing the success of the support center with underprepared students is tutor availability [2]. A center's hours of operation should be comparable to students' schedules, meeting the needs of the students.

A majority of students who lack math study skills tend to memorize formulas and steps in solving problems, instead of understanding the process involved. In addition, many of these students suffer from math anxiety which leads to frustrations and a loss of interest. To address these issues,

QMaSCs should offer workshops to help students learn how to study math, defeat anxiety, and set them on the right track for success. Finally, the center should provide useful handouts and the links to websites for additional support.

3 Returning Adult Students

Many higher education institutions define adult learners as college students over the age of 25. These students are often referred to as nontraditional students. Reasons why adults may enter college include job loss, denied promotions due to the lack of a college degree, discharge from the military, divorce, or just wanting a better job.

A large number of adult students enter college unaware of the expectations and are underprepared for college-level course work. Generally, these students lack basic math skills, and get anxious at the thought of taking required math classes. In addition to being underprepared for college course work, adult learners often have extremely busy schedules due to job obligations, family responsibilities, health issues, and other outside factors that reduce the time available for education. Since these students face many challenges and obstacles, they may need additional support to achieve academic success in mathematics. Effective tutoring services may help adult students succeed in pursuing their degrees.

In an effort to adequately assist adult learners, directors have to be aware of the fact that adult students are significantly different from students coming directly from high-school. According to Nunn, adults learn best when the content and learning process is personally relevant to their life experiences or present concerns [6]. Malcolm Knowles identified four common characteristics of adult learners: (a) adult learners are self-directed and take responsibility for their own learning, (b) adult learners have an extensive life experience, (c) adult learners want to learn and they are internally motivated, and (d) adult learners need to know the reason for learning something [7]. The following shows how Knowles' Adult Learning Principles could be applied to tutoring for the adult student:

1. **Adult learners are self-directed and take responsibility for their own learning** [7]. The tutor should guide the student through a process of solving a problem and ask leading questions that would help the student understand each step. According to Fidishun, "adult learners resist learning when they feel others are imposing information, ideas or actions on them" [8]. Tutors should also allow the student to have control of the tutoring process, encourage independent learning and help the student gain confidence in his/her own ability. During the tutoring session, adult learners should be actively involved in the learning process for learning to be meaningful [9].

2. **Adult learners use life experience as a foundation for their learning** [7]. Adult learners may have had numerous life experiences upon reentering college, compared to the

typical first-year student, and consequently, learn in different ways. These students want to apply their life experiences to learn new material [7]. In order to successfully tutor returning adult students, it is essential to know their learning styles. Thus, the director should encourage these students to take a learning style inventory test to help them identify their learning strategies preferences, strengths, and weaknesses. Once the learning style is identified, tutors should use the appropriate tutoring techniques that work best for the student [9]. Furthermore, tutors should find out about the student's interests and past experiences and try to relate problems at hand to the students past skills whenever possible.

3. **Adult learners want to learn and they are internally motivated** [7]. Adult students are in college because they want to learn, not because their parents made them attend, which may be the case for some traditional students. Unlike traditional underprepared students, adults will seek assistance to achieve their academic goals. When they struggle with material, they will go to a support center if the environment in the center suits their learning preferences. In order to effectively assist adults, the QMaSC has to provide a relaxed and motivated environment; the staff must be supportive, respectful and good listeners. Due to age differences, adult learners experience some difficulties connecting to younger students [10]. As a result, adult students would prefer to receive assistance from tutors closer to their age, so being conscientious as a staff to have the older mentors working with these students would provide a more comfortable working experience. Additionally, most of the adult students work and may not be able to come for tutoring during the day, so hours of operation should include evening and weekend hours.

   When working with adults, it is essential to provide a comfortable, supportive environment that encourages learning. As with traditional, underprepared students, the tutor should start a session with a friendly greeting and a casual conversation to set the stage for a relaxed and attentive atmosphere. During a tutoring session, the tutor should establish rapport with the student, ask open-ended questions, and encourage the student to participate in the session as much as possible.

   Knowles also argues that adult students are motivated to learn and stay in college by some external factors, such as promotions, a salary increase, and obtaining a better job. However, internal motivators such as improving self-esteem and desire to grow intellectually have greater impact on adult learners [8]. Having this in mind, the tutor should learn the student's possible motivations, and try to relate the material to the motivating factors.

4. **Adult learners need to know the reason for learning something** [7]. Most students desire to know why they are learning certain material and how will retaining the material personally benefit him or her; adult students are no different. "[Adult students] are less interested in learning for learning's sake than in learning to achieve some immediate or not

too far distant life goals" [6]. They can be impatient with long discussions on theory, and like to see theory applied to practical problems. Therefore, whenever possible, the tutor should relate theory to real world applications.

4 English as a Second Language (ESL) Students

In most colleges, the English as a Second Language (ESL) student population includes a diverse profile of immigrants, refugees, and international students. ESL students face numerous challenges and obstacles in higher education. Besides the language barriers, the ESL student may have difficulty adapting to the demands of different cultures [11]. Furthermore, many ESL students may not have the background to understand textbooks that are written for students who are educated in the U.S. Consequently, these students may need additional support to achieve their academic dreams.

Despite the fact that mathematics is a universal language, many ESL students may struggle with math and science courses. They may know a mathematical concept, but it is the language and cultural background within math problems that they may not understand. This student population may need individualized attention provided by a tutor. In an effort to adequately assist ESL students, QMaSCs should offer a wide range of tutoring services, staff the centers with diverse tutors, provide an extensive tutor training on cultural differences, and constantly collaborate with ESL teaching faculty members and offices for international students.

Centers should provide a wide variety of academic support to meet the needs of the ESL student population. From my own experience as a math tutor, math instructor, and a former ESL student, I have learned that tutoring in a one-on-one setting is most effective and provides the best results in assisting ESL students. This is especially true for those enrolled in developmental math courses. Furthermore, ESL students may benefit when tutored in a small group consisting of a diverse student population. In order to better assist ESL students, staff should also provide various handouts written in simple English covering different math or quantitative topics, and a description of commonly used formulas and terminology.

Another important factor influencing the success of a center with ESL students is tutor diversity. The QMaSC should be staffed with tutors who have different cultural backgrounds whenever possible. ESL students feel more comfortable when they walk into a center with students who also speak their language or relate to them on a personal level. In addition, it is a good idea to assist ESL students in their primary language whenever possible. Some math terms, word problems, and abstract ideas might not make sense until directly translated into the student's native language [12]. Having a tutor who can explain the concept in the student's primary language will make tutoring more productive.

At the beginning of each semester, the director should provide extensive training on how to tutor ESL students. Before the training, directors should find out the number of ESL students on the campus and the language they speak. The directors also should learn basic information about linguistic, cultural, and academic backgrounds of these students and share the information with tutors. For example, tutors should be aware that in some cultures female students may not want to shake hands with a male, or in some cultures it may not be acceptable for a student to ask for additional help. Furthermore, the math center director could include ESL faculty members into the training process to provide additional information and useful tips for assisting ESL students.

When it comes to actual tutoring, the center staff should be aware of the fact that quantitative material could be more challenging for some ESL students. Besides the conceptual understanding of mathematics, ESL students often have problems with math linguistics. As with adult students, it is essential to create a comfortable atmosphere during a tutoring session. The first tutoring session should take place in a one-on-one setting. Tutors should speak slowly, write down formulas, draw diagrams or pictures if appropriate, and paraphrase questions whenever the tutor sees confusion on the student's face. When asking a question, the tutor should wait for the student to respond. Increasing wait time will give students a chance to process what they have heard and formulate answers. ESL students could flourish in their quantitative classes with adequate academic support.

5 Students with Disabilities

An individual with a disability is a person with a physical or mental impairment which substantially limits one or more major life activities, such as walking, seeing, hearing, speaking, working, or learning [13]. According to Federal Law, Section 504, "...colleges and universities may not discriminate in the recruitment, admission, educational process, or treatment of students" [13]. Due to this, the number of individuals with a wide range of disabilities in higher education has significantly increased.

Students with disabilities face a wide range of academic and personal challenges while on the college campuses. Faculty and support staff are legally obligated to provide necessary accommodations for these students to help them overcome many obstacles to achieve their academic goals [14]. Students with disabilities are often less academically prepared, and have lower self-confidence and self-esteem than students without disabilities [14]. As a result, these students may need extra support to thrive academically. The college support center plays an essential role in providing the necessary tools for these students to succeed.

In order to effectively accommodate students with disabilities, all tutoring locations must be wheelchair accessible, and tables should be accommodating for a wheelchair. In addition, tutoring centers should be staffed with well-trained tutors who could combine content knowledge with empathy. Lastly, the QMaSC director should regularly collaborate with the office of services for students with disabilities to become informed about each disabled student that seeks assistance within the center.

As mentioned above, students could be physically or mentally disabled. One of the major mental impairments is a learning disability. A learning disability is a neurological condition that interferes with a person's ability to receive, process, store and respond to information [15]. Learning disabilities can affect one's ability to read, write, speak, spell, compute math, reason, and also affect a person's attention, memory, coordination, social skills and emotional maturity [14]. Typical learning difficulties include dyslexia (problems with reading), dyscalculia (problems with math concepts), and dysgraphia (writing problems) [16]. In order to successfully assist a student with a learning disability, the tutor should be aware of the particular learning disabilities the student has. Since a student with a disability has a right to confidentiality, tutors may not be able to see the actual reports about the student's disability. Instead, the math center director could contact the office of services for students with disability to obtain the information about a particular student's strengths and weakness and possible strategies for assisting them. Once strengths and weaknesses for the disabled student are understood, tutors should use techniques and strategies that work best for the students.

When working with a college student with a learning disability, it is essential to find a location where the student feels comfortable to work on problems without fear of being embarrassed. Students with learning disabilities need to feel comfortable sharing their frustrations with the tutor. Based on my experience, students with learning disabilities learn best when they are tutored in a one-on-one setting by professional tutors. The majority of these students have a hard time disclosing their educational obstacles to their peers. Tutors should be extra careful when assisting a student with dyscalculia. These students have difficulty performing simple algebra and solving word problems. "Dyscalculia makes it impossible for learners to function with story problems that present math facts in sentences or paragraphs" [17]. When a tutor suspects that the student may have dyscalculia, the tutor should be extremely patient, repeat and paraphrase the question whenever there is confusion on the student's face, draw diagrams and charts, and encourage the student to use a calculator to perform any calculations when needed. Well-trained tutors who are able to combine content knowledge with empathy could help these students successfully complete their math requirements.

6 Student Veterans

With the end of the wars in Iraq and Afghanistan, many colleges and universities are seeing a significant increase of student veterans. Entering a college environment after serving in the military may be a difficult transition for the majority of veterans. Typically, student veterans tend to be older and are considered non-traditional students. Like other non-traditional students, veterans face many challenges and obstacles. A significant number of veterans have physical and/or psychological injuries. "A study conducted by the Veterans Administration reveals that nearly 30% of its patients who served in Iraq and Afghanistan have PTSD Post Traumatic Stress Disorder]" [18]. As a result of their injuries, veteran students have different needs than traditional college students and may need unique support to achieve their academic goals.

Research shows that veterans were less likely to seek academic support and getting them come to a QMaSC is quite a challenge [19]. For these reasons, the QMaSC director needs to use proactive efforts to reach out to these students and encourage them to utilize available support. The director should establish a close working relationship with the Veterans Office on campus (if one exists), or in town and ask their staff to make veterans aware of additional support. If possible, the director should invite a representative from the Veterans Office to participate in the tutor training and to inform tutors about veterans' obstacles and challenges. Furthermore, research found that academic success and persistence of student veterans depends on veterans' connection with students and faculty with military experience [19]. In an effort to attract student veterans to use the QMASC's services, the director should staff the center with tutors and support staff that have some military experience whenever possible. By working closely with the Veterans Office and hiring tutors with military experience, the director of a QMaSC could improve tutoring effectiveness and increase the number of veterans in the QMaSC.

7 Creating Welcoming Atmosphere in QMaSCs

In an effort to attract and motivate different types of students to utilize the services provided by the QMaSC, the director should staff the center with diverse and well-trained tutors and make sure that the atmosphere in the center is relaxed and non-threatening.

Studies show that qualified tutors constitute a successful QMaSC. Hiring qualified tutors is one of the main challenges facing the QMaSCs administrators, and whenever possible, tutors in the QMaSC should represent the cultural diversity that exists on campus. Close working relationships between the director and other offices (international students, ESL, students with disabilities, veterans office, student affairs, etc.) can be helpful in recruitment of potential tutors.

The effectiveness of QMaSCs with a diverse student population is strongly influenced by the tutor training. Directors could train their tutors in many different formats to ensure the quality of tutoring services. At Mohawk Valley Community College, two-day tutor training classes are offered at the beginning of each semester to help maintain the quality of tutoring services. The main goal of the tutor training is to ensure tutors understand the learning process in relation to content knowledge and to review and clarify the policies and procedures. Tutors should be trained to create a relaxed atmosphere in the tutoring session and to encourage independent learning. The director should invite the representative from the Office for Students with Disabilities to provide guidelines for tutoring students with disabilities. It would also be beneficial to include staff from the Office for International Students and ESL faculty members to address cultural differences during the tutor training. In addition, the director could ask faculty members teaching quantitative and math intensive classes to share their experience with tutors. By doing this, tutors will have a chance to meet the faculty and be aware of what faculty expect from tutors. Each semester, the director should observe and evaluate tutors and make appropriate suggestions for improvement.

QMaSCs should be conveniently located, and include adequate space and furnishings. The facilities should include: tutorial rooms, private offices and reception. All facilities must be wheelchair accessible and in compliance with relevant building codes. The physical appearance of each location in the center plays a significant role in getting students into the center. An attractive décor, such as seasonal and holiday decoration, posters of scientists from different countries, and student artwork, will enhance learning environment. For example, having a poster that says Albert Einstein had a learning disability may attract students with learning disabilities to utilize the tutoring service more often. Tutorial rooms need to be comfortable and quiet. Basic furnishing in tutorial rooms should include study carrels, tables, and chairs. In addition, every QMaSC should have several private offices for academic advising and individual tutoring for students that may not be comfortable sharing their frustration in larger groups. Also, the center should have a receptionist station to help students with scheduling and navigating the center. The reception area should also be decorated to promote diversity. By implementing the above recommendations, the director of a QMaSC could significantly improve the center success with diverse learners.

8 Conclusion

Today's higher education institutions are characterized by a diverse student population. College students differ in ability, academic preparation, age, ethnicity, language and cultural background. A significant number of these students enroll in college underprepared for rigorous college course work. One of the major obstacles in their educational adventure is mathematics. A large number of students do not learn math from lectures even when taught by excellent instructors. In order to successfully complete math requirements, many students will seek additional support. QMaSCs play a significant role in helping college students achieve academic success in quantitative and math-intensive science courses. The success of the QMaSCs with diverse student populations may be positively influenced by several factors. First of all, tutors should be selected based on their academic achievements and empathetic abilities, and represent the student diversity on the campus. Second, all tutors must be trained in tutoring techniques, learning styles, interpersonal skills, and self-esteem development. The learning environment in the center should be relaxed and non-threatening, so that students feel comfortable asking questions and expressing their frustration. Tutor availability must be comparable with students' schedules. Finally, the staff must publicize tutoring services and collaborate with faculty and other student support services on the campus. By executing the previous steps, the center director could significantly improve academic support for ever-increasing diverse student populations.

9 Bibliography

[1] N. Heckart, "Equal access: Universal design of tutoring and learning centers." DO-IT, University of Washington, 2012. Accessed from http://www.washington.edu/doit/equal-access-universal-design-tutoring-and-learning-centers June 2016.

[2] H. Baril, "Addressing diverse audience: Tutoring the underprepared student," in *Tutor Training Handbook* (S. Deese-Roberts, ed.), pp. 81–83, College Reading and Learning Association, 2003.

[3] W. J. McKeachie, *Teaching and Learning in the College Classroom: A Review of the Research Literature.* 1986.

[4] M. Hurley and M. Gilbert, "Basic supplemental instruction model," in *Supplemental Instruction: Improving First Year Student Success in Higher Risk Courses, Monograph No.7* (M. E. Stone and G. Jacobs, eds.), (Columbia, SC), 2008. http://eric.ed.gov/?id=ED354839.

[5] "Mohawk Valley Community College Learning Center Survey," Spring 2013.

[6] D. Nunan, *Learner-Centered English Language Education: The Selected Works of David Nunan.* 2012.

[7] M. Knowles, *Andragogy in Action: Applying Modern Principles of Adult Learning.* 1984.

[8] Queensland Occupational Therapy Fieldwork Collaborative, "Adult learning theory and principles." Accessed from http://www.qotfc.edu.au/resource/?page=65375 on May 13, 2013.

[9] K. J. Rodrigues, "It does matter how we teach math," *Journal of Adult Education*, vol. 41, pp. 29–33, November 2012.

[10] S. L. Hagedorn, "Second chances: Older students in community colleges." Accessed from http://www.academia.edu/2650227/Second_Chances_Older_adults_in_community_colleges on May 22, 2013.

[11] M. S. Sylvain, *The Language of Success: A Case Study of the Academic Achievement of ESL Students Who Thrive in Spite of Language Barriers.* PhD thesis, Capella University, 2013. Accessed from http://eric.ed.gov/?id=ED522266 on June 2016.

[12] B. Jasper, J. Huber, S. Taube, F. Taylor, M. Klespis, M. Swarthout, C. Long, E. Avalos, L. Gerardo, and S. Boha, "Teachers guide to teaching mathematics for english language learners," November 2005. Accessed from http://txcc.sedl.org/events/previous/092806/8ExplorStrats/mell-teacher-guide.pdf on May 23, 2013.

[13] U. S. Department of Health and Human Services, Office for Civil Rights, "Your Rights Under Section 504 of the Rehabilitation." Accessed from http://www.hhs.gov/sites/default/files/ocr/civilrights/resources/factsheets/504.pdf June 2016.

[14] Academic Skills Center University of Wisconsin , "Assisting students with disabilities." Accessed from http://www.uwec.edu/ASC/resources/studentswithdisabilities.htm.

[15] National Association of Special Education Teachers, "Introduction to learning disabilities." Accessed from http://www.naset.org/2522.0.html on May 29, 2013.

[16] Disabilities, Opportunities, Internetworking, and Technology Center at the University of Washington, "Learning disabilities." Accessed from http://www.washington.edu/doit/Faculty/Strategies/Disability/LD/ on May 29, 2013.

[17] D. R. Jordan, *Teaching Adults with Learning Disabilities.* Malabar, Florida: Krieger Publishing Company, 1996.

[18] J. Reno, "Nearly 30% of Vets Treated by V.A. Have PTSD," 2013. Accessed from http://www.thedailybeast.com/articles/2012/10/21/nearly-30-of-vets-treated-by-v-a-have-ptsd.html.

[19] D. DiRamio, "Transition 2.0: Using tintos model to understand student veteran persistence," *ASHE Higher Education Report*, vol. 37, no. 3, 2011.

[20] M. Snyder, "About MVCC." Mohawk Valley Community College website. Accessed from http://www.mvcc.edu/about on June 3, 2013.

Strategic Planning and Budgeting

© Gail Hilyard,
Holyoke Community College

1 Introduction

Why do we strategically plan for quantitative and math support centers? Strategic planning is like a road map. It helps determine what direction we want to take our services by setting goals and objectives, and the goals help to standardize services and assess programs. The budget should align with the institution's strategic plan. It is crucial to identify diverse funding sources to be able to offer important future services. In determining goals, it is important to involve stakeholders such as senior administration, faculty, staff and students to ensure widespread support for the program. Once the strategic planning process has clarified the goals of the center, the director or coordinator must decide whether to implement a particular program to achieve them.

2 Goals and Objectives

All QMaSCs should have goals and objectives. Goals help to create the vision, and objectives help to break the plan down into manageable parts. Goals and objectives must be connected to the mission of the institution and its values, visions, and goals. For example, one institution's mission may be to provide comprehensive and high quality educational programs that will meet the intellectual and practical needs of a diverse student population [1]. The QMaSC must then attempt to align its goals with the institution's goals as stated in its mission. In addition, the

Suggested Citation: Gail Hilyard (2016), "Strategic Planning and Budgeting", `http://dx.doi.org/10.5038/9780977674435.ch5` in G. Coulombe, M. O'Neill, M. Schuckers (Eds.) *A Handbook for Directors of Quantitative and Mathematical Support Centers*, Neck Quill Press, `http://scholarcommons.usf.edu/qmasc_handbook`.

This material is based upon work supported, in part, by the National Science Foundation under Grant DUE-1255945. Any opinions, findings, and conclusions or recommendations expressed in this material are those of the author(s) and do not necessarily reflect the views of the National Science Foundation

institutional mission may be to encourage and support a variety of instructional strategies [1]. Thus, the center at this institution might offer an assortment of contemporary instructional and supplemental support programs for students, including programs like the math learning coach, math mini prep, or e-tutoring. Likewise, the center services would try to support the lessons instructors are teaching in their classrooms. For another example, if the goal of the institution is to serve an economically, educationally, and linguistically diverse population and the student body reflects this goal, it is important for the director or coordinator to hire a diverse professional and peer tutoring staff that reflects the college's student profile. By adapting this hiring approach, this center may gain support for new and existing programs as well as being authorized to hire more tutors.

3 Effect of Data on Budget and Strategic Planning

Why collect center data? Because data give accountability and evidence of the need for financial support, data collection is vital for the survival of current and future support services in the center; therefore, the director/coordinator must collect student data. If the college has a long-range planning committee (LRPC), the center must tie its goals to the committee's goals: This helps show which center programs and activities are helping to achieve those goals set by the committee. It then becomes practical for the college to support the QMaSC budget for current and future services because center goals are aligned with LRPC and college goals. Some colleges collect student data manually — for instance, through a "Tutor Log" — while some may use a student tracking system software package — e.g. AccuTrack — and some may collect manual and electronic data at the same time to build in a backup system.

What types of data might be collected? The answer might depend on the characteristics of the students coming to the college: What are their learning gaps? What are their ethnic, socioeconomic, and linguistic backgrounds? And what type of support services are needed to help them succeed? The center should also collect data such as student identification numbers, appointment dates and durations, attendance, and grades (if the institution permits). The math center could even design its own survey to collect whatever data are needed. Collecting the above data will be useful in examining student retention, which tends to be an institutional goal at most colleges. If the data can show increased student achievement, which influences institutional retention, then it will be in the best interest of the institution to financially support programs and activities and future services in the center.

4 Identifying Diverse Funding Sources

It is important to target one's own college as a funding source by consulting with the development office, alumni office or grant writing division. There are also private funding sources like the

Bill and Melinda Gates Foundation that support initiatives in education.

Several key funding sources can be used to help create a new center or support an existing one. Title III and Student Support Services (TRIO) grants are from the U.S. Department of Education. These federal grants are used to improve higher education in areas such as academic development and student support services. Another great source is STEM (Science, Technology, Engineering and Mathematics) funded by the National Science Foundation (NSF).

In Massachusetts, another funding source is the Vision Project Grant from the Department of Higher Education (DHE). This grant examines the role of public higher education in Massachusetts in comparison to other states, including factors such as graduation rates, student learning, and workforce development. The grant seeks to strengthen these areas, and other states may offer similar programs.

First Year Experience (FYE) Programs are funded by Strengthening Institutions Grants from the U. S. Department of Education. This program provides support for unique student subpopulations: underrepresented ethnic groups, specific genders, students with learning disability, etc.

5 Involving Key Stakeholders in Program Development

The involvement of stakeholders in program development is key to the success of present and future programs, projects, activities and initiatives.

How are stakeholders identified? Certain faculty or professional staff may ask the director/coordinator to do workshops with their students in the classroom or meeting room, or may bring their students to tour the QMaSC. They may also communicate their own ideas and concerns about center services. One way these genuinely interested stakeholders could serve is through an Advisory Committee; in this way, they will be able to enhance the services, vision and mission of the center. These individuals can help develop interdisciplinary strategies with other departments and clarify needs for the center. It can be especially helpful to have senior administration or other influential people serve on the Advisory Committee. The committee should be very diverse and have members from every aspect of the campus if possible.

In addition, it is very important to meet with the Vice President of Academic Affairs and other senior administration individually. The senior administration is faced with the difficult decision of what services, projects, programs, initiatives or activities to financially support. Therefore, it is imperative that the QMaSC director/coordinator show them how the center's goals are consistently meeting the goals of the administration. For example, student retention and course completion rates tend to be great concerns – keeping students in college is a huge issue for senior administration. The proposed goals of the center can become stakeholders' goals as they gain experience with the center's services and see that their goals are in alignment with the center's goals. These stakeholders, such as interested math department instructors, will then share the goals with other math instructors and the overall institution.

Specific programs could be developed through the strategic planning/budgeting process. Although not every institution will need any specific program, the following discussion will describe how a center director/coordinator could ask key questions to help decide whether to establish a particular type of program.

6 One Example of How to Tie a Program to Budgets and Strategic Planning

Some QMaSCs have developed a program called the Math Learning Coach to assist students and give them specialized one-on-one instruction. The program assists in breaking down the content and helping the student develop critical thinking skills and problem solving. The Math Learning Coach is often a professional mathematics specialist who identifies an individual student's learning styles and preferences, and develops an individual plan for each student to enhance her learning and success. See a more complete description of the program in the appendix.

What data should be examined by the director or coordinator to determine if this new initiative would be suitable for the institution? One important element to examine would be the characteristics of the student body at one's institution. For instance, a college might have a growing number of students who are coming to college with learning disabilities:

> Nearly nine out of 10 of the nation's two- and four-year colleges enroll students with disabilities, and of the 86% of those that enroll students with learning disabilities, only 24% say they can help disabled students 'to a major extent' says an Education Department report published in June. Of about 67% of such students who pursued college or vocational school after high school, only 24% disclosed their impairment to college officials, a necessary first step toward getting accommodations, says a federally funded study that has tracked students from their teen years into young adulthood [2].

Likewise, a large segment of the student population may represent the first generation in their families to attend college. Many of these students have a difficult time succeeding in college. There could also be students who come with poor mathematical skills and a poor concept of their math ability that affects their learning [3].

How would the director/coordinator find out if the Math Learning Coach would be a good use of the center's budget? One might pilot the program for a year to see how beneficial it would be to students and whether an effective program could be aligned with LRPC goals.

If the pilot program indicates that the Math Learning Coaches are needed, then the data collection on the various student populations using this service would help show the need for this position: the data could show the benefit it has had for achievement, retention and completion rates, which are typically the goals of most institutions. The director or coordinator would need to track and collect the students' grades, attendance, subjects, and student comments on the coaches. The director or coordinator in that case would need to advocate for this position by meeting with all stakeholders, such as senior administration and the vice president from academic affairs, in order to secure financial support for this position.

7 Summary

First, one establishes goals and objectives connected to the mission of the institution. Then one needs to align one's budget with the goals and to use data collected to establish accountability for the center's programs. Seeking diverse funding sources from the college, governmental and private foundations may be helpful, and one cannot overestimate the importance of collaboration with key stakeholders. Every college has its own situations and priorities, so actual goals and objectives will vary.

8 Bibliography

[1] Holyoke Community College, "Mission statement." http://www.hcc.edu/about-hcc/welcome/mission. Accessed June 14, 2013.

[2] M. B. Marklein, "Learning-disabled students get firmer grip on college." http://usatoday30.usatoday.com/news/education/story/2011-10-17/college-and-learningdisabilities/50807620/1. Accessed June 14, 2013.

[3] A. F. Gourgey, "Tutoring developmental mathematics: Overcoming anxiety and fostering independent learning," *Journal of Developmental Education*, vol. 15, no. 3, pp. 10–14, 1992.

9 Further Reading

J. Cullinane and P.U. Treisman, "What Policies and Practices Work for Students?," *Prepared for the NCPR Developmental Education Conference Teachers College*, Columbia University, Sept. 2010. Retrieved from http://www.postsecondaryresearch.org/conference/PDF/NCPR_Panel4_CullinaneTreismanPaper_Statway.pdf

Massachusetts Community College Executive Office (MassCC). "100% Math Initiative: Building a foundation for student success in developmental mathematics," 2006. Retrieved from http://www.masscc.org/sites/massc.drupalgardens.com/files/mathinitiativefinal.pdf

10 Appendix

The Math Learning Coach assists the student with:

- critical thinking skills
- reading comprehension
- time management
- study skills
- problem-solving
- note-taking
- test anxiety

Students set-up a schedule to meet with Math Learning Coaches on a weekly basis — sometimes 1, 2 or 3 times a week — depending on the need.

As part of the assessment and planning, the Math Learning Coach:

- Fills out a personal interview form with the student (15 to 20 minute)
- Administers an untimed diagnostic test to determine strengths and weakness, if needed
- Explores students' learning styles — LASSI (Learning and Study Strategies Inventory) and VARK Questionnaire (Visual Auditory Reading Kinesthetic)
- Recommends an academic plan in order to address both strengths and weaknesses
- Offers other resources to address weaknesses (videos, book, software, etc.)

Incorporating Technology

© Maria Belk,
Bard College

1 Introduction

This chapter will discuss the ways in which computers and other technology can be used in a Quantitative or Mathematics Support Center (QMaSC). Section 2 will consider computer tools that can help with the administrative tasks involved in running a tutoring center, such as keeping track of the number of students who attend and getting feedback from students on how the sessions went. Sections 3 through 5 will look at computer tools that can be helpful for students and tutors during tutoring sessions.

2 Computer Tools to Help with Running a QMaSC

Computers can help with many of the organizational tasks involved in running a QMaSC. Here are some suggestions for using computers to help a QMaSC run more smoothly.

1. **Create an online application form for tutors.** An online application is easier to distribute and collect than a paper form, and the responses can be automatically stored in a spreadsheet, making it easier to look through the tutor applications. See the information on online forms below for examples of web applications that make it easy to create an online form.

Suggested Citation: Maria Belk (2016), "Incorporating Technology", `http://dx.doi.org/10.5038/9780977674435.ch6` in G. Coulombe, M. O'Neill, M. Schuckers (Eds.) *A Handbook for Directors of Quantitative and Mathematical Support Centers*, Neck Quill Press, `http://scholarcommons.usf.edu/qmasc_handbook`.

This material is based upon work supported, in part, by the National Science Foundation under Grant DUE-1255945. Any opinions, findings, and conclusions or recommendations expressed in this material are those of the author(s) and do not necessarily reflect the views of the National Science Foundation

2. **Have students sign in electronically when they enter the QMaSC.** There are many advantages to having an electronic sign-in system instead of simply using paper and pen. With an electronic system, it is easier to compile data on how many students are making use of the QMaSC, and it is possible to search the data to see whether a specific student has been attending. There are commercial visitor management systems that can be used for this purpose, and there are also commercial programs such as AccuTrack or TutorTrac that include an electronic sign-in system along with many other capabilities.

3. **Keep a log of activity for the QMaSC.** This could include the number of students who attend, what topics they need help with, and the results of their tutoring sessions. The most obvious software to use for this would be a spreadsheet such as Excel or a spreadsheet on a cloud storage service (see below for information on cloud storage services). Alternatively, tutors could fill out an online form after each session, and the information would be automatically stored in a spreadsheet (see below for information on online forms). Commercial programs such as AccuTrack or TutorTrac can also keep track of all of this information.

4. **Schedule tutor appointments online.** Some QMaSC's offer individual tutoring, and computers can be helpful for scheduling these appointments. One option is to have an online form for students fill out to request tutoring appointments. Alternatively, a course management system can be used to schedule appointments (see the information below on course management systems). Commercial programs such as TutorTrac and AccuTrack also have ways to schedule appointments.

5. **Create an online tutor evaluation form.** An online evaluation form can be sent to students at the end of every semester, or there can be an evaluation form on the web page for the QMaSC, so that students can provide feedback throughout the semester. See the information on online forms below.

6. **Share information with tutors and staff members online.** For example, the tutor schedule, the contact information for all the tutors and the tutor training materials can be stored online in a cloud storage system (see below) or a course management system (see below).

7. **Create an online math assessment.** Some QMaSC's require students to take a math placement test or quantitative assessment. Others may find an online assessment helpful in identifying areas students need help with. Online homework systems (see below) can be used to create online assessments (in addition to online homework), and some course management systems (see below) include ways to create online quizzes or tests.

Many tools can help with these tasks:

Cloud Storage Service: A cloud storage service such as Google Drive or Dropbox is a way to store documents online and share them with others. It can be used to share information with tutors, or simply to share files between several staff or faculty members.

- Google Drive (`drive.google.com`) is a free service that includes a simple web-based word processor and spreadsheet. Documents on Google Drive can be shared by many users and edited by multiple people at the same time.

- Dropbox (`www.dropbox.com`) is a free service for file sharing and backup for any computer. A downloadable program creates a folder on the computer in which files can be stored; the files are also accessible online and on other computers on which the Dropbox folder has been downloaded. Because the files are stored on the computer, the files can be accessed without an internet connection. It can be a problem if multiple people edit a Dropbox file at the same time; Dropbox will save both edits, but as separate files.

- Additional cloud storage services include Microsoft One Drive (`https://onedrive.live.com`), iWorks for iCloud (`www.apple.com/iwork-for-icloud`), and Amazon Cloud Drive (`www.amazon.com/clouddrive`). These services all work in a way similar to Dropbox.

Online Forms: Google Forms and SurveyMonkey are two web applications that make it easy to create forms that students can fill out online. Online forms can be useful for many purposes such as tutor applications, tutor evaluation forms, and requests for tutoring appointments. If a QMaSC offers one-on-one tutoring, it could use an online form for students to request one-on-one tutoring appointments. The responses would go to a spreadsheet, and someone would need to check the spreadsheet regularly and match tutors with the requests. Online forms can also be used to keep track of information about the students attending the QMaSC. The tutors can fill out a form with this information after each tutoring shift, and then the information can be stored in an online spreadsheet.

- Google Forms is one of the document types on Google Drive. The responses to a Google form are stored in a Google Drive spreadsheet, which can be downloaded as an Excel file. Spreadsheets on Google Drive can have up to 200,000 cells, so the maximum number of responses is 200,000 divided by the number of questions.

- SurveyMonkey (`www.surveymonkey.com`) is another website for creating online forms. It allows more question types than Google forms; however, the free version allows only 10 questions per survey and 100 responses per form. This limit on the number of responses makes it less useful for some tasks.

Course Management Systems: Most colleges and universities use a course management system such as Blackboard or Moodle. These systems allow the college to create web pages accessible only to certain students and faculty, such as the students enrolled in a specific course or the faculty members in a certain department. A QMaSC could have a page that is accessible only to the tutors, and the page could be used to share information with them.

Course management systems can be used to create online quizzes or tests, or to create an online math assessment. Moodle allows you to use LaTeX commands when writing quizzes, which makes it easy to include math symbols in the questions.

In addition, Blackboard and Moodle both include scheduling tools that can be used by tutors to schedule one-on-one appointments with students. Using the scheduling tool on the course web page, tutors can indicate what times they are free to meet, and then students can sign up for appointments. This system works best if each course is assigned a designated tutor who can coordinate with the professor.

Electronic Visitor Management Systems: Visitor management systems are intended for businesses to use as electronic sign-in forms for visitors to an office. At a QMaSC, software such as this can be used to keep track of which students attend. Students can sign in when they enter the QMaSC and sign out when they leave. In case a student forgets to sign out, the software can be set to sign students out automatically at the end of the day. Many visitor management systems are made for tablets (such as iPads or Android tablets) the touchscreen is convenient for an electronic sign-in form, and a tablet is less expensive than desktop computer. An example program is SIGN IN, which works on the iPad and costs a one-time fee of $54.99 (as of October 2013). The data from SIGN IN can be exported as a .csv file for use in Excel or another spreadsheet program.

Online Homework Systems: There are a variety of online homework systems. These are intended to be used along with a course, but they can also be used to create an online math assessment. One such online homework system is WebAssign (`https://webassign.net`). WebAssign contains a library of pre-made problems, and it also allows instructors to design their own problems. It is free for instructors but requires a small fee for each student. Another online homework system is Maple T.A. (`https://maplesoft.com/products/mapleta`), which is a commercial program by the company that makes the computational algebra system Maple. Maple T.A. has pre-made math placement tests which have been approved by the Mathematical Association of America.

AccuTrack and TutorTrac: AccuTrack (`www.engineerica.com/accutrack`) and TutorTrac (`http://www.go-redrock.com/products/tutortrac/`) are commercial software packages that can keep track of information such as how many students are attending the tutoring center and what subjects they are receiving help in. TutorTrac and AccuTrack are fairly expensive, but either is probably worth the cost for a large college or university, where the tasks of keeping track of the visitors to the tutoring center and scheduling tutor appointments might otherwise be unmanageable. As an example, the following information shows how TutorTrac works (AccuTrack works similarly):

- Using TutorTrac, students sign in when they enter the tutoring center. When they sign in, the system can automatically ask them to record what class they need help with.
- When students leave the tutoring center, they sign out, and the TutorTrac can survey them on how the session went. Alternatively, it could send an e-mail to the student with the survey, so that the student can fill it out later.
- If a QMaSC offers one-on-one tutoring in addition to drop-in tutoring, TutorTrac can also be used to schedule individual appointments. Using TutorTrac, tutors who are available for one-on-one consultation can indicate which times they are free, and then students can sign up for one of the available the time slots. TutorTrac can automatically e-mail the students and the tutors a reminder about the scheduled tutoring session.
- It is easy to look up all the data that TutorTrac has stored on the students visiting the tutoring center. For example, it is possible to look up how many students attended the center over a given period, or how often a specific student attends the center.
- TutorTrac can be used with multiple tutoring centers on a campus (such as a QMaSC and a writing center).

All of the tools considered above are programs that help with collecting data, organizing data, and sharing information with others. As technology changes, some of the specific programs suggested may become obsolete; however, similar tools are likely to continue to exist, with functionality improving over time. In particular, there are likely to be an increasing number of programs for the web and for tablets that make use of the internet and cloud storage. It can be helpful to search online every few years for new tools that help with collecting and organizing data.

3 Mathematical Review Tools

One of the challenges faced by a QMaSC tutor is how to help students who are struggling with weak backgrounds in mathematics. Such problems can be hard for tutors to deal with although tutors are adept at helping students with homework problems, they can have difficulty creating practice problems on material from previous courses. Fortunately, there are many resources online for reviewing mathematics. Tutors can direct students to these resources, and then the students can either try learning the material on their own, or they can work through the material with the tutors' help.

The following web pages offer review materials. The tutors could be told about these web pages, and it may also be helpful to include links on the QMaSC website.

- Khan Academy (`www.khanacademy.org`) provides many free short videos on a wide range of topics, not all mathematical. In math, it has videos on topics ranging from arithmetic to college-level courses such as calculus, statistics, and linear algebra. The videos vary in quality, but they can be a great resource to help students review material they may have forgotten. One disadvantage of the Khan Academy is that they don't provide many exercises for students to practice with after watching the videos.
- Kuta Software (`www.kutasoftware.com`) sells several applications for automatically generating worksheets on algebra and calculus. Their website also boasts a large number of free, downloadable worksheets.
- Visual Calculus (`archives.math.utk.edu/visual.calculus`) is a website that has several tutorials, demonstrations, and drills on topics from precalculus and calculus courses. The problems in the drills have complete solutions, which can be very helpful for students trying to learn the material.
- ALEKS (Assessment and LEarning in Knowledge Spaces, `www.aleks.com`) is a commonly used commercial program to help with learning mathematics. It covers material from basic mathematics to precalculus, as well as statistics. ALEKS can be purchased by an individual as well as by an institution, so it could be recommended to individual students even if the college or university has not purchased ALEKS.
- Catchup Math (`http://catchupmath.com`) is another commercial program that helps students learn mathematics. Catchup Math focuses on mathematics taught in high school, including pre-algebra, algebra, and geometry. It can be useful for a student who needs to review these topics.

In addition to the resources listed above, many textbook publishers have online resources for use with the textbook.

4 Other Online Resources

In addition to those listed above, there are several other online resources useful to tutors and students. The tutors can be told about these resources, and it may also be helpful to include links on the tutoring center's web page.

Tutors and students may find it helpful to look up online a math topic that they have forgotten. Wolfram MathWorld (`http://mathworld.wolfram.com`) and Wikipedia (`www.wikipedia.org`) are useful websites for looking up math topics, including more advanced math topics.

Wolfram|Alpha (`www.wolframalpha.com`) is a "computational knowledge engine" created by Wolfram Research, the company that sells Mathematica; it will do mathematical computations such as solving equations and computing derivatives and integrals. Unlike a computer algebra system, Wolfram|Alpha has a natural-language interface, so it is easy for students to use to check their answers. It has a free version, or for a monthly fee it offers more capabilities, such as step-by-step solutions to math problems. There are also Wolfram|Alpha apps for iPads and Android tablets.

Tutor Universe (`www.tutoruniverse.com`) is a website where tutors are available to answer questions in many subjects, including mathematics. The tutors on Tutor Universe charge money, but it could be useful to students who are off campus or who otherwise have difficulty making it to the QMaSC.

Mathematics Stack Exchange (`http://math.stackexchange.com`) is a question and answer website where anyone can post a math question, and then others will respond with a detailed answer. Many of the participants on the site are mathematics professors or graduate students, and the questions range in level from precalculus to graduate-level mathematics. Students can use this website to get help with homework, and are encouraged to include a "homework" tag if they are only looking for a hint rather than full solutions.

5 Computers in a QMaSC

Previous sections have discussed various online tools and software packages that might be helpful for running a QMaSC. This section will consider the role of desktop computers and other technology in the tutoring center itself.

Many of the computers in the tutoring center may belong to the students themselves. Increasingly students in college have laptops — in 2010, eighty-eight percent of college students had laptops [1]. Many students bring their laptops to the QMaSC, and they also bring tablets and smartphones. They might use these devices to complete online homework assignments, to access course web pages, or to read electronic textbooks. To accommodate the students using the internet on their devices, a QMaSC should have wireless internet available. It may also be helpful to have extra power outlets, and possibly a charging station for USB devices.

Although many students have laptops, desktop computers are still useful to have in the tutoring center, for students who either do not have laptops or do not bring their laptops to the tutoring center. Also, mathematical software (such as Mathematica or Maple) that may not be available to students on their laptops can be installed on the desktop computers.

In addition to having desktop computers, some colleges have been experimenting with using tablets (such as iPads or Android tablets) in their QMaSCs. Tablets work as well as desktop computers for many web applications and can also be used as graphing calculators and as e-readers for digital versions of textbooks. As of 2013, neither Mathematica nor Maple is available for tablets — though that is likely to change in the coming years — but both companies currently offer downloadable apps that can run existing Mathematica or Maple files. Also, as was mentioned in Section 2, tablets can be used as electronic sign-in forms when students enter the QMaSC.

Many colleges and universities use computational software in lower-level mathematics courses, so it can be helpful to have this software installed on the desktop computers in a QMaSC. Though some software options are expensive, it may be possible to install this software on the computers in a QMaSC as part of the site license at the college or university (check with computing services). Listed below are some examples of computational software that may be available on QMaSC computers for use in lower-level mathematics courses :

- Mathematica, Maple, Matlab, and Sage are computational software often used in mathematics courses. Mathematica, Maple, and Matlab are commercial software, whereas Sage is free and open source.
- LaTeX is a program used for typesetting mathematics. It is sometimes used in advanced math courses.
- IBM SPSS (Statistical Package for the Social Sciences) and Stata are two commercial programs commonly used in statistics courses. R is a free program sometimes used in statistics courses.
- Microsoft Excel or other spreadsheet software is often used for data analysis and visualization in low-level mathematics courses as well as in science and economics.

Ideally, any computational software that is used in the courses at a college or university would be available on the computers in the QMaSC. If this is not possible, there could instead be regular drop-in tutoring sessions in a computer lab for students who need to learn the software or who must use the software to complete their homework.

Since students will have questions about using the computational software, it will be helpful for some of the tutors in the QMaSC to know how to use it. One solution is to determine which tutors know how to use the software and advertise this information so that students know when they can get help. Another solution is to hold a tutor training meeting devoted to learning the computational software. Even if the tutors used this software in a previous course, they may not remember how to use it very well, so they may appreciate a quick review.

Finally, a tutoring center can also benefit from certain other technology such as a computer projector and projector screen, or an interactive whiteboard. A computer projector can be useful for a tutor working with a group of students. For example, the tutor can put on the projector a list of problems for the group to work on. A projector can also be helpful during tutor training, and it can be used by students to practice presentations. An interactive white board can save material that has been written on it. This can useful during a tutoring session — information that the tutor writes on the board can be saved and given to the student, so that the student can concentrate on understanding the material and not writing it down.

6 Conclusions

This chapter has considered ways in which computers and technology can be helpful for a QMaSC. There are several computer tools that can help with the administrative tasks of running a QMaSC — online forms, cloud storage services, and course management systems can be useful for collecting data and sharing information among tutors and staff members. There are also many online resources and commercial software programs to help students learn mathematics; these can be used by a student who is working with a tutor, or they can be helpful resources for when a tutor is not available. Finally, computers in a QMaSC can be used by students for a variety of tasks, including completing online homework or accessing course web pages. The computers can also contain the computational software that is used in the lower level mathematics courses at the college.

7 Bibliography

[1] A. Smith, L. Rainie, and K. Zickhur, "College students and technology." Pew Internet and American Life Project: http://pewinternet.org/Reports/2011/College-students-and-technology.aspx, 2011.

Math Anxiety and the Impact on QMaSC Services

© Gail Hilyard,
Holyoke Community College

1 Introduction

How can a QMaSC support math anxious students? Math anxiety is a major barrier to student success in learning mathematics, especially developmental mathematics. In order to provide sufficient services, QMaSC directors must be fully informed about math anxiety, its causes, and the impact of anxiety on learning.

2 Understanding Math Anxiety

Research demonstrates that learning mathematics is often emotionally charged, evoking strong feelings of hatred and fear of failure, thereby causing math anxiety. Chinn believes that most math anxiety is due to a learning disability called dyscalculia [1]. Shields, along with some educators, believes that teachers need to address the issue of math anxiety as soon as it appears in the classroom [2]. Because students with math anxiety typically assign blame to teachers for its cause, educators must analyze the causes of the math anxiety and address the anxiety immediately instead of letting it fester in the learning environment. The article "Coping with Math Anxiety" discusses

Suggested Citation: Gail Hilyard (2016), "Math Anxiety and the Impact on QMaSC Services", http://dx.doi.org/10.5038/9780977674435.ch7 in G. Coulombe, M. O'Neill, M. Schuckers (Eds.) *A Handbook for Directors of Quantitative and Mathematical Support Centers*, Neck Quill Press, http://scholarcommons.usf.edu/qmasc_handbook.

This material is based upon work supported, in part, by the National Science Foundation under Grant DUE-1255945. Any opinions, findings, and conclusions or recommendations expressed in this material are those of the author(s) and do not necessarily reflect the views of the National Science Foundation

various definitions of math anxiety, social and educational causes, math myths that lead people to math anxiety, how to take control over anxiety, and strategies for succeeding at math [3].

Other students will choose or have chosen to avoid taking mathematics courses to protect their self-worth and to avoid embarrassment in front of others. Jackson and Leffingwell argue that an investigation of different types of instructors (K through college) and their associated behaviors verifies that some instructors may create and exacerbate anxiety in students. The authors determined that math anxiety first occurs in these students in the third grade [4].

Chapter Four of *Student Learning and the Learning Environment* discusses how mathematics faculty and their institutions need to create a student-based environment to maximize student learning in mathematics in and outside of the classroom by enabling students to be partners in the learning process [5].

Marchewka criticizes the approaches of many researchers and educators who only address ways to ease the symptoms of math anxiety through humor, coping strategies, and specialized instruction, but do not examine the main reasons for the math anxiety. He argues that math anxiety is a *bona fide* anxiety that needs to be addressed because it causes students difficulties with learning mathematics [6].

3 How Students Bring Math Anxiety to Developmental Mathematics

Woodard explains the nature of anxiety for developmental students and suggests ways to alleviate such anxiety [7]. She sees herself as an expert after having taught developmental mathematics for several years and having witnessed the effect math anxiety can have on developmental math students. Woodard's study explores the differences in math anxiety scores based on gender or age, and if there is a relationship between math anxiety scores and achievement scores. She used the Mathematical Anxiety Rating Scale (MARS) to measure math anxiety levels and to determine the relationship between the anxiety levels of the students and their achievement scores. Her results indicate there is no significant difference in math anxiety among traditional (less than 25 years of age) and non-traditional (over 25 years of age) students.

Likewise, Smith's study examines the psychological predictors of academic achievement in students enrolled in self-paced developmental math courses [8]. Smith used online instruments to gather data from students in order to measure the psychological predictors such as procrastination, optimism, self-esteem, anxiety, locus of control, need for achievement, and fear of success. Smith concludes that an estimated 95 percent of college students have procrastination issues, that self-esteem and anxiety are related to procrastination, and that a relationship exists between procrastination and age. Both Smith and Woodard found a higher level of anxiety in women than men.

Guevara's investigation of the Fall 2005 implementation of the Achieving the Dream initiative at Vista College found no significant differences in achievement between Hispanic female and male students enrolled in their first developmental math courses [9].

Knowing how math anxiety impacts developmental math students can help the QMaSC director/coordinator better design support services and programs to help these students.

4 Resource Literature on Math Anxiety

The QMaSC director may choose to provide a collection of books on math anxiety for students to borrow. The self-help handbook *Math! A Four Letter Word*, written by Sembera and Hovis, gives suggestions for helping to change negative math attitudes and to develop study habits [10]. Researchers Kogelman, Warren and Smith offer students a series of workshops to address negative attitudes towards math, to debunk math myths, to identify the causes of and strategies for decreasing math anxiety, and to suggest approaches for working on certain math problems [11][12]. *Mastering Mathematics*, by Smith, gives strategies to improve learning and math grades, offers a diagnostic test that contains numerous suggestions for activities, and includes a section in the appendix titled, "Study Habits Improvement Check" [12].

Nolting's *Winning at Math* focuses on math reform and the future requirements of the American Mathematical Association of Two-Year Colleges (AMATYC). The goal of the book is to improve the nation's math literacy for all students, especially minority students and those with learning disabilities. Students can use the book as a study guide to learn math study habits, while instructors can use the book as a part of any math course or as an independent course [13].

Shelia Tobias, a pioneer in the study of math avoidance, wrote *Overcoming Math Anxiety* to convince women and men that their fear of mathematics is the result, and not the cause, of their negative experiences with mathematics. She encourages readers to give learning mathematics another chance. Tobias argues that when students are not successful learning mathematics, some blame themselves for failing instead of looking at the instructor as a cause of the problem [14].

5 Instruments for Assessing Math Anxiety

Many surveys and instruments exist that can be used to identify students with math anxiety. The Mathematical Anxiety Rating Scale (MARS) is commonly used to measure math anxiety levels and to determine the relationship between the anxiety levels of the students and their achievement scores. The list below in Table 1 identifies many of the available anxiety instruments categorized by those instruments that assess anxiety in general and those instruments that specifically assess math anxiety.

Table 1: Comparison of Math Anxiety and General Anxiety Instruments

Math Anxiety Instruments	General Anxiety Instruments
Dreger and Aiken, 1957 – **Math Anxiety Scale** This is a 3-item scale used to assess emotional reactions to mathematics.	Beck, 1982 – **Situational Anxiety Checklist (SAC)** The SAC is an experiential measure of the severity of somatic and cognitive symptoms of anxiety.
Fennema-Sherman, 1976 – **Mathematics Attitude Scales (MAS)** The MAS is designed to measure attitudes toward the learning of mathematics by women and men. It consists of nine instruments of which one measures math anxiety.	Beck, Steer and Brown, 1985 –**The Anxiety Checklist (ACL)** The ACL is designed to assess the severity of anxiety symptoms in depressed patients. The 21 items selected reflect somatic, affective, and cognitive symptoms that were common to anxiety and not of depression.
Richardson and Suinn, 1972 – **Mathematics Anxiety Rating Scale (MARS)** The MARS is 98-item scale designed to assess anxious reactions to using mathematics in ordinary life and academic situations.	Hamilton, 1959 – **Hamilton Rating Scales for Anxiety (HAM-A)** The HAM-A is one of the first rating scales developed to measure the severity of anxiety symptoms. The scale consists of 14 items, each defined by a series of symptoms, and measures both cognitive and somatic anxiety.
Suinn and Edwards, 1982 – **MARS-A** The MARS-A is the counterpart to MARS, which is used for adolescents.	Spielberger et al., 1970 – **State- Trait Anxiety Inventory (STAI)** The inventory measures anxiety and differentiates it from depression.
	Taylor, 1953 – **Taylor Manifest Anxiety Scale (TMAS)** The TMAS is designed as a device for selecting subjects for inclusion in psychological experiments on stress, motivation, and human performance. TMAS is used as a general indicator of anxiety as a personality trait, it is not intended as a specific measure of anxiety as a clinical entity.
	Zung, 1971 – **Self-Rating Anxiety Scale (SRAS)** The scale is designed to measure anxiety and differentiate from depression.

The appendix includes a sample math anxiety instrument for those QMaSC directors interested in developing an informal in-house survey to examine a student's level of math anxiety

6 Possible Strategies for Addressing Math Anxiety

QMaSC directors may want to consider one or more of the following strategies to address math anxiety: employing a professional math tutor specializing in math anxiety; introducing a math anxiety workshop series; facilitating professional development and training for staff and faculty; and providing additional resources and services.

Employing a Professional Math Tutor Specializing in Math Anxiety

If resources allow, the director might hire a professional math tutor with a specialty in working with students with math anxiety. This professional tutor might explore and purchase literature on math anxiety, find methods for reducing the anxiety, and offer appropriate strategies to students. After making an initial contact with the center, which might include completing an intake form (see appendix for sample), students with math anxiety would be referred to the Math Specialist. The Specialist might begin with setting up an appointment to find out what math-related issues are going on with the student. More generally, the Math Specialist may also consider the big picture to find out why many students are anxious, consult with other learning specialists, and consult with faculty. For QMaSCs with limited budgets, the director could assume this specialized position. In addition, the campus counseling services could be of some assistance with staff training related to anxiety.

The Math Anxiety Workshop Series

The QMaSC director could develop a series of workshops on issues related to math anxiety addressing the following topics in succession: an introduction to math anxiety, strategies for reducing math anxiety, and developing a plan for success. The director or Math Specialist could conduct or lead the workshops. Such workshops might be offered outside of the math classroom or in class at the request of a faculty member. One benefit of these workshops is that students in attendance will realize that they are not alone when dealing with math anxiety. See the appendix for a sample outline of topics included in such a workshop series.

Training and Tutoring of Staff and Faculty

At a minimum, the training of staff on dealing with math anxiety should include when to refer students to the appropriate services in the center. Role-playing activities can be used to show new staff members how to recognize math anxiety, what strategies to employ, and how to refer students to more in-depth services and resources. Likewise, workshops, like those described in the previous section, might be adapted to train faculty on identifying math anxiety and strategies for addressing it. See the appendix for a sample outline for a faculty workshop entitled, "What Faculty Need to Know about Math Anxiety."

Additional Resources and Services

For those students who may not have the time or inclination to work with the center over an extended period of time, a library of additional resources could be available. These could include handouts, textbooks, and videos, or online presentations for students to use in the center or at home. For example, there are a number of YouTube videos on math anxiety such as `http://www.youtube.com/watch?v=0w3_7xytvmU` that can help identify math anxiety and how to go about reducing it [15].

7 Conclusions

QMaSC directors need to understand that math anxiety is a legitimate concern. It is beneficial for directors, tutoring staff, and faculty to become more informed about math anxiety, its causes, and its effect on student learning, such as preventing students from demonstrating their math ability on some forms of assessment. A QMaSC needs to be able to assess math anxiety in students and provide a variety of strategies and programs to address this student concern.

8 Bibliography

[1] S. Chinn, *The Trouble with Maths.* New York: RoutledgeFalmer, 2004.

[2] D. Shields, "Taking math anxiety out of math instruction," *National Association for Developmental Education Digest*, vol. 1, no. 1, pp. 55–64, 2007.

[3] W. H. Smith and B. S. Smith, "Coping with math anxiety." `http://www.mathacademy.com/pr/minitext/anxiety/`. Accessed November 2009.

[4] C. D. Jackson and R. J. Leffingwell, "The role of instructors in creating math anxiety in students from kindergarten through college.," *The Mathematics Teacher*, vol. 92, no. 7, pp. 583–586, 1999.

[5] Beyond Crossroads Writing Team, "Student learning and the learning environment," in *Beyond Crossroads: Implementing mathematics standards in the first two years of college* (R. Blair, ed.), pp. 17–28, American Mathematical Association of Two-Year College, 2006. `http://beyondcrossroads.matyc.org/doc/PDFs/BCAll.pdf`.

[6] E. Marchewka, "Math anxiety with adult learners." `http://www.imagine-america.org/online-journal/math-anxiety-adult-learners`, 2010.

[7] T. Woodard, "The effects of math anxiety on post-secondary developmental students as related to achievement, gender, and age," *Inquiry*, vol. 9, no. 1, 2004.

[8] D. L. Smith, "Personal characteristics related to developmental math courses." http://www.webclearinghouse.net/volume/1/SMITH-PersonalCh.php, 1998.

[9] H. Guevara, *Achieving The Dream: Success Of Hispanic Students In Developmental Math Courses In A Texas Community College*. PhD thesis, Baylor University, 2007.

[10] A. Sembera and M. Hovis, *Math! A Four Letter Word*. Wimberley, Texas: The Wimberley Press, 1990.

[11] S. Kogelman and J. Warren, *Mind Over Math*. New York: McGraw-Hill Paperbacks, 1978.

[12] R. M. Smith, *Mastering Mathematics*. Pacific Grove, CA: Brooks/Cole Publishing, 3rd edition ed., 1988.

[13] P. Nolting, *Winning at Math*. Bradenton, Florida: Academic Success Press, 1997.

[14] S. Tobias, *Overcoming Math Anxiety*. Boston: Houghton Mifflin, 1978.

[15] Community Colleges of Spokane, "Math anxiety." http://www.youtube.com/watch?v=0w3_7xytvmU, 2011.

9 Appendix A

Sample Instrument to Determine Math Anxiety

Name: ______________________________

Date: ______________________

Please answer questions as honestly as possible:

1. Do you have math anxiety? *Please circle* YES or NO or Uncertain

2. How do you know when you are experiencing math anxiety?

3. Can you recall any negative experiences in your prior education, please describe?

4. If you had to measure your level of math anxiety on a scale from 1 to 10 where 10 is the highest, what would be your level, please circle.

   1 2 3 4 5 6 7 8 9 10

10 Appendix B

Personal Intake Form

Date__________________ Time__________________ Telephone Number __________________

Name__________________ Major__________________

Address ____

____

Student Number __________________

Affirmative Action (***optional***): Race or Ethnic Identification

______ Caucasian ______ Native American ______ Black ______Cape Verder

______ Spanish-Surname______ Asian ______Other

MATH CONCERNS:

Do you like math?

Math Experience (positive or negative)

What math course are you taking this semester, if none, then why?

FOR OFFICE USE ONLY

COMMENTS:

EDUCATIONAL BACKGROUND DATE:__________________

	school/college	city	year graduated
HIGH SCHOOL			
COLLEGE			
OTHER			

COURSE	YEAR	GRADE	COURSE	YEAR	GRADE
Algebra I			Business Math		
Algebra II			Other		
Plane Geometry					
Trigonometry					

CHECK THE MATH COURSES COMPLETED (High School or College)

Indicate the problem(s) you have experienced in the past in your study of mathematics

What are your expectations upon the completion of this course?

11 Appendix C

Outline of Math Anxiety Workshop Series for Students

Workshop 1: Introduction to Math Anxiety

1. Explore the nature of math anxiety with students.
2. Have students share their personal stories about math anxiety by writing Math Autobiographies, as described by Shelia Tobias [14].
3. Explore the two common types of anxiety: Cognitive and Somatic
    - Cognitive occurs in the mind.
    - Somatic occurs in the body.
        a) *Have students identify whether they have experienced one or both of these types of anxieties.*
        b) *Briefly describe appropriate techniques to reduce these types of anxieties.*

Workshop 2: Strategies for Reducing Math Anxiety

1. Explain that math anxiety is a learned response, which can be unlearned. People are not born with math anxiety.
2. Discuss "The 12 Myths About Test Anxiety."
3. Have students practice the Short-Term Relaxation Techniques, as described by Paul Nolting [13].
    - *The Tensing and Differential Relaxation Method*
    - *The Palming Method*
4. Advise anxious students to follow up with the professional math tutor who specializes in math anxiety in the QMaSC.

Workshop 3: Developing a Plan of Success

1. Explore the time commitment needed for success in learning math.
2. Investigate proven methods for reading a math textbook.
3. Demonstrate and set-up how to organize a math notebook

12 Appendix D

Sample Faculty Workshop

Workshop on Faculty Training: What Faculty Need to Know about Math Anxiety

- Discuss how an instructor can recognize when a student is experiencing math anxiety.
    1. Students show signs such as sweating, crying, overreacting, breathing change, fidgeting, looking away, breaking out in hives, lacking eye contact, facial expressions, physically tense, emotional, or body language.
    2. Students have shut down, turned off learning, stopped doing homework or are not engaged in classroom activities.
    3. Students show fear or being afraid around quizzes and tests.
    4. Students take longer on quizzes and tests.
    5. Students self disclose, self-identify, or tell you that they have math anxiety or are experiencing math anxiety.
    6. Students avoid class, extra help, and tests.
    7. Students are nervous before and during quizzes or tests. And students do poorly on them.
- Explore what an instructor might do when a student tells the instructor he or she has math anxiety.
    1. Instructors can calm students by listening, giving pep-talks, or by walking them through the problem.
    2. Instructors can offer to spend extra, or one-on-one, time with the student to help the student learn the material.
    3. Instructors can break down material step by step to help the student learn the material.
    4. Instructors can suggest that students use resources and academic support services like the QMaSC.
    5. Instructors can suggest students visit a counselor or the office for that serves students with disabilities.
    6. Instructors might be flexible with assessments such as giving extra time on tests or using an alternate testing place.
    7. Instructors can give students strategies for taking exams and techniques to reduce math anxiety.
    8. Instructors can suggest that student visit during office hours.
    9. Instructors can motivate and encourage students to persist with learning the course material.
    10. Instructors can explore students' past experiences to try to understand their math anxiety better.

Part III

Community Interactions

Coordinating With Institutional Administration

© Frederick L. Van Swearingen,
Winston-Salem State University

1 The "Politics" of Coordination

Little has been written about how the director of a Quantitative and Mathematics Support Center (QMaSC) should interact and coordinate with other institutional groups. In fact, little has been written about this topic for directors of other centers (such as Writing Centers) on campus. For a fortunate few, this information is passed down from a mentor, perhaps a previous director. However, in smaller organizations, the previous director might have left before a replacement is hired and the knowledge is lost. Then, the new director has to learn the hard way – by learning from his or her own mistakes.

It takes time to develop relationships and allies on campus. Most center directors have not been trained in how to negotiate campus politics, and fewer still enjoy the process. Directors who neglect these relationships do so at their programs' peril. A center's health may depend on its director's ability to implement or resist change. Thus, these relationships and this coordination not only require that the director let others know about his or her program, they also mean that a director has to know what is important to others.

Suggested Citation: Frederick L. Van Swearingen (2016), "Coordinating With Institutional Administration ", `http://dx.doi.org/10.5038/9780977674435.ch8` in G. Coulombe, M. O'Neill, M. Schuckers (Eds.) *A Handbook for Directors of Quantitative and Mathematical Support Centers*, Neck Quill Press, `http://scholarcommons.usf.edu/qmasc_handbook`.

This material is based upon work supported, in part, by the National Science Foundation under Grant DUE-1255945. Any opinions, findings, and conclusions or recommendations expressed in this material are those of the author(s) and do not necessarily reflect the views of the National Science Foundation

2 Coordinating with Upper Administration

Most directors of a QMaSC do not report directly to the uppermost levels of administration. Although some may report to a provost, most will report to a dean, department chair, or some other individual. But regardless of reporting lines, many requests will end up on an upper administrator's desk. It is important that upper administration knows about the good work that the QMaSC does, and to be most effective, the director also needs to know what is important to the upper administration.

Upper administrators base their decisions on institutional priorities. These priorities may be established by boards, legislatures, or accrediting bodies [1]. They may result from the current strategic plan. They may also result from the vision of a president or chancellor [2]. The director should know that the highest level of administration may report to a board or system administration. Since decisions that these administrators make set precedents, those decisions may come slowly and with limits.

A director should also realize that upper administration will have limited knowledge about the QMaSC. Even if they are former faculty members of the institution, upper administrators do not really know everything that the center does. They do not need to be experts about the center; they just need to know the basics. Although the director might feel protective of his or her center, to upper administration it is just another unit of the institution. It has assigned space, a defined reporting structure and leadership, and an established budget. And, it is in competition with other units for scarce resources.

Having regular communication with administrators or the director's liaison in administration will be important. This may be accomplished through face-to-face meetings, regular reports, or other means. To be valued by administrators, a director must use the administrators' time efficiently, be prompt in supplying information, and work toward the institution's goals. The director must follow proper procedures, without demanding special consideration. He or she must also be flexible and able to handle QMaSC-related problems. In short, upper administrators value directors who are skilled, informed, and professional.

Directors who have problems or requests should prepare carefully before bringing those issues before an administration. They should gather convincing data and prepare possible options. Fortunately for a director of a QMaSC, many administrators value quantifiable data more highly than anecdotal information. Directors should also align their needs with institutional priorities. As an example, a new initiative could be offered as an answer to a strategic plan initiative. Perhaps an online tutoring program could support the expanded online and distance learning offerings mentioned in the strategic plan. Or, a student success objective might relate to increased retention and persistence initiatives. In this case, collecting data to show that center attendees have reduced drop/fail/withdrawal (DFW) grades might be an appropriate response. Additional information can be found in the chapter on "Internal Evaluation and Reporting."

Regardless of the strength of the data, a director should prepare for a "no" answer. Whining is not an option. For many administrators, the answer to a whine is "no." The smart director will realize that the groundwork has been laid to support a "yes" answer at a later time.

Going around the chain of command to get more favorable treatment at a higher level is also not an option. The director who attempts that will make enemies and lose the trust of administrators.

However, following the chain of command can sometimes be difficult in an academic setting. The QMaSC may be seen as a campus-wide resource, and its director may be approached by other departments for help with their initiatives. For example, a department might have gotten a grant to provide tutoring help for a special population of students, even though the QMaSC is set up to help all students. The director might have to help by following the department's (or the grant's) reporting structure. This does not mean that this request should be turned down. After all, it can be an opportunity to build relationships and, perhaps, get something in return. In such a case, it is imperative that the director keeps his or her own supervisor informed of these situations.

To prepare for coordinating with upper administration, data gathering is an important task. To assist the director, the university's institutional assessment or institutional research group can be a powerful ally. It can help a director access and analyze the center's data along with other institutional data needed to show how the center is supporting the university's objectives. The research group can also point out useful information from national and regional databases, as well as data from peer institutions.

A member of the institutional research group is usually part of higher administrative committees and may even be present at some Board of Trustees meetings. It can be of immense benefit to a center if the Board of Trustees knows it has a growing program that is helping the student population to succeed.

3 A QMaSC's Support Structure

Like Writing Centers, Quantitative and Mathematics Support Centers are institution-specific in both structure and function. Some centers are responsible just for math courses; others add support for science courses, while others add help for other areas (business, nursing, computer and social sciences). Some just offer tutorial help, others add online resources, and others offer additional services. As a result, all Quantitative and Mathematics Support Centers do not have the same reporting structure.

Many Math Support Centers start out in a traditional mathematics department. The director is likely to report to the department chair. Quantitative Support Centers may be aligned with a larger tutorial unit, a faculty development center, or student services center. The director's supervisor may be a middle- or upper-level administrator of various ranks. Still other QMaSCs are in non-departmental academic units headed by deans or provosts. Regardless of the configuration, the director should have a well-defined structure for both reporting and budgetary support.

The director will interact with his or her immediate supervisor for many of the same reasons that they might interact with upper administration. These include making long-range and day-to-day policy decisions, handling personnel decisions, budgeting, and short-range (annual) and long-range (3-5 years) planning. Ultimately, the director will have to work with his or her supervisor to get the data and write the reports needed by upper administration to make their decisions. To maintain communication and also go beyond dealing only with urgent issues or requests, regular meeting times should be established.

In some cases, the director may have both faculty and non-faculty status. This can create problems for the director if the faculty status is within one department while the center is under the supervision of someone from another area (such as a faculty development center or dean of another college). However, directors having faculty status in the departments they serve may create natural ties to the department through this status, which may help to minimize politics and communication problems.

What happens if a director's supervisor is not a"collaborator?" People can refuse to collaborate, and this can include one's supervisor. "Refusing to collaborate" can be through overtly avoiding collaboration, such as acting hostile to new ideas. It can also be through inactivity or neglect, such as a supervisor who sits on ideas or never responds to requests for action. This leaves the director with few options. It may mean that the director has to make collaborative alliances elsewhere. This has the potential of "going over the head" of the supervisor if the alliances are within the unit, which is risky and frowned upon by upper administration. To avoid this situation, the director should document requests, discuss strategies with other directors, and seek to communicate clearly and openly with all involved.

In addition to direct supervisors, some institutions also set up an advisory board or advisory committee for the center (or all of the support centers of the institution). Most of these committees are for advisory purposes only. Committee members may include faculty, staff, and students. Members usually come from the areas the center serves (e.g., the chemistry department, biology department, mathematics department). If advisory, the board's goal is to bring additional viewpoints to issues of importance to the director and the center, which can be a valuable communication vehicle. To enhance these relationships, the director should coordinate regular meetings with this board, bring agendas to meetings to help direct conversations, and share positive aspects, statistics, and ideas related to the center.

4 Coordinating with Other Academic Support Centers

For many universities, a Quantitative and Mathematics Support Center is not the institution's only academic support center. Some institutional administrators prefer to define various academic services as the responsibility of "home" departments. Thus, a math lab might report to a mathematics department, a writing center might report to the English department, and a speech lab might report to a journalism department. Although all provide services to help students, it is generally easier for administrators to structure and assess centers if they are within existing academic departments. However, it is becoming more common for several of these centers to be located within one entity, such as a faculty development center, student academic support services, or a University College. So, how should a director of a QMaSC coordinate with the directors of these other centers?

The answer depends on the situation. All directors have to remember that their mission is to benefit the students, which engenders collaboration between the centers. Having this common goal and recognizing similarities in center structures can be helpful in developing assessment strategies, long-term planning, working with administration, and dealing with shared or common issues and practices. Although this collegiality can go a long way, budgetary issues and contesting for resources may create competition. If the centers are within different departments with different budget structures, this competition might not be as acute as if the centers are within one group competing for one pot of money. In either case, additional initiatives might be in competition if they have to be funded by the same upper administrator.

Usually with this shared vision for helping student learning, relationships between directors are positive and alliances can be created. In areas where all center directors do similar things, they can work together and help each other on projects. For example, they can create a master schedule of services each semester. Instead of each group competing to get its information out to a smaller group of students, they all work together to get a complete master schedule out to all students. They can also jointly develop public relations materials. Writing center faculty and staff have many clever ideas that could be used to promote a QMaSC program. Conversely, often the QMaSC director has the math background to help all centers analyze data each semester to assess their programs on a consistent basis. Since this group of directors is a very valuable resource for one another, regular meeting times should be established with this group.

5 Coordinating with Academic Departments

Department chairs and faculty can be powerful allies of the director of a QMaSC. But they usually do not initiate contact with the QMaSC, not unless there is a problem. The smart director needs to reach out and connect with departments.

A Quantitative Support Center usually provides academic support for students taking courses in multiple departments. This offers the director a unique advantage. Having the ability to work with many departments, the director is frequently seen as "neutral." This can make it easier to establish collaboration and coordinate service.

The director of a QMaSC can usually establish relationships with individual faculty. Some will be more interested in "going the extra mile" for their students than others. They will put in more effort to get their students help. The director must make sure, though, that his or her program is seen as working collaboratively with faculty, not doing their work for them. Tutors do not do the faculty member's teaching.

To be more effective, the director will need to develop good relationships with department chairs. Although it is important to work with individual faculty, failure to work with the "important people" (department chairs) will cause problems. Chairs are protective of "their turf," so they will want to be included (or at least notified) if the QMaSC director works with their faculty.

As an administrator, a department chair will respond to data. A director must be willing to share center successes with department chairs. For example, if a director creates a report to show upper management how helped students have lower DFW rates than unhelped students, that information would also be of interest to a department chair. It gets a chair's attention knowing that students are passing a course in his or her department at higher rates if the students are coming to QMaSC tutoring or supplemental instruction (SI) sessions. Some chairs might even pass this information along in a faculty meeting to encourage their faculty to send students for additional support.

Department chairs and faculty can help a director in several ways. They can recommend good students as peer tutors. If they are working on an educational grant, they can ask for tutorial support funding. If they are already paying for tutors out of departmental or grant money, they might be willing to let the director advertise their tutoring services along with the center's services. This guarantees that word about the QMaSC program gets out to students. Some faculty might even give incentives to their students who attend tutoring sessions, an effective method for increasing center attendance. Encourage these efforts, and acknowledge any support that the departments give the QMaSC program.

When working with departments, the director must be careful to communicate clearly with each party served, keep department chairs informed about services and issues, and keep his or her supervisor abreast of any issues that may arise. Sometimes navigating these relationships can be tricky. For example, a department chair may recognize that a faculty member in their department could use the center's help for their course. However, the faculty member may see this as a challenge to their abilities and might resist help. To handle this situation, a director might begin by thanking the department chair for their confidence in the service that the QMaSC supplies. Next, the director might ask the chair to schedule a joint meeting with the faculty member where all three can discuss implementation of the chair's suggestion. It can also be beneficial if the director makes his or her own supervisor aware of this meeting. Whenever possible, keep communication open and seek to maintain a neutral position on any department politics.

6 Coordinating with Other Campus Groups

The director of a QMaSC needs to interact and coordinate with other campus groups. A few of these include:

1. The Honors Program/Graduate School Departments – These programs can provide additional sources of well-qualified student workers for the center. Many of these students look for extra sources of income. An added benefit is that they are often active leaders in many programs, so their fellow students know them. They provide added visibility to the center. Graduate students, if available, can add an even higher quality of peer tutoring because of their experience. Assistantship programs with a Graduate School program can provide ongoing graduate student positions that can ensure a consistently high caliber of student assistance for the center.

2. Athletics – Many athletic programs are under increasing pressure to graduate their students. A director might be able to coordinate with the athletic program to set up a satellite center in the gym. Although its proximity would benefit the athletes, it could be open as an additional site that is conveniently available to all students.

3. Library – Some centers are physically set up in the library. If not, a small satellite location within the library should be considered. Library administrators are usually agreeable to coordinate joint programs that help students and bring more "customers" to their location.

4. Residence halls – Directors of residence halls are also increasingly being asked to do more with the residential students. This offers the QMaSC director additional opportunities to set up satellite service areas. As always, it helps to know who the QMaSC "customers" are. For example, a freshman residence hall might need a tutor with different capabilities than a residence hall housing nursing students, athletes, or honors students.

When coordinating to supply service to different locations, the director should recognize that not all efforts will be immediately successful. However, "failures" can represent opportunities. It might not be that the location was wrong; the problem could be that the timing of the services was not optimal. The director should continue to work with directors of these other programs to find ways to further benefit their students.

Finally, tutors can be a uniquely qualified group, as they may be candidates for serving as either paid or volunteer tutors in off-campus situations. Tutors appreciate added income (if any), and the institution can benefit from the positive relationships developed through this arrangement. To avoid potential problems, the director should work with the appropriate organization (such as Student Affairs) that coordinates the volunteer services group within the institution. This group will be more familiar with issues related to transportation, safety, and appropriate contact information between student tutor and outside agency. With a few constraints in mind, positive relationships can be developed and the center's tutors can continue to grow in their tutoring capabilities through these situations.

7 Conclusions

The director of a Quantitative and Mathematics Support Center will find it necessary to coordinate and collaborate with many different groups on campus. This will require the forming of new relationships. The director should expect not only to tell others about the QMaSC program, but also to give them the information that they need to know in ways that they need to know it. It also requires the balancing of these multiple relationships. But this should not be anything new to the director of a QMaSC. Service, coordination, and relationships lie at the heart of what QMaSCs do, whether internally or externally.

8 Bibliography

[1] J. Simpson, "Managing encounters with central administration," in *The Writing Center Director's Resource Book*, pp. 199–214, 2006.

[2] W. B. McLeod and J. M. Young, "A Chancellor's Vision: Establishing an Institutional Culture of Student Success," *New Directions for Institutional Research*, no. 125, pp. 73–85, 2005.

Outreach and Promotion

© Laurie Wern Overmann,
Southeast Missouri State University

1 Introduction

A Quantitative Mathematics Support Center (QMaSC) can be viewed as a learning community within the larger college or university campus milieu. A literature review prepared for the U.S. Department of Education with the direct aim and title of *Strengthening Mathematics Skills at the Postsecondary Level*, emphatically states: "The importance of communities in learning cannot be emphasized strongly enough with adult learners" [1]. The writer elaborates, suggesting that adult learners benefit from both interaction with peers and interaction with "knowledgeable others." Outside the college classroom, learning community-centered QMaSCs — typically open to the entire campus community and staffed with mathematically well-trained individuals – provide these beneficial interactive opportunities. Thriving QMaSCs are attractive learning facilities where students from a wide variety of backgrounds can congregate and interact.

Effectively developing a learning community requires outreach and promotion. A QMaSC director can develop and promote awareness of the benefits of QMaSCs within the larger campus community through targeted publicity and outreach. Outreach and promotion will, of course, be wasted effort unless the QMaSC is staffed by individuals truly devoted to assisting students. Thus, the paramount goal of anyone responsible for a QMaSC is to foster and promote a welcoming, threat-free learning community.

Suggested Citation: Laurie Wern Overmann (2016), "Outreach and Promotion", http://dx.doi.org/10.5038/9780977674435.ch9 in G. Coulombe, M. O'Neill, M. Schuckers (Eds.) *A Handbook for Directors of Quantitative and Mathematical Support Centers*, Neck Quill Press, http://scholarcommons.usf.edu/qmasc_handbook.

This material is based upon work supported, in part, by the National Science Foundation under Grant DUE-1255945. Any opinions, findings, and conclusions or recommendations expressed in this material are those of the author(s) and do not necessarily reflect the views of the National Science Foundation

Community outreach has two components: providing service to diverse populations and raising awareness of existing services. Providing excellent service to the student population has the reciprocal benefit of raising awareness of existing services. Increasing awareness of existing services can result in a thriving, vibrant QMaSC learning community. This chapter suggests ways of developing a learning community through outreach and promotion to a diverse campus community. Effective, continual outreach will bring in reluctant students, and the vibrancy of the QMaSC learning community will keep learners returning.

2 Types of Consumers

Experience indicates that a QMaSC caters to broadly two types of consumers. One type uses a QMaSC as a place to congregate and interact with like-minded peers. This student group tends to be mathematically inclined, comprised of majors in mathematics, mathematics education, or other STEM fields. The second type of QMaSC consumer seeks academic support from mathematically knowledgeable others. Both groups benefit from the interaction opportunities provided by a QMaSC. Mathematically proficient students often come to "hang out," benefitting from interaction with like-minded peers; outreach to this group of students is minimal, for they will come despite promotional efforts. The less mathematically inclined student consumers seek interactions with hired support staff or with their more mathematically minded peers. The less mathematically proficient students are the more reluctant consumers, requiring QMaSC promotion by the director, faculty, staff, or student peers.

To attract QMaSC consumers, a director engaged in community outreach works long term in ever-widening circles, creating an across-campus network of faculty and staff to identify and direct students to a QMaSC. Over time, a director can contact and communicate across campus with individuals who have direct or indirect access to students in need of mathematics assistance. Directors must interact not only with mathematics faculty, but with faculty and staff in academic support, freshman experience, minority student affairs, disability service offices, and administration; in short, any institutional faculty and staff members who interact with students. Creating a network of individuals across campus to funnel students to a QMaSC is a strategic role of the director.

3 Promotion in the Innermost Circle

Research by Solomon, *et al.* suggests mathematics faculty are effective at reaching mathematically inclined students seeking a place to congregate with like-minded individuals [2]. Interestingly, their research indicated that too many mathematics majors flounder in their first year and change majors, not due to lack of mathematics preparation, but because of a lack of support and interaction with mathematically like-minded peers. Their article claims that the "colonization of [a] mathematics learning space" by mathematically inclined students can be a major determinant in these students' continuing study in mathematics. QMaSCs provide a place to congregate and collaborate with like-minded mathematically inclined individuals, facilitating students' mathematical maturation and consequently continued interest in mathematical studies. For a QMaSC director, this student cohort is a desirable and easy-to-access part of a thriving QMaSC learning community. They not only enjoy and benefit from the camaraderie of mathematically like-minded students but also often act as tutors to their less mathematically capable peers.

The challenge, then, for a QMaSC director is identifying and reaching the less mathematically inclined and typically more reluctant students. A QMaSC director's outreach begins by enlisting the aid of mathematics faculty and institution staff that have direct access to mathematically needful students. To this end, a QMaSC director must establish and maintain an ongoing positive relationship with the institution's faculty, advisors, and staff. Frequent face-to-face interaction and conversation with these colleagues will build relationships. These relationships are maintained through follow-up e-mails or newsletters that provide information about QMaSC services and events.

Directors can enlist the aid of mathematics faculty and university staff in advertising the QMaSC by e-mailing a template announcement or link that includes center hours and policies that can be copied and pasted for inclusion in a syllabus or course management system (e.g. Moodle or Blackboard) site. The advertisement can include pictures of the center or maps to the location so that an interested student gets a "feel for the center," perhaps reducing perceived barriers for coming to the center. Providing faculty and staff with colorful, inviting classroom posters or small hand flyers with the QMaSC schedule and policies to be distributed to students is also helpful.

Provided faculty are willing to give up a few minutes of class time, some QMaSC directors and staff make classroom visits early in the term or at other strategic times in the term (before or immediately following exams) in a direct appeal to the hesitant student. Perhaps outside the classroom milieu, a QMaSC staff member armed with an inviting attitude and a colorful handbill can tempt a student to seek tutoring services when a classroom professor cannot. In larger departments with many classes, directors can and should bring tutorial student staff along on classroom visits to model an inviting attitude before the student staff member undertakes classroom visits "solo." Student-staff tutor presentations can be especially effective because some students receive information more readily from peers.

Directors can send reminders to faculty at key moments of the semester to promote QMaSC services when mathematically needful students are reluctant to seek help. Key moments include those times when semester progress reports are sent out or exams occur. Although faculty can target students — for example those who receive D's or F's on the first exam — all students, even those who might want to improve grades from B to A, benefit from the reminder of QMaSC services. Directors can encourage faculty to advertise the QMaSC by providing faculty a template e-mail to send to students.

Sometimes mathematics faculty members spend one or more office hours in the QMaSC. Over time, faculty time spent in a QMaSC can be a very effective means of directing mathematically needful students to a QMaSC. Anecdotal experience indicates that some students view the QMaSC as a more comfortable and less threatening atmosphere than a private office and may be more likely to seek out their mathematics professors in a QMaSC.

Once mathematics department outreach is underway or established, a director of a QMaSC can engage in further promotion through networking with offices across campus.

4 Promotion in Widening Campus Circles

Although the name of the office may differ, many campuses have a broadly based academic support office. Such an office provides academic support services to any requesting student, including traditionally underrepresented groups, minorities, students with disabilities, and first generation college students. Oftentimes, one of the functions of this office is to give academic counseling or grant access to a subsidized one-on-one tutor to any requesting student. One goal of the one-on-one tutoring is to help a student become a more independent learner. To this end, some academic support offices require a designated number of times for lab attendance before individual tutors are assigned. Communicating and sharing resources with these academic counseling and support individuals serves as another means of outreach, directing mathematically needful students to the QMaSC.

Academic support offices often also give group counseling by way of campus presentations on topics of general interest. As another means of campus community outreach, directors can make presentations on such topics as mathematics study skills or mathematics anxiety. These sessions are often well attended, especially when scheduled strategically. A session on mathematics study skills can serve to help struggling mathematics students to understand the importance of preparation and persistent study in overcoming mathematics anxiety and succeeding in a quantitative course. The goal of such a session is to assist students in becoming independent learners, directing them to the campus QMaSC for mathematics content help. A director can collaborate with mathematics faculty to provide well-timed math anxiety and study skill sessions, an excellent means of outreach to students in need of QMaSC services.

At many institutions, something akin to a "learning fair" is held to highlight the many campus resources available to students. Having a booth that has colorful inviting posters or distributing small hand flyers with the QMaSC schedule is an effective means by which to initiate contact with students. Working in a booth at this fair will also serve to alert and inform a QMaSC director about the needs of diverse student populations. For instance, although a campus is deemed handicap accessible, students with mobility issues are more likely to use a QMaSC when equipped with information to help them navigate access to the building and room in which the QMaSC is housed, e.g., on which side of the building the ramp and door are located, where the elevator is located with respect to the QMaSC, etc.

Many campuses also have an office of first-year experience, perhaps with a different name, which serves as an outreach and information clearinghouse for the newest and most naïve students. A first-year experience office and staff can be very effective at advertising the existence of a QMaSC.

Freshman seminar-style courses are quite common at institutions across the country and are usually coordinated by a first-year experience office. Such courses are designed to guide and support students through their first, sometimes daunting, semester, to disseminate correct or dispel incorrect information about the institution, and to direct the most vulnerable students to offices and campus resources. Designed to help students develop communication and critical thinking skills in a non-threatening atmosphere, these courses can be intensely student discussion- and activity-based, serving to develop camaraderie among the group. Further, these courses are often taught by experienced faculty, who are familiar with campus offices, resources, personnel, and most importantly are sympathetic to the plight of first year students. QMaSC directors can use these freshman seminar courses as a means of outreach to students and an opportunity to network with seasoned faculty. Because these courses are typically not driven purely by academic content, professors are often willing to allow QMaSC staff, again at strategic times in the semester, to "pitch" their programs. Connection with the freshman experience office and a network opportunity with like-minded colleagues provide another avenue of outreach and promotion.

Notwithstanding the Family Education Right to Privacy Act of 1970 (FERPA) , anecdotal evidence indicates that, for better or for worse, parents are directly contacting mathematics faculty more and more often about a young adult child's progress in a mathematics course [3]. According to FERPA, before an employee of education institution can discuss a student's status with anyone apart from the student, that student must sign a waiver granting an institution representative permission to discuss the student's status or progress with a given individual, typically a well-intentioned parent. Proactively addressing the increase in mathematics faculty and parent interactions, a QMaSC director can schedule a presentation to both students and parents at first year student and parent orientation programs held before the onslaught of fall semester. New student orientation sessions can provide another means of directing students and alerting parents to academic assistance available in a QMaSC before the semester gets underway. Especially refreshing to first year student parents at this time is learning that QMaSC services are offered free of cost.

Academic advisors across campus generally need to be apprised of the location, schedule, and policies of a QMaSC. Inviting these stakeholders to meet periodically is an effective means of promoting an institution's QMaSC. Currently many institutions nationwide are undertaking course redesign in lower-division first-year mathematics courses. At the time of redesign, often campus academic advisors need information not only about course placement, but also about the policies of a redesigned course to better advise students. A QMaSC director meeting and interacting with academic advisors at this juncture provides answers to advisors' multitude of questions about courses and the QMaSC's mission. Inviting advisors to the campus QMaSC familiarizes advisors with the physical location, mission, and policies of this campus asset. If a face-to-face meeting with advisors is not feasible, sending an email or newsletter with schedules, resources, links to policies or procedures, template emails for advisors to send to struggling advisees, and a color picture of the center can help make this connection.

Finally, at times when campuses nationwide are faced with budget constraints, a director will eventually be required to promote the existence of a QMaSC to a campus administration or budget review committee, likely requiring data to justify funding for a QMaSC. Thus, QMaSC directors are advised to maintain records over time of student usage patterns. Data collection usually comprises some subset of the following: student user name, ID number, course, instructor, date, time in/out, and QMaSC staff contact. Information may be collected via paper/pencil, swiping a student ID card, or electronic sign-in. In addition to usage data, a director may wish to collect attitudinal data through surveys or a complaint/compliment box, again either traditionally or electronically. In addition to promotion efforts, student usage data can be used to improve QMaSC customer service and satisfaction.

5 Promotion and Outreach in Even Wider Circles: Traditional and On-line

Networking with staff and colleagues across campus is perhaps the single most important way to attract and funnel students to a QMaSC. However, distributing newsletters, posters or handbills and an on-line presence provide an efficient means for students to access scheduling information. Key to effective newsletters, posters, handbills, or websites, is an economy of, but welcoming, language, and a schedule of hours. Enlisting the aid of an on-campus graphic design student or group in need of a project can result in an attractive, inviting poster, newsletter, handbill, or website.

Many post-secondary institutions' QMaSCs have an on-line presence as evidenced by a quick Google search of "Mathematics Tutoring Lab." All on-line websites for QMaSCs include hours of operation; some websites include names of tutors often with their particular mathematics expertise (statistics, differential equations, etc.). Several websites include information about group tutoring sessions or mini-lectures, access to on-line mathematics practice software, and links to old exams. Some websites have an on-line application for a student to schedule an appointment directly with a lab tutor. Some have information about private tutors and their fees in addition to the free-of-charge

QMaSC. Some websites include links to other familiar mathematical websites (The Mathematics Forum with Ask Dr. Math, Cut-the-Knot.org, The History of Mathematics, and others).

Less common among these websites, but worth the consideration of QMaSC director implementing an online presence, is the promotion of responsible learning. Naïve students desperate for mathematics help too often arrive at QMaSCs with unrealistic goals, leading to frustration for themselves and their tutors. Proactive directors promoting QMaSC services should include a list of what to expect and tips for a successful tutoring session which can serve to assist in the academic maturation of the learner. To address some common issues, more than one website includes a warning about how busy a lab can be, suggesting that students "plan for success" and manage study time leading up to an exam. All websites should offer tips about a successful tutoring session, including reminders to a prospective tutee to bring both learning tools (book, notes, paper, writing implement) and to come with specific questions. All of these tips can also be made available in hard copy within the QMaSC.

A cursory tour of various websites yields some singular suggestions for a QMaSC director contemplating an online presence. One website has a YouTube video of a walk-through tour of the lab. The short tour includes the physical space in addition to some short out-takes of one-on-one tutor student conversation and groups of students working together. This is a thoughtful, welcoming gesture, providing viewer a virtual glimpse of what to expect at a lab without raising anxiety by requiring any commitment on the part of a curious student. Another QMaSC offers Skype conferencing with tutors.

6 Full Circle

The paramount goal of a QMaSC director is to foster and promote a welcoming, threat-free learning community. A quality learning experience includes the provision of a learning space and of resources within that space that facilitate student and peer interaction. Such a thriving learning community is developed over time through tireless outreach, networking, and promotion.

7 Bibliography

[1] U.S. Department of Education, Office of Vocational and Adult Education, "Strengthening mathematics skills at the postsecondary level: Literature review and analysis." `www.eric.ed.gov`, 2005.

[2] Y. Solomon, T. Croft, and D. Lawson, "Safety in numbers: Mathematics support centres and their derivatives as social learning spaces," *Studies in Higher Education*, vol. 35, no. 4, pp. 421–431, 2010.

[3] U.S. Department of Education, Family Policy Compliance Office, "Family Educational Rights and Privacy Act (FERPA)." http://www.ed.gov/policy/gen/guid/fpco/ferpa/index.html, 2013.

Building Effective Quantitative and Math Faculty Development

© Matthew Salomone, Kathryn Bjorge,
Bridgewater State University

1 Introduction

Faculty development — in the form of elective programs to encourage effective teaching — can transform a campus from within, and quantitative and math support centers (QMaSCs) are well positioned to lead such programs. In order to have the most positive influence, these programs must be responsive to specific campus needs, sustain partnerships across organizational boundaries, and, over time, be assessed regularly for continuous improvement.

Support for student learning and student success is central to the mission statement of any QMaSC, and the staff of such support centers is likely to spend the majority of their time working directly with students. These students, however, may often report that it is not only the content of their course with which they struggle, but the student's preferred learning style and needs may not align with the style of instruction and level of support provided within their course. Students' experiences with math teachers have tremendous power to shape their attitudes toward mathematics in profound and persistent ways. Unfortunately, these experiences may affect students negatively as well as positively. The effectiveness of teaching styles has been shown to be a factor in the development of math anxiety, which hampers student learning and fosters innumeracy at all levels of education [1].

Suggested Citation: Matthew Salomone, Kathryn Bjorge (2016), "Building Effective Quantitative and Math Faculty Development", http://dx.doi.org/10.5038/9780977674435.ch10 in G. Coulombe, M. O'Neill, M. Schuckers (Eds.) *A Handbook for Directors of Quantitative and Mathematical Support Centers*, Neck Quill Press, http://scholarcommons.usf.edu/qmasc_handbook.

This material is based upon work supported, in part, by the National Science Foundation under Grant DUE-1255945. Any opinions, findings, and conclusions or recommendations expressed in this material are those of the author(s) and do not necessarily reflect the views of the National Science Foundation

Matthew Salomone is the corresponding author for this chapter.

Support centers can facilitate the use of more effective methods and manners of instruction in the classroom not only by providing a positive environment for tutoring *and* supplemental instruction, but more indirectly, through investing in both student and faculty excellence. Through faculty collaboration and provision of professional development opportunities, support centers can see their investment multiply, as each instructor they reach may in turn reach dozens more students. Centers that participate in faculty development may also find benefits from being better integrated into their campus community and culture, because they provide valuable service to both the student affairs and academic affairs divisions of their institution.

As with effective teaching approaches for students, faculty development programs can vary in their effectiveness. At best, faculty development programs can be transformative for both faculty members and their students, succeeding by both drawing upon and reinforcing positive faculty culture and institutional values [2]. Strong faculty development programs, like strong curricula, are outcome-driven and research-based. They have their own cycle of assessment and are valuable parts of their participants' self-assessment and professional growth processes. They seek partnerships, both horizontally across academic departments and vertically across instructional class, supporting senior and junior faculty, part-time lecturers, graduate teaching assistants, and undergraduate tutors and mentors. Faculty development programs foster a culture of continuous improvement among their participants and use campus-wide events to issue challenges and celebrate achievements.

Above all, though it may take many different forms, successful faculty development is intentional, even scientific. It proceeds from needs to hypotheses, experiments, conclusions, and reflection. It happens most effectively within a community of faculty, staff, and students, in which to "build the commonalities and connections so essential to our education and our society" [3, p. 92]

2 What is Faculty Development?

Faculty members hold multifaceted jobs, and so the term "faculty development" suffers from a multiplicity of definitions [4]. Most commonly, "faculty development" refers to programs that focus on the role of faculty as teachers. These programs provide support for faculty designing course syllabi and assignments, evaluating their students' learning, improving their classroom management and practice, and improving their students' experience. Other faculty development programming focuses on the role of faculty as scholars, providing support for their scholarly writing, grant writing, and administration activities. "Teacher-scholar" programs that target the intersection of these two primary faculty roles can be particularly effective; for example, encouraging faculty to bring their scholarly interests into the classroom provides for both more authentic teaching and more inclusive scholarship. Finally, faculty development programming may also focus on supporting the individual's personal professional development, wellness, and leadership, all of which can help to transform faculty's service roles on campus from onerous to effective. The particular focus of faculty development in an institution is driven largely by institutional culture and associated contexts, such as those that emerge during a needs assessment.

Whatever faculty development is, however, it is crucial to understand what it is not. Faculty development is formative, not summative: it is an elective process of a faculty member's personal professional growth. It is not a job requirement or a part of the evaluative process by which promotion and tenure is granted. As such, faculty development programs are typically not coordinated by those faculty or administrators with direct supervision over the participants. Instead, it is initiated and supported by campus teaching and learning centers who maintain participants' confidentiality [5]. Teaching and learning center staff, replete with expertise in good teaching practice, are invaluable partners in the establishment of teaching-focused faculty development programs, on which this article shall focus.

In addition to the organizational and pedagogical support provided by a teaching and learning center, a strong faculty development program requires an active facilitator to coordinate its efforts. The QMaSC director could be well positioned to facilitate such a program, particularly if he or she is also well connected with the faculty that would participate. Because support centers occupy a unique position between student affairs and academic affairs, this affords the QMaSC access to both the unvarnished perspective of struggling students and the faculty who work with these students on a daily basis.

The kinds of events and programs that faculty development offers to its participants can also vary. Offering public presentations or colloquia on teaching practices is a natural first step. Just as with students, however, faculty members learn best through active engagement with facilitators and with one another. Faculty development programs that offer targeted skill or subject workshops, intensive teaching institutes, mentorship programs and open classrooms, and teaching consultations have been shown to be most effective in both supporting good teaching and improving faculty culture [6].

Due to its elective nature, faculty development programs sometimes find it necessary to incentivize participation either in a tangible fashion with stipends, teaching release, internal grants, or professional memberships, or in a less tangible fashion by helping participants to showcase the results of their participation in their academic portfolios. Whether or not these external incentives exist, a faculty development program's success and sustainability depends on its ability to draw on the felt significance of both its participants and its institution. Locating this significance — defining an answer to "what's in it for me?" for both the faculty that participate and the administrators that support faculty development is the crucial job of a needs assessment.

3 Why Faculty Development? Defining a Scope and Assessing Needs

How will a quantitative or mathematical faculty development program benefit faculty, students, and the institution? Establishing specific rationales for programs will serve both to recruit the right coordinators and participants and justify its continued existence to administrators who support it. Moreover, effective academic development is academic in approach, grounded in research, and driven by data [7]. Motivation for better teaching, and for the role of a support center in helping to bring it about, can come from both the perspectives of the academic affairs and the student affairs divisions of the institution. These motivations can inform both initial and ongoing assessments of campus needs, and progress toward meeting them.

What Counts? Supporting Students' Academic and Social Development

As with everything a QMaSC does, the ultimate goal of any faculty development program is to improve the experience of students at the institution. Fortunately (and unfortunately), students' college lives are sufficiently multi-faceted that their experience may be affected in a wide variety of ways. Effective faculty development, then, is focused so as to do the most good for the most students.

Among the most obvious targets for faculty development are the academic lives of students. Outwardly, these manifest as course grades and GPA, persistence and graduation rates, and post-graduate and workforce placements. Inwardly, these are reflected in students' acquisition of particular skills and development in higher-order thinking. These are the aspects of students' college lives with which faculty members are most intimately familiar, and with which their superiors — from department chairs to deans and provosts — are primarily concerned. Even so, they make up only one piece of the student success puzzle.

Students are most likely to succeed academically when their environment is conducive to success, and this environment includes not only the academic milieu but also the social. How engaged they are in their education, their level of access to supportive educational experiences, their integration into the campus community, and their one-on-one connections with faculty members all correlate positively with students' academic achievement. Student affairs divisions are primarily concerned with these aspects of students' lives, and since they are the aspects with which faculty are least familiar, student affairs professionals are a valuable resource in faculty development.

A truly student-centered approach to faculty development addresses each phase of, and each transition within, students' college experience, from input factors (what are students' backgrounds and educational goals?) to environmental factors (how and how well are they meeting those goals while in college?) to output factors (where do students go when they leave?). Measuring student outcomes against these goals, and the mission of the institution, is essential [8] and seminars that acquaint faculty with their students' and institution's particular goals have been shown to maximize faculty involvement in development [9].

Case Study: "Streams" Retention Enhancement[10] (Bridgewater State University, 2012)	
Identified Need:	Students leaving their math/science majors due to high rates of attrition in introductory courses
Key Metric:	Failure rates in introductory courses, initially 30-35%
Initiative:	Structured learning assistance (SLA), assigning a dedicated peer tutor to each introductory course section and creating required small-group tutorial sessions
Faculty Development:	Train faculty to design tutorial activities integrated with lecture material and to expect students to make the most use of their peer assistant
Support Center Role:	Recruit, hire, train, and oversee peer SLA assistants
Result:	Failure rates in supported courses decreased to 15-25%

Student success can be challenged most during transitions into, within, and out of college. Transition points such as students' declaration of major, enrollment in their first college math course, their first failing exam grade or course grade, and their beginning upper-level coursework, represent crucial moments in which a students' situation, self, support, and strategy are all tested [11]. QMaSCs provide an important network of peers, faculty, and resources that can provide more, and deeper, points of contact for struggling students. Moreover, faculty development is a valuable tool for reinforcing these contacts, creating shared purpose and message, and giving faculty strategies to help their students cope with difficult transitions.

Effective faculty development, then, must not miss either the academic or the student-support perspective, and the institutions with the most successful students actively attend to both aspects of their students' college experience [12]. Collaboration that includes both faculty and student affairs staff affords the best opportunity for holistic, student-centered programming enriching both students' academic and social lives.

What Can We Accomplish? Mission- and Needs-Driven Faculty Development

Effective (and sustainable) faculty development is reflective of its institution's mission, and responsive to its campus' needs. Formulating both long- and short-term goals and intended outcomes for a program will help demonstrate its effectiveness, and justify its continued existence. In this, a quantitative and math support center – of all campus units! – must be driven by data. The choice of one or two key measures of student success to address through faculty development, and track over time, is vital to a focused and impactful program.

Measures of student success come in many forms, and most institutions already collect a wide variety of them. Academic measures, which are typically collected by institutional research departments, may include the success rates of students in particular courses; persistence, retention, and graduation rates; outcome- or exam-specific assessments; or academic portfolios. Such data are often used in departmental program evaluations, and department chairs and deans are responsive to them. Due to their visibility among faculty, these are often the first measures of student success which faculty development seeks to target.

Students' social support factors, while less obvious to measure, are no less important to their success, as discussed above. There are several widely-used instruments to measure these factors, already in place on many campuses. On the input side, the CIRP Freshman Survey is a valuable tool for acquainting a campus with its incoming students' backgrounds, expectations, and behaviors [13]. The environment can be measured using instruments such as the Inventory for Student Engagement and Success (ISES) and the National Survey of Student Engagement (NSSE). These assess, respectively, the conditions present for student success on a campus and the extent to which students engage and thrive in those conditions [14, 15]. Output measures, such as exit surveys and graduating senior focus groups give students the ability to look back upon their whole college experience and provide feedback on the experiences they found most significant to their success.

The particular metrics chosen to assess a campus' needs and gauge the impact of a faculty development program will also have the effect of defining its future participants. For example, a program whose goal is to reduce the failure rate in a particular calculus course is likely to invite participation from faculty who teach pre-calculus and calculus courses, while a program whose goal is to strengthen the numeracy content of the liberal-arts curriculum as a whole is likely to invite participation from faculty across departments. Each choice comes with a unique set of opportunities and challenges.

Development Programs for Mathematics Instructors

Programs focused specifically on mathematics courses often have the luxury of taking their participants' content knowledge as given, which enables participants to focus on the pedagogy of syllabus and assignment design, classroom practice, and student mentorship and support. The Mathematical Association of America regularly shares articles on best practices in teaching undergraduate mathematics and activities for faculty development [16]. The MAA also coordinates a special interest group around research in undergraduate mathematics education. The Math Teachers Circle model, in which in-service teachers gather regularly for problem-solving and enrichment activities, may also be useful in engaging college math faculty in developing classroom strategies [17].

Faculty development focused on achievement and retention of underrepresented students in mathematics courses have been shown to be successful in improving these students' outcomes [18]. Inclusive teaching methods, such as the incorporation of active and project-based learning, feature heavily in these programs and are especially effective in increasing the engagement and performance of diverse populations of students.

Development Programs Across Disciplines ("QuAC")

Programs aimed more broadly at strengthening quantitative reasoning in liberal-arts core curriculum courses will need to engage all faculty who teach those courses, thereby establishing a "conspiracy" across disciplines [19]. Both the MAA and the Association of American Colleges & Universities [20] have developed educational standards for cross-curricular quantitative reasoning. Likewise, the National Numeracy Network [21] strongly supports numeracy-oriented faculty development through their organization's journal *Numeracy*. "Quantity Across the Curriculum" (QuAC) programs have begun to do for the basic skill of numeracy what similar Writing Across the Curriculum (WAC) programs have succeeded in doing for student writing in the postsecondary curriculum. In fact, QuAC programs have found success emulating and partnering with established WAC programs to reach broader segments of their institutions [22].

Numeracy-oriented faculty development across disciplines entails a unique set of challenges. Faculty in disciplines that are not traditionally quantitative often do not know how to incorporate numeracy content in their teaching. Moreover, some such faculty struggle with their own math anxiety in a similar way as pre-service elementary educators do [23], and in the U.S., even highly educated adults struggle with numeracy skills [24]. Assisting faculty in locating "sophisticated reasoning using elementary mathematics" in their discipline [25] is important to place numeracy into the context of their teaching, in order that they can connect their students with those contexts in turn.

After the institutional needs and objectives of a faculty development program have been identified and made public, and a group of participants have come forward, the work of building a learning community begins.

4 What's It Look Like? Learning and Partnerships

Because they are detached from formal evaluative processes, faculty development programs often forgo hierarchical organization and thrive in small, decentralized groups. Where the need for these programs is less urgent, teaching circles, book clubs, "brown bag" lunch groups, and seminar series, may be sufficient to carry on the work. Where specific, urgent needs, and a timeline in which to meet them, exist, these informal structures may not be insufficient. In cases that call for more intensive, focused work, faculty learning communities (FLCs) have been shown to be effective platforms for achieving program goals.

Faculty Learning Communities

Faculty learning communities are focused versions of "communities of practice," groups of practitioners bound together by shared expertise, interest, and purpose [26]. In an FLC, a relatively small group of 4–10 faculty, instructors, and/or staff engage in an active, often year-long, collaboration. Frequent activities, workshops, and seminars are held, designed to accomplish a well-defined teaching and learning development goal. Participants in an FLC "determine both their individual and collective outcomes and ways they would achieve them" [27] and may share the responsibility of facilitating the group [28]. The establishment, facilitation, and assessment of FLCs have been extensively documented [29]. Due to their size, FLCs make smaller the daunting task of launching a new faculty development initiative and can foster close interpersonal relationships between their participants on campuses of any size. Over time, FLCs can develop into networks of support across campus that foster not only faculty and student learning, but organizational learning and progress.

There is also evidence to support the inclusion of undergraduate students, and especially undergraduate tutors, as participants in a faculty learning community. Not only does this enfranchise students in the "co-creation" of teaching approaches and course design, which is of particular value to students who plan on careers in education, but the collaboration often uncovers new insights and encourages shared responsibility for learning [30, 31].

Integrating Tutor Training and Faculty Development

One innovative approach to faculty development is to offer workshops that bring together tutors and instructors. QMaSCs with well-established programs for tutor training may find it useful to build faculty development atop their training program. The goals of tutor training programs — for instance, to increase student engagement in learning activities and help students to develop positive study habits — are goals both tutors and instructors can help to address in complementary ways [32].

"Vertically-integrated" programs for improving both teaching and tutoring are in some ways analogous to corresponding vertically-integrated programs for undergraduate, graduate, and postdoctoral research, such as the National Science Foundation's Vertical Integration in Graduate Research and Education (VIGRE) program, and the benefits of shared expertise and cross-difference dialogue are similar.

Case Study: "Beyond the Formula" Combined Tutor/Faculty Workshop [33]	
Objective:	Give faculty and tutors tools to encourage students' quantitative reasoning using non-algebraic problem solving techniques
Format:	Half-day workshop, using a combination of presentations and small-group inquiry activities; each group included both tutors and faculty
Attendance:	8 faculty/staff members; 20 undergraduate tutors

5 Is It Working? Assessment for Continuous Improvement

Faculty development programs can require substantial investments of institutional funding and faculty time to sustain. Thoughtful assessment of the program not only illustrates its value to administrators, it also highlights its own strengths and weaknesses to its facilitators and participants.

But what, and whom, should be assessed? The foremost answer to this question is linked to the needs identified when the program is established (see Section 2, above). If the mandate for a faculty development program comes from deficiency in a key metric, such as a calculus course pass rate, assessment can and should begin with this metric. However, one data point rarely paints a complete picture of the impact of a faculty development program, especially if it reflects its impact only on students. The program's impact on its faculty participants is equally deserving of measurement.

The domains of assessment for faculty learning communities [29] are applicable to other faculty development models as well. A program may assess the development of its faculty participants to demonstrate the program's reach, the efficacy of its individual program components to demonstrate its efficiency, and student learning in participants' courses to demonstrate its impact. Of these, only student learning may be assessed purely via institutional metrics. The first two must be assessed by the program itself, or by the teaching and learning center to which it reports.

Insofar as faculty development resembles other organizational training programs, its impact on participants and the effectiveness of its individual components may be evaluated on four levels [34].

1. **Reaction:** How well did participants enjoy their experience?
2. **Learning:** To what extent did participants acquire knowledge and skills?
3. **Behavior:** Did participants use this knowledge and skill to change their practice?
4. **Results:** Did participants' experiences lead to a tangible outcome?

These four levels need not be assessed with equal frequency. Participants' reaction and learning may be assessed at each workshop, institute, or event through brief exit surveys. These quick evaluations provide immediate feedback to facilitators that can help guide future programming on an ongoing basis. Assessments of participants' behavior and results are likely to be infrequent — biannual or annual — and more circumspect. For example, a key behavioral assessment tool for a teaching institute or learning community is the collection of its participants' pre- and post-workshop syllabi or the exhibition of novel assignments or pedagogical experiences they implemented as a result of their participation. Assessments of results may be tied to institutional metrics, but may also include student work or works of faculty scholarship.

Thorough assessment can also provide a faculty development program with its own best publicity. Not only is an effective program likely to secure continued funding from administrators, it is also likely to attract wider and more energetic participation from faculty.

6 Conclusions

QMaSCs are uniquely positioned on campus to facilitate dialogue and establish partnerships that lead to more effective teaching in quantitative and mathematics courses. Faculty development programs that assess and address specific departmental, student, or campus needs; sustain partnerships in focused learning communities that bring together participants from across boundaries of department and instructional class (including students); and engage in ongoing assessment have considerable influence in improving the experiences, effectiveness, and overall success of both students and faculty in the educational process [11].

7 Bibliography

[1] M. H. Ashcraft, "Math anxiety: Personal, educational, and cognitive consequences," *Current Directions in Psychological Science*, vol. 11, pp. 181–185, October 2002.

[2] A. E. Austin, "Faculty cultures, faculty values," *New Directions for Institutional Research*, vol. 68, pp. 61–74, 1990.

[3] F. Gabelnick, J. MacGregor, R. S. Matthews, and B. L. Smith, eds., *Learning Communities: Creating Connections Among Students, Faculty, and Disciplines*, vol. 6. San Francisco: Jossey-Bass, 1990.

[4] Professional and Organizational Development (POD) Network in Higher Education, "What Is Faculty Development?," 2013. Accessed from `http://podnetwork.org/about-us/` on June 28, 2016.

[5] C. E. Cook and M. D. Sorcinelli, "The value of a teaching center." Accessed from `http://podnetwork.org/about-us/what-is-faculty-development/the-value-of-a-teaching-center/` on June 1, 2013.

[6] W. C. Nelsen, *Renewal of the Teacher-Scholar: Faculty Development in the Liberal Arts College*. Washington, DC: Association of American Colleges, 1981.

[7] R. Miller, "Connecting beliefs with research on effective undergraduate education," *Peer Review*, vol. 11, pp. 4–8, Spring 2009.

[8] J. H. Schuh and M. Upcraft, *Assessment Practice in Student Affairs: An Application Manual*. San Francisco: Jossey-Bass, 2001.

[9] J. A. Centra, "Types of faculty development programs," *Journal of Higher Education*, vol. 49, no. 2, pp. 151–162.

[10] R. Matheson, "A Rich Stream," 2013. Accessed from http://bridgew.edu/news-events/news/rich-stream on June 1, 2013.

[11] N. J. Evans, *Student Development in College: Theory, Research, and Practice.* San Francisco: Jossey-Bass, 2009.

[12] G. D. Kuh, J. Kinzie, J. H. Schuh, and E. J. Whitt, *Student Success in College: Creating Conditions that Matter.* San Francisco, CA: Jossey-Bass, 2nd ed., 2010.

[13] J. H. Pryor, K. Eagan, L. P. Blake, S. Hurtado, J. Berdan, and M. H. Case, "The American Freshman: National Norms Fall 2012." Accessed from http://www.heri.ucla.edu/monographs/theamericanfreshman2012.pdf on June 1, 2013.

[14] National Survey of Student Engagement, "Promoting student learning and institutional improvement: lessons from NSSE at 13," tech. rep., Indiana University Center for Postsecondary Research, Bloomington, IN, 2013. Accessed from http://files.eric.ed.gov/fulltext/ED537442.pdf on June 1, 2013.

[15] A. S. Green, E. Jones, and S. Aloi, "An exploration of high-quality student affairs learning outcomes and assessment practices," *NASPA Journal*, vol. 45, no. 1, pp. 133–157, 2008.

[16] Mathematical Association of America, "Teaching and learning," 2013. Accessed from http://www.maa.org/programs/faculty-and-departments/curriculum-department-guidelines-recommendations/teaching-and-learning on June 28, 2016.

[17] T. Shubin, "Math circles for students and teachers," *Mathematics Competitions*, vol. 19, no. 2, pp. 60–67, 2006.

[18] J. Handelsman, D. Ebert-May, R. Beichner, P. Bruns, A. Chang, R. DeHaan, J. Gentile, S. Lauffer, J. Stewart, S. M. Tilghman, and W. B. Wood, "Scientific Teaching," *Science*, vol. 304, no. 5670, pp. 521–522, 2004.

[19] D. Hughes-Hallett, "Achieving numeracy: The challenge of implementation," in *Mathematics and Democracy: The Case for Quantitative Literacy* (L. Steen, ed.), pp. 93–98, Princeton, NJ: Woodrow Wilson National Fellowship Foundation, 2001.

[20] Association of American Colleges & Universities, "Liberal Education and America's Promise (LEAP)." Accessed from http://www.aacu.org/leap/ on June 1, 2013.

[21] National Numeracy Network, "Teaching resources." Accessed from http://serc.carleton.edu/nnn/teaching/index.html on June 1, 2013.

[22] C. Rutz and N. D. Grawe, "Pairing WAC and Quantitative Reasoning through Portfolio Assessment and Faculty Development," *Across the Disciplines*, vol. 6. Accessed from `http://wac.colostate.edu/atd/assessment/rutz_grawe.cfm` on June 1, 2013.

[23] G. Gresham, "A study of mathematics anxiety in pre-service teachers," *Early Childhood Education Journal*, vol. 35, no. 2, pp. 181–188, 2007.

[24] Organization for Economic Cooperation and Development, "OECD skills outlook 2013: First results from the survey of adult skills." Accessed from `http://www.oecd.org/skills/` on June 28, 2016.

[25] L. A. Steen, ed., *Achieving Quantitative Literacy.* Washington, DC: Mathematical Association of America, 2004.

[26] E. Wenger, *Communities of Practice: Learning, Meaning, and Identity.* Cambridge: Cambridge University Press, 1998.

[27] M. C. Petrone and L. Ortquist-Ahrens, "Facilitating faculty learning communities: A compact guide to creating change and inspiring community," *New Directions for Teaching and Learning*, vol. 97, pp. 63–69, 2004.

[28] L. Ortquist-Ahrens and R. Torosyan, "The role of the facilitator in faculty learning communities: Paving the way for growth, productivity, and collegiality," *Learning Communities Journal*, vol. 1, no. 1.

[29] M. D. Cox and L. Richlin, eds., *Building Faculty Learning Communities.* San Francisco: Jossey-Bass, 2004.

[30] C. Bovill, A. CookSather, and P. Felten, "Students as co-creators of teaching approaches, course design, and curricula: Implications for academic developers," *International Journal for Academic Development*, vol. 16, no. 2, pp. 133–145, 2011.

[31] T. M. Redd and C. E. Brown, "Using students to support faculty development," in *To Improve the Academy* (J. Miller and J. Groccia, eds.), vol. 29, pp. 233–245, San Francisco: Jossey-Bass, 2010.

[32] Utah State University, "Tutor training and certification course packet," 2011. Accessed from `http://sunykorea.ac.kr/sites/default/files/Utah%20State%20Training%20and%20Certification%20Course%20Packet.pdf` on June 28, 2016.

[33] M. Salomone, "Beyond the formula," 2012. Accessed from `http://webhost.bridgew.edu/msalomone/quac/beyondformula.pdf` on June 1, 2013.

[34] D. L. Kirkpatrick, *Evaluating Training Programs: The Four Levels.* San Francisco: Berrett-Koehler, 2nd ed., 1998.

[35] T. Kling, "Student retention enhancement across mathematics and science." Accessed from `http://microsites.bridgew.edu/streams` on March 8, 2014.

[36] M. Salomone, "Quantity across the curriculum." Accessed from `http://www.quacprogram.com` on March 8, 2014.

[37] M. Nathan and K. Koedinger, "An investigation of teachers beliefs of students algebra development," *Cognition and Instruction*, vol. 18, no. 2, pp. 209–237, 2000.

8 Acknowledgements

The authors wish to acknowledge Roben Torosyan and Ann Brunjes for their energetic support of faculty development across the curriculum at Bridgewater State University; Stacey Sheriff, now director of Colby College's Writing Program, for her partnership in launching and co-facilitating the Quantity Across the Curriculum faculty development initiative at BSU; and Grace Coulombe and the QMaSC team for putting together this valuable project.

9 Appendix

The following are materials used in the two faculty development case studies of this article: a faculty program to encourage high-impact educational practices in the sciences, and a combined faculty and student program to develop strategies for communication through formula.

"STREAMS" course development program [10, 35] (Bridgewater State University, 2012.)

Student Retention Enhancement Across Mathematics and Science (STREAMS) was a series of programs supported by the National Science Foundation's Science Talent Expansion Program (NSF-STEP). The primary goal of STREAMS was to increase the number of students graduating with a math or science degree.

STREAMS supported redesigning both curriculum and pedagogy in introductory courses with high attrition rates. As part of this effort, the faculty who coordinate and teach these courses received course development grants and attended a faculty development event. In its call for grant applications, STREAMS listed its expectations that grant recipients:

1. Introduce new course elements and commit to the implementation of these elements for several years,
2. Define or clarify course learning objectives,

3. Develop and implement an assessment plan for learning objectives,
4. Share their approach with faculty colleagues at a STREAMS workshop, and
5. Present their work at a campus-wide symposium.

In the application and in subsequent faculty development workshops, grant recipients were engaged in discussion around the following questions:

1. Describe the aspects of your current course that you believe are least conducive to student learning / achieving learning outcomes.
2. Describe the pedagogical changes you believe will best address those weaknesses.
3. What literature exists that is relevant to the changes you propose?
4. How will you know whether the changes are working? What is your assessment plan?

"Beyond the Formula" faculty/student quantitative reasoning workshop [33, 36]

As part of an interdisciplinary quantitative reasoning faculty development program, the workshop entitled "Beyond the Formula" was offered to both faculty and peer tutors. The workshop's objective was to develop techniques for conveying quantitative concepts and reasoning without relying upon formulas. The workshop was offered first as a half-day event; it was abbreviated to two hours when repeated in a subsequent semester.

The program for the half-day event included both presentations and interactive, inquiry-based discussions in mixed groups of faculty and tutors:

1. Discussion: The Trouble With Formula
   Objective: To challenge participants' assumptions about whether students are more successful reasoning within an applied context, or using a formula, using a study [37].
2. Presentation: Meaning and Context: Reclaiming Formula
   Objective: To introduce participants to their "expert blind spot" and connect formal (exact, algebraic) with informal (approximate, numerical) modes of reasoning.
3. Discussion: Telling Formula Stories in the Classroom
   Objective: To engage participants in developing a narrative around selected formulas to explain their "meaning."
4. Discussion: Peer Assisted Learning (PAL) in Formula-Rich Courses
   Objective: To invite student participants to reflect on their own tutoring practice, summarize takeaways from the workshops, and develop individual statements of teaching philosophy.

For all discussions, participants were provided with worksheets of prompts with space to write their responses. Handouts and resources from this and a variety of other quantitative reasoning faculty development workshops have been made available online [36].

Course Collaboration Models

© Eric Gaze,
Bowdoin College

1 Introduction

Quantitative and Mathematics Support Centers (QMaSCs) come in many different shapes and sizes and the services they provide are just as variable. Ranging from the most basic model of drop-in tutoring help for certain introductory math classes, to teaching and learning centers that provide assistance to course faculty at multiple points throughout the semester and in multiple disciplines. Options for course collaboration can include:

- Course placement and advising based on assessment administered and evaluated by QMaSC directors.
- Just-in-time course support with drop-in tutoring and online tutorials, including support for software packages (Mathematica, SPSS, Excel, etc.).
- Creation of ancillary materials to support and scaffold background quantitative and mathematical knowledge.
- Ongoing structured support in study groups and supplemental instruction including targeted workshops.
- Expert support of quantitative projects in non-quantitative fields, including in-class workshops and presentations on writing with numbers and database navigation.
- Follow-up assessment to measure outcomes at the course and institution level.

Suggested Citation: Eric Gaze (2016), "Course Collaboration Models", http://dx.doi.org/10.5038/9780977674435.ch11 in G. Coulombe, M. O'Neill, M. Schuckers (Eds.) *A Handbook for Directors of Quantitative and Mathematical Support Centers*, Neck Quill Press, http://scholarcommons.usf.edu/qmasc_handbook.

This material is based upon work supported, in part, by the National Science Foundation under Grant DUE-1255945. Any opinions, findings, and conclusions or recommendations expressed in this material are those of the author(s) and do not necessarily reflect the views of the National Science Foundation

2 Assessment

The topic of assessment might not seem to belong in a chapter on course collaboration models, but more and more QMaSCs are involved with the administration of quantitative and mathematical assessments and the subsequent evaluation of collected data. Many QMaSC directors share this information with faculty and advisors, and often participate in the accurate placement of students in courses. Likewise, directors funnel students with weak incoming skills into targeted support opportunities. Diagnostic tools include made-to-order instruments such as Accuplacer, and in-house creations that seek to assess algebra skills, calculus readiness, and overall quantitative reasoning (QR) abilities. Algebra skills and calculus readiness tests are straightforward to develop and are utilized at institutions of all sizes to accurately place students in appropriate algebra, pre-calculus and calculus courses. Relying on high school GPA is problematic, as variability in rigor in high schools throughout the country can lead to significant differences in interpreting high school preparation. Community colleges struggle to place students in the current "layer-cake of algebra dominated mathematics" [1]. As of Fall 2010, the developmental math programs at 2-year colleges centered upon algebra, in particular, 61% of all math enrollments consist of some form of algebra. This statistic is even higher given that 30% of 2-year schools have their pre-college level math programs offered outside of the math department in developmental (remedial) programs: Table 1: 2-year course enrollments, *note that 30% of developmental math programs at 2-year schools are outside of math departments and thus not listed here [2].

Table 1: 2-year course enrollments, *note that 30% of developmental math programs at 2-year schools are outside of math departments and thus not listed here (CBMS 2010) [2]

Two-year College Course Enrollments (in 1000's) In Math* Courses					
	Pre-College Level				
	1990	**1995**	**2000**	**2005**	**2010**
Pre-Algebra	45	91	87	137	226
Elementary Algebra (HS level)	262	304	292	380	428
Intermediate Algebra (HS level)	261	263	255	336	344
	Pre-Calculus Level				
College Algebra	153	186	173	206	230
College Algebra + Trigonometry	18	17	16	14	11
Total	**1272**	**1425**	**1347**	**1696**	**2024**

Theoretically, the justification for so much algebra is to prepare students for Calculus, despite the fact that quantitative reasoning (QR) courses are steadily increasing in number and attracting more and more students. Enrollments in Math for the Liberal Arts and Finite Math courses rose 63% from 1995 to 2010; whereas mainstream Calculus I enrollments rose only 20% over the same period [2]. The total enrollment in these two general education math courses exceeded mainstream

Calculus I enrollments 318,000 to 300,000 in fall 2010:

Table 2: 2-year course enrollments, *note that 30% of developmental math programs at 2-year schools are outside of math departments and thus not listed here (College Board of Mathematical Sciences (CBMS) 2010) [2]

College Board of Mathematical Sciences 2010 Survey (Enrollments in 1,000's)				
	1995	**2000**	**2005**	**2010**
Finite Math (2-year)	**24**	**19**	**22**	**18**
Liberal Arts Math (2-year)	**38**	**43**	**59**	**91**
Calculus I (2-year)	58	53	51	65
Finite Math (4-year)	**59**	**82**	**94**	**62**
Liberal Arts Math (4-year)	**74**	**86**	**123**	**147**
Calculus I (4-year)	192	192	201	235

Similarly, enrollments in *Liberal Arts Math* courses, those most often designated as QR, went up 113% over this period. This movement of "math for everyone else" is in part a recognition that the majority of college students will not major in a Science, Technology, Engineering, or Math (STEM) field. Of the 1,303,050 college-ready students in Fall 2005 (from the 4,012,770 cohort of 2001 9th-graders) only 277,550 went on to major in STEM fields, and only 166,530 graduated with a STEM degree in May 2011 [7]. This illustrates why institutions are providing an increasing number of QR course offerings to their students. Colleges and universities strive to provide a mathematical experience that will directly benefit their students' QR abilities. This new emphasis requires assessment of both traditional algebra and Calculus readiness alongside assessment of QR. The latter is more problematic for QMaSC directors because QR is relatively new on the academic landscape and traditional college-readiness tests, like the SAT, are not designed for QR. The Mathematical Association of America's Special Interest Group in Quantitative Literacy (SIGMAA-QL) conducted a survey in 2009 of 1,554 institutions and found that while 87% of respondents had some form of quantitative requirement [8] only about 23% of them have pre-post assessment of quantitative literacy and reasoning (QLR) skills. Most of these internal assessment tools have no national norms with which to compare. Furthermore, the actual construct of "quantitatively literate" remains undeveloped [9]. A QLR test is available through a National Science Foundation (NSF) funded project, QLR Assessment (QLRA), in the Transforming Undergraduate Education in STEM (TUES) program. The QLRA project created a non-proprietary nationally normed, valid, reliable QLR test[1] , which was piloted at over 30 institutions across the country from 2-year schools to public universities and liberal arts colleges. To learn more about this assessment, contact the

[1] `http://serc.carleton.edu/qlra/index.html`

principal investigator, Eric Gaze (egaze@bowdoin.edu). The 2012 pilot data for this 23 question multiple-choice test shows a nice distribution with no ceiling or floor effects:

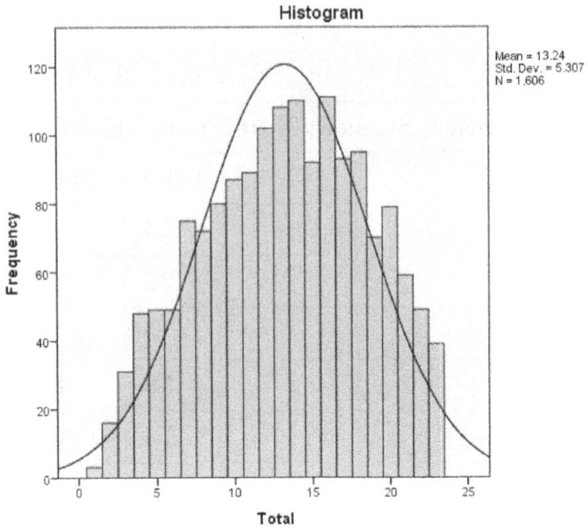

Figure 1: QLRA 2012 test data

Bowdoin College administers a QR test to all incoming students (initiated in 1995) and uses the scores for placement and advising purposes in all math and science courses. At Bowdoin, a student's Q-score is among the best predictors of GPA with the following correlation coefficients derived from 6 years of data (N $\sim$ 3,000):

- R = 0.39 Q-score vs. Overall GPA
- R = 0.48 Q-score vs. Math/Science GPA
- R = 0.48 Q-score vs. First year overall GPA

Accurate placement in math/science coursework is an incredibly important part of the QMaSC director's job. Before directors of QMaSCs can collaborate with the courses themselves, it is crucial that students be placed into the correct level giving them the best chance of a successful start. QMaSC directors are often concerned mainly with Calculus placements and the QLRA test may be a valuable part of this advising process, especially for students with no AP scores.

3 Tutoring and Study Groups

Once students have been placed into an appropriate math or science course, QMaSC directors will often develop and oversee a broad range of ongoing academic support activities. Drop-in tutoring for quantitative and mathematics courses is the most obvious way in which a QMaSC can support instruction. Directors and student tutors hold "office hours" in a dedicated tutoring space.

Students looking for help can either sign up for a block of time to work one-on-one with a tutor, or they can just drop-in to work collaboratively with tutors and other students. Course specific study groups, coordinated by the QMaSC director, provide another option for students seeking assistance. Working closely with course faculty, the director trains and assigns student tutors to lead the study groups. Study group leaders meet their assigned class at the start of the semester to introduce themselves and explain the purpose and value of attending study groups. The leader emphasizes the importance of forming regular study habits, developing academic peer support structures, and working productively under the guidance of trained peer tutors. Individualized attention is also available at many institutions with QMaSC directors coordinating individual tutorials. These arrangements for one-on-one tutoring provide students with a more private experience, which is especially helpful in cases of students with severe math anxiety or learning differences. QMaSC directors can also work with faculty to develop tutorials on specific math concepts for students in math and science courses. These tutorials can take the form of worksheets, workbooks, or online modules. The following list provides a subset of available resources on the internet that QMaSC directors may find useful in an attempt to avoid reinventing the wheel:

- Khan Academy: Definitely the giant in this arena and has expanded beyond simple math to almost all disciplines. `https://www.khanacademy.org/`
- Center of Math Organization: Great resource with lots of tutorials and digital materials. `http://www.centerofmath.org/`
- Hippocampus: Nice collection of different resources including video tutorials. `http://www.hippocampus.org/?gclid=CKK18vzPh7oCFUWd4AodxG8AZA`
- Cool Math: The name says it all, you can be the judge. `http://www.coolmath.com/algebra/`
- Math Planet: Videos and problems available. `http://www.mathplanet.com/`.
- Calculus Help: Fun problems in addition to online tutorials. `http://www.calculus-help.com/`

These "just-in-time" activities free up course faculty from spending valuable class time on review material. QMaSC directors can also create tailored assessments for course specific skills to identify weaknesses among students taking the course.

4 Workshops

QMaSC directors may find workshops to be particularly useful in providing supplemental support. The instructor of a science course that explores exponential growth can ask for a workshop on this topic led by either the director or a trained student tutor. Discipline specific content can also be reviewed using a similar format but led by student tutors who have previously and successfully completed the course. Workshops on the basics of logic or proof by induction can be offered for students in a Foundations of Mathematics course. In addition, a series of workshops can be designed to provide technological support for software packages being used in courses. For example, statistics instructors can request workshops on the use of R while mathematics instructors can request supplemental instruction in typesetting using LaTeX or in using programs such as Mathematica, Maple, MatLab, or SPSS. Likewise, instructors of humanities courses that use databases can request that the QMaSC director lead a class or workshop on working with data or writing with numbers. Follow-up instruction can then take place as needed using the QMaSC tutoring resources.

See the appendices for detailed examples of a calculus workshop designed for a physics class and a faculty development workshop on reasoning from evidence. These examples might be particularly useful for QMaSC directors when thinking about creating their own materials.

5 Conclusion

QMaSCs are more than just a place where students can go to get help factoring quadratics and integrating trigonometric functions. QMaSC directors are involved throughout the semester helping to place students in appropriate courses using well-developed assessment tools and providing supplemental instruction and tutoring to assist students as the course progresses. QMaSC directors can assist faculty with identifying weaknesses in students' background knowledge and coordinating resources to address such deficiencies. QMaSC directors can lead the campus community in an effort to incorporate QR across the curriculum, raising awareness for the value and utility of reasoning from all forms of evidence, including quantitative information. In the process QMaSC directors positively impact all shareholders on campus, from faculty and students in math and quantitative courses to the students who provide support as tutors and study group leaders.

6 Bibliography

[1] J. M. Shaughnessy, "Endless algebra the deadly pathway from high school mathematics to the college mathematics," February 2011. Accessed from `http://www.nctm.org/News-and-Calendar/Messages-from-the-President/Archive/J_-Michael-Shaughnessy/Endless-Algebra%E2%80%94the-Deadly-Pathway-from-High-School-Mathematics-to-College-Mathematics/` on June 28, 2016.

[2] Conference Board of Mathematical Sciences, "Statistical Abstract of Undergraduate Programs in the Mathematical Sciences in the United States," 2013.

[3] Curriculum Renewal Across the First Two Years MAA/CUPM Subcommittee of the Committee on the Undergraduate Program in Mathematics (CUPM), "College Algebra Guidelines," 2007. Accessed from `http://www.maa.org/cupm/crafty/` on June 11, 2013.

[4] B. Madison and L. A. Steen, "Evolution of numeracy and the national numeracy network," *Numeracy*, vol. 1, no. 1, 2008. Accessed from `http://services.bepress.com/numeracy/vol1/iss1/art2` on June 11, 2013.

[5] National Council of Teachers of Mathematics. Reston, VA: National Council of Teachers of Mathematics, 1989.

[6] L. Steen, *Why numbers count: Quantitative Literacy for Tomorrow's America.* New York, NY: The College Board, 1997.

[7] D. Kaplan, "The STEM Challenge," *Fortune*, June 14, 2010.

[8] M. Schield, "Quantitative graduation requirements at us four year colleges," 2010. Presented to the SIGMAA-QL board at the 2010 Joint Math Meetings.

[9] J. Bookman, S. Ganter, and R. Morgan, "Developing assessment methodologies for quantitative literacy: A formative study," *American Mathematical Monthly*, vol. 115, no. 10, pp. 911–929, 2008.

7 Appendix

A. Calculus Workshop Designed for Use in Physics Course

The following was created for a workshop on calculus in a physics class. The professor wanted to make sure students understood the "kinematic chain," an expression that is unfamiliar to many mathematicians. Kinematic chain refers to the relationship between the position, velocity and acceleration functions. In particular, students in the physics class were expected to be able to work backwards from graphs of acceleration to velocity and position. Discipline specific vocabulary and notation is often a huge stumbling block for students in transferring their knowledge in a math class to an applied domain. This workshop starts by reviewing basic concepts of the derivative and then moves on to discussing the kinematic chain, first using notation from a calculus class then comparing with the notation from physics.

Calculus Workshop on Kinematic Chain for Physics Class

There is a fundamental relationship between the signs of the 1st and 2nd derivatives and the shape of the function. The function is **increasing/decreasing** when the 1st derivative is positive/negative, and **concave up/down** when the 2nd derivative is positive/negative. **Extrema** occur when the 1st derivative changes sign (zeroes) and **inflection points** occur when the 2nd derivative changes sign.

Intervals	...-5	-5 ... 2	2 ... 18	18 ... 34	34...57	57 ...
1st derivative $f'(x)$ or dy/dx	Positive (+)	Negative(-)	Negative(-)	Positive(+)	Positive (+)	Negative(-)
2nd derivative $f''(x)$ or d^2y/dx^2	Negative (-)	Negative(-)	Positive (+)	Positive (+)	Negative (-)	Negative (-)
Original function $f(x)$	Increasing	Decreasing	Decreasing	Increasing	Increasing	Decreasing
	Max		Min		Min	
		Inflection Point		Inflection Point		

When applying this table to graphs of position $x(t)$, velocity $v(t)$, and acceleration a(t) the following guidelines will help:

1. A function is CONSTANT/HORIZONTAL ($\leftrightarrow$) derivative is ZERO.
2. A function is DECREASING $\leftrightarrow$ derivative is NEGATIVE.
3. A function is INCREASING $\leftrightarrow$ derivative is POSITIVE.
4. A function is QUADRATIC $\leftrightarrow$ derivative is LINEAR.
5. A function is LINEAR $\leftrightarrow$ derivative is CONSTANT/HORIZONTAL.
6. A function is CONCAVE UP/DOWN $\leftrightarrow$ derivative is INCREASING/DECREASING

Thus for the following graph of the acceleration we can reconstruct the velocity as follows.

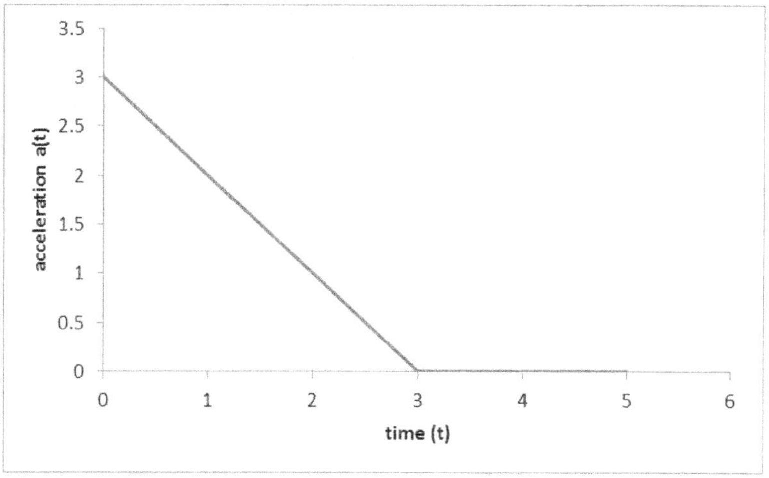

Let's first write down what we know and what this implies using the 6 equivalences above:

- a(t) the derivative is LINEAR from 0 to 3 ↔ v(t) the function is QUADRATIC from 0 to 3
- a(t) the derivative is POSITIVE from 0 to 3 ↔ v(t) the function is INCREASING from 0 to 3
- a(t) the derivative is DECREASING from 0 to 3 ↔ v(t) the function is CONCAVE DOWN from 0 to 3
- a(t) the derivative is ZERO from 3 to 5 ↔ v(t) the function is CONSTANT from 3 to 5

In order to draw a graph having these characteristics we need a starting point or initial condition. Assuming the initial velocity $v0 = 0$ we start our graph at the origin, changing this **initial condition** simply raises or lowers (shifts vertically) this shape determined by the conditions above:

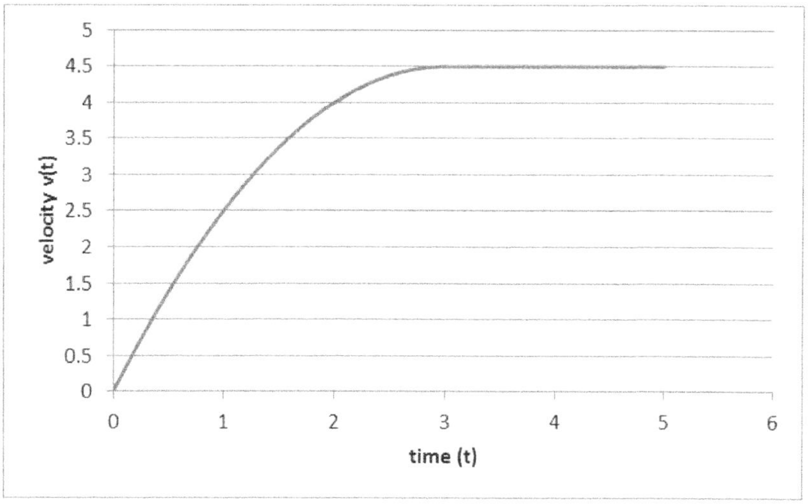

B. Faculty Development Workshop on Reasoning with Evidence

The following agenda gives a sense of what is covered in a workshop on Reasoning from Evidence, aimed at engaging the campus community in a discussion on how QR can permeate the curriculum.

Reasoning from Evidence Workshop Agenda

1. Welcome and Introductions
2. Lunch Conversation about our group and this workshop
3. Participant descriptions of evidence and argumentation in their disciplines
    a) Conversation
    b) White board notes in 3 columns:
        i. **R**easoning
        ii. **E**vidence
        iii. **A**rgumentation
4. Sample papers
    a) English Poetry and Dependency Ratios
    b) How do our categories for REA apply to these papers?
    c) Differences and Similarities between disciplines
5. Sample Rubrics
    a) AACU Critical Thinking Rubric [2]
        i. How does this relate to our notes?
    b) Carleton's Quantitative Engagement[3] rubric
        i. Simple yet powerful in changing culture!
    c) Bowdoin's 6 Questions in First Year seminars
        i. Looks familiar!
6. Wrap Up ☺

The link given to Numeracy Infusion Course for Higher Education (NICHE) is another great resource for QMaSC directors.

---

[2] http://www.aacu.org/value/rubrics/index_p.cfm?CFID=46004977&CFTOKEN=58030724
[3] http://serc.carleton.edu/NICHE/ex_qr_assessment.html

Virtual Presence

© McKayla Nuffer,
St. Lawrence University

1 Introduction

Computers and information technologies have been a growing interest throughout society and, particularly, the area of education over the past fifty years. Modern, high-speed computers, and, more specifically, instant access to the internet, produces a dependency within the global economy and scientific age in which we live. The launch of Sputnik in 1957 prompted national efforts to reform education with new technological practices and an improved philosophy to educate the masses [1]. Since then, the National Defense Education Act has supported educational reform and the incorporation of technology within school curricula [1]. More recently, the use of the World Wide Web has increased at a prominent rate and become responsible for the success of businesses, organizations, and educational institutions.

Access to information and services through the "click of a button" is appealing to many, specifically college students. Students find comfort in accessing assistance and academic support at their convenience outside of the classroom, especially during late night cram sessions before quizzes and exams. A quantitative and mathematics support center (QMaSC) provides services and information to students that will aid the progression in their studies, and QMaSCs aim to provide quality assistance in a convenient manner. In addition to its physical location, a QMaSC must also be virtually present in meeting the needs of students. This chapter addresses how to be virtually

Suggested Citation: McKayla Nuffer (2016), "Virtual Presence", http://dx.doi.org/10.5038/9780977674435.ch12 in G. Coulombe, M. O'Neill, M. Schuckers (Eds.) *A Handbook for Directors of Quantitative and Mathematical Support Centers*, Neck Quill Press, http://scholarcommons.usf.edu/qmasc_handbook.

This material is based upon work supported, in part, by the National Science Foundation under Grant DUE-1255945. Any opinions, findings, and conclusions or recommendations expressed in this material are those of the author(s) and do not necessarily reflect the views of the National Science Foundation

present and the kinds of information and services to provide virtually in order to market effectively the services provided by the center.

2 What is Virtual Presence?

Before deciding what to include on a center's website or what devices to use to become virtually present as a center, one should know what it means to be virtually present. According to Heiner Wolf (2005), virtual presence is about mutual awareness and synchronous communication on the web [2]. An individual can be virtually present at a specific web location, but virtual presence on the World Wide Web can range from viewing a website to being present in an artificial 3D space. Differing from tele-presence, virtual presence does not result in looking into real world settings, but only artificial space [2]. These types of "presences" are linked because 3D models of realistic space can be captured artificially and transferred into virtual representation [2]. Tele-presence and virtual presence can be used jointly to create options for students to seek the maximum assistance, fully demonstrating all that the web has to offer.

Given the prevalent internet usage, specifically social media — email, Facebook, Twitter, Instagram, Skype, iChat, or Google+ Hangouts — as part of everyday life, the QMaSC's ability to be virtually present is critical for public access to resources. Allowing on-the-go users access to information, updates, and course-related materials from portable devices such as laptops, smart phones, or tablets enhances the center's ability to serve more students. Acting as a catalyst for academic growth and support is rewarding for a center. The implementation of a website and other virtual systems will often increase student engagement in using resources, thereby allowing the center to act as a catalyst for academic growth and support.

Many different types of services and tools are offered within the realms of virtual and tele-presence. Tools that can be incorporated into the center's website to supply students with as many virtual resources as possible include social networking forums, such as Facebook, Twitter, and personal blogs, or online tutoring and direct communication forums, such as wikis, Skype, iChat, Microsoft Lync, Lynda.com, video streaming, YouTube videos, and eBeam. Successful sites often include links to data and software resources. Likewise, learning management systems, known as virtual learning environments (VLEs), can provide virtual assistance and tutoring to students through the posting of course syllabi, course materials, and texts, in addition to discussion forums for the courses supported by the QMaSC. Popular VLEs include Moodle, Sakai, Angel, Blackboard, Claroline, and ATutor; any of these can serve as an additional resource that can be developed outside of the standard website [3]. Directors can learn more about these tools by talking to other QMaSC directors, talking with the information technology departments at the parent institution, or by additional online research. Ultimately, all of these virtual tools, organized through the creation of a website, will enhance the virtual presence of the center.

3 Beginning a Center Website

While many avenues exist in beginning to develop a website, the most common, and strongly encouraged, path is to contact your institution's information technology and web services department. Most institutions enforce policies regarding the use and public access of institutional computing resources and the campus network. At some institutions, these departments issue strict guidelines and standards for the creation of webpages. Standards may be stylistic in nature governing color, font selection and size, or guidelines may concern content such as hyperlinks, text, images, and navigation.

At St. Lawrence University, standards are established for all webpages for academic and administrative departments addressing required links, required content, and limits on design. Examples of these guidelines include the following instructions: "have a link to the St. Lawrence homepage on them," "have your department or section name on them," and "avoid using varied design elements, such as different background colors or fonts on different pages" [4].

Not all directors actively create the center website but many, if not all, supply the content for and assist in developing the ideas for the website. By contrast, some directors create the entire website under the review and guidance of a web team. It is essential for directors to understand the various processes and procedures required by the information technology and web services department; therefore, directors must start by setting up meetings with the appropriate departments and offices.

Building a website involves a number of major decisions, four of which include: choosing the type of website, selecting a web host, establishing a domain, and identifying and designing key features for the website. In turn, these decisions influence future decisions such as content, design, organization, and additional features that make a user-friendly website. For those who will build a web site from the ground up, and those without guidelines mandated by their institution, sections 4 – 6 address guidelines for the first steps in creating a web site. For those with institutionally mandated guidelines, much of the discussion in sections 4 – 6 will be handled completely by the web services department or in partnership with that department.

4 Static vs. Dynamic Websites

For directors creating a website from scratch, the first step is *deciding on the desired type of website:* static or dynamic. Static websites have "content embedded in HTML code" while "the content and layout [are moderately] independent" [5]. There are no special requirements on the web host (see section 5 on web hosts) and entering content involves writing HTML code, this is similar to computer programming [5]. Static HTML is used to display pages with no comments, blogs, daily additions, or special features that require advanced functionality [6]. The files are loaded, then the browser retrieves and displays images, item links, video, and other materials. Static websites effectively establish a center's virtual presence because they are clean, secure, load quickly, and are easy to manage [6].

In comparison, dynamic websites allow content to be retrieved from a database, and content can be arranged by date, category, or similar organizational structures. Dynamic websites are ideal for blogs or chronological entries. The content and layout of dynamic websites are independent, and entering content is similar to word processing. Dynamic websites present a desirable alternative for those unfamiliar with writing HTML [5]. Dynamic websites are also popular because social media, surveys, and feedback are easily integrated into the pages; this allows centers to have both a virtual and a tele-presence. The advanced functionality of dynamic websites allows direct interaction between the creator and the user; however, dynamic sites are slower to load and more expensive than static websites [7, 6].

Both types of websites are suitable for a QMaSC, but the choice depends on the creator's willingness and time, knowledge of and skill level with the different content management systems or existing website building systems possibly supported by the parent institution, and the associated institutional constraints and support.

5 Choosing a Web Host

After selecting the type of site to be designed, *choosing a host* is the next, and most crucial, step. Knowledge and skill come into play because many different types of web building systems exist that are crafted for either the static or the dynamic website. Those building a website from scratch or those under the assistance of a web services team must select a host compatible with the institution's server, this ensures the website's approval and associated support of the university communications and public relations office.

Manually setting up a host requires the knowledge of code and script and requires special web design software [6]. Such website creators must know how to code and save files in HTML and how to use cascading style sheets (css) which maintain consistency of fonts, colors, and styles throughout the website [5]. While manual web hosts provide templates, knowledge of HTML coding is essential to enable the page creator flexibility in customization. Both Dreamweaver and Joomla, tools for setting up manually created sites, require the knowledge of HTML code content management systems to organize and create static (Dreamweaver) or dynamic (Joomla) websites.

Using Dreamweaver enables creators to design pages that exhibit pictures of staff, schedules and other documents in PDF format, links to other websites and data search engines, among other features. Advertisements, widgets, and other gadgets can be added to the QMaSC website using this web design software.

Joomla is ideal for building small to medium sites and is known to be fast and accessible in many places [2]. The inclusion of dynamic features, such as image galleries, YouTube videos, guest books, polls, calendars, file managers, and forms [2], combine multimedia, communication, collaboration, and data management all within the center's website.

Other mechanisms exist for manually creating a website, many are designed to use preexisting templates and style options for creating customized work with little to no experience in either coding or programming. Artisteer and WordPress are web hosts that create both static and dynamic websites. The features they provide make the process of creating and organizing content easier and more user-friendly [6]. With Artisteer, coding can be used to edit and customize a site, but be advised that the learning curve is steep [5]!

Google Sites provide another web hosting option requiring no coding knowledge. Unlike the aforementioned web hosts, Google Sites is free. Google Sites users need only have a Google account in order to create pages using the website templates. Other free options exist for hosting but most carry the risk of losing your entire site at the will of the web host. To ensure this does not happen, choose a host so that the center URL does not include the host name — a free web host gains publicity by including its name in the domain name of its patrons. Google Sites provide a safe and free option for centers with limited funds for producing a website.

6 Establishing a Domain

The selected web host influences the last step, *establishing a web domain*, in preparation of designing the website. Establishing a domain is the same as identifying a *web address for your site.* Registering a domain involves serious thought and consideration, despite seeming like a minor task in the creation of a website. While determining a name for the site may at first appear to be an elementary task, there are many ways to go about deciding a web address.

In naming the website, consider both the future QMaSC users and the keywords that best describe the center. Because the web address will be frequently advertised and used continually, it should be catchy and appealing to the user. Consider the length and characters within the domain. Avoid discouraging users by using a long address or one that is embellished with varying capitalization, numbers, or symbols [6]. Finally, avoid using any preexisting trademarks or names that may inadvertently lead a user to the wrong website [6]. (A name that is closely related to another website may discourage users who have difficulty locating the site and may eventually require a major and costly name change.) Hence, initial research into the possible domain names for ease of use and uniqueness will be worth the effort.

These three steps form the necessary preparation for the more enjoyable aspect of creating a website: design and organization. As previously stated, most of these steps are typically mandated by your institution, but having the background knowledge in what it takes to create a website can help directors when assisting a web team. Once a domain and a web host have been established, the focus shifts to content and features.

7 Recommended Features for Websites

Each center is managed and run in its own particular manner, but those commonalities in services that most centers provide for students should be displayed on the website. Whether the page employs hyperlinks, text, images, or PDF files, the content and material described below can be included to enhance to the center's virtual presence.

General Information and the Home Page

General information about the center, including *the mission statement, brief history, and overview of services* are possible items for display within the site. The overview can identify the courses that are supported by the center and may be accompanied by the names of professors and instructors for the courses; however, more general descriptions are more practical because maintaining updates for this level of detail is time consuming. Typically, general information is located on the "Home" page or an "About" page that can lure users into reading further and sets the stage for what the entire website and center have to offer. The home page should contain an opening message, welcoming and inviting users to continue reading the website.

The "Home" page is often the first space that users see, so the selection of user-friendly colors and fonts is crucial for this visual representation of your center. Web page creators should choose a color scheme that is cohesive and complementary if the color scheme is not mandated by the center's parent institution. Maintaining stylistic consistency throughout the site is critical; for example, make headers, text, and links follow a set color scheme. Not only will this make the site look professional but will help users in recognizing key features and information through color coordination. Likewise, avoid fancy, cursive or symbolic fonts. Font should be sized large enough so that all viewers are able to read the page. Always think from the professional and student perspectives when deciding on the aesthetic style of the website. Make it "eye-popping" and clear. Unless the site is visually stimulating and simple, users may not see all the details. Attractive and inviting pictures of the center help students become virtually familiar with the space and welcome the users to visit the center.

Tutor and Staff Directories

Before receiving or considering services, students and faculty appreciate a clear and pictorial directory of mentors, tutors, directors, assistant directors, and other personnel. This offers users a sense of comfort and familiarity for they already have a name, picture, or both to refer to upon entering the center. Many sites have "*Staff*" or "*People*" pages which group the staff based upon their roles within the center. A possible design format for the directory entry for a mentor or tutor might include his or her name, class year, major(s), minor(s), course support, software support, and photo. Similarly, the site might contain short biographical blurbs for each of the directors and other staff members of the QMaSC.

Expectations and FAQ Pages

Informative pages often include a description of *what to expect from a tutoring session* or other services. Such information can be discussed and organized in more than one way. For example, on the website for the Peterson Quantitative Resource Center at St. Lawrence University, two pages provide information on: "What to Expect" and "What Not to Expect" when coming to the center. Having both pages clearly distinguishes the expectations and definition of a standard mentoring session, which explains what mentors do for students and gives a clear definition of the mentor's role in assisting students. Such concerns may also be addressed on a "*FAQ*" (Frequently Asked Questions) page. FAQs often provide descriptions of a typical tutoring session, directions for entering the center, the location of the center, and how to contact a tutor for individual sessions.

Schedules

Hours of operation and appropriate *schedules* should be posted on the website. These can be attached as PDF files, Excel spreadsheets, or presented in a table. Centers may have multiple schedules for a variety of services, but the most important schedule is often the drop-in tutoring schedule. At any given time, students might want to access information about when specific tutors are working. Similarly, a schedule organized around course coverage and giving a list of supported courses, along with the hours of operation for each course, is useful to students. Depending on how the center is structured, whether open hours are offered, or a specific set of hours is set aside for certain courses, providing schedules is vital information to be displayed on the website.

Attendance Tracking

Directors need to keep track of the flow of traffic in and out of the QMaSC for data collection purposes. Some center websites include a *log-in* page. This can be accomplished by using a commercially packaged attendance tracker or by developing an in house system using a database. Regardless of the method used, directors can manage and gather data for their own personal records or, at any given time, present statistics and figures to faculty and administrators. Demographic data about students can be gathered by requesting the following information at login: name, class year, whether this is a first time visit to the center (newly opened centers might want to ask how the visitor first learned about the center), the course for which assistance is sought (this may vary depending on the center's services), and other pertinent information sought for tracking purposes. Check with the Registrar's Office or Dean's Office regarding FERPA laws related to data collection and use. Some institutions allow students to use services confidentially or have students check a box stating that the data can be shared with administrators and faculty. If clearance is given, the resulting login statistics may be useful to faculty or an advisory board.

Similarly, a *tutorial report form* and *tutor evaluation form* can also be established to help monitor the activities and performance of a tutor. Because the primary goal of a resource center is to meet the needs of the students through providing quality services, satisfaction surveys can be used to seek information on the quality of the overall session, the tutor's presentation, and additional comments. This feedback will help in addressing major concerns of the students through improvements that can be made. Online surveys are always available to visitors whether completed on site or online after students leave the center. The surveys can be easily generated using the attendance tracking systems, Google forms, Qualtrics surveys, or other online forms. In this way, the virtual presence of the center helps to monitor the success and perception of the center.

Use of Social Media

The popularity of social media suggests that methods for students to *contact* and communicate with the center is an important area of focus. According to Eric Chapman (2011), "between 85 percent and 99 percent of college students use Facebook ... [and] students spend an average of 106 minutes on Facebook per day" [8]. University and college admission programs use social media to contact students and provide information and materials to prospective students at low costs through Facebook, Twitter, and YouTube. The use of such platforms is advantageous to a QMaSC in reaching students through these heavily utilized media. Centers can use social media sites to deliver efficient updates and important messages to students about upcoming exams, tutorials offered by the center, and changes in hours of operation and scheduling. Depending on the chosen web host, calendar attachments and templates can be used and easily adapted to meet the needs of the QMaSC in addressing similar updates to students or through certain email accounts, such as Gmail and Microsoft Outlook (Google's calendar template can be attached directly to the center's site when Google Sites is used as the web host). Some colleges or universities use course-management systems (CMS) such as Moodle, Blackboard, or Sakai. Directors may wish to set up a page on the institutionally supported CMS site to manage tutors or establish "sub-boards," a forum-type interface, for updating students. Center mentors and staff can form a Facebook group in which members can post a status in order to find a sub for a given drop-in shift and respondents can post a comment when they are able to take the open shift. Similarly, listservs or email groups also meet this need.

Mixed reviews exist on whether to use social media for the purposes of finding tutor/mentor coverage, networking, and supplying students information and updates on the website. In a recent roundtable discussion, some directors argued that these areas were already being addressed without the need of social media [9]. Some tried it and then deleted the social media accounts because they duplicated what they were already doing on their main website [9]. While placing information in more than one area is not necessarily excessive, maintaining multiple virtual forums can get complicated. Conversely, placing an advertisement in multiple forums gives more visibility and may result in more patrons coming to the center. Being mindful of particular student populations and how they communicate can help determine where information is placed and assist in making the necessary choices to become virtually present.

External Resources

Because tutors and mentors cannot staff drop-in hours 24 hours per day, providing students with web access to *extended resources* such as web addresses for scholarly research, data search engines, video tutorials, fact sheets, or other materials expands the effectiveness of the center beyond the physical services of the QMaSC staff. Students will appreciate the time and energy used to create these resources and will be more receptive and willing to explore what these pages offer. Outside online resources can disappear or change, so it is important to regularly manage and verify links to ensure they continue to offer appropriate resources.

Beyond providing links and other preassembled resources, directors can post the same help sheets that are provided at the center or distributed during workshops or tutorials throughout the year. Segmented video recordings of workshops can also be posted for students to view at their convenience. While this may reduce the number of students who visit the center or attend workshops, it provides students with alternate options for getting help. Some students prefer to work with an actual person rather than looking at a worksheet, but those who are fearful of using services will appreciate the online option. Most students like having multiple options such as working with others when they are having difficulty understanding a concept, taking the personal instructional route, seeking assistance of the center's staff, and using the online resources for quick references.

Placing these resources on the QMaSC website, coupled with previously mentioned components, creates a strong virtual presence for the center. Other possible additions to the site include applications for becoming a tutor, links to all departments supported by the center, YouTube videos for tutor training, pictures of the center, photo albums of events, and more. Depending on factors such as the availability of the center's staff for specific courses, the hours of operation, and the size of institution, online tutoring options that use virtual whiteboards and other commercially developed, or free, systems are alternative components for meeting student needs.

8 Conclusion

Being virtually present and building a website is a progressive route any QMaSC director can employ. Many technologies are used within education. Knowing and continuing to seek assistance from other centers and experts, within the education and technological fields, will aid directors in providing the best support opportunities for eager students. Because many colleges and universities have a quantitative or mathematical literacy requirement for graduation, more and more students will need assistance. QMaSCs can guide the way! As computers, tablets, and other portable devices become more prevalent, assembling virtual locations for students to access the skills, knowledge, and materials designed to meet their academic needs becomes essential. The QMaSC website can showcase the work and assistance provided to the future mathematicians, scientists, engineers, professionals, and humanities students of the world. Become virtually present and join the information age, because no one wants to "byte" the dust!

9 Bibliography

[1] A. Molnar, "Computers in education: A brief history," *The Journal*, 1997. Accessed from http://thejournal.com/Articles/1997/06/01/Computers-in-Education-A-Brief-History.aspx?Page=1 on March 14, 2013.

[2] H. Wolf, "Home page." Accessed from http://www.virtual-presence.org on June 6, 2013.

[3] I. Clement, "Virtual learning environment comparison," 2003. Accessed from http://atutor.ca/atutor/files/VLE_comparison.pdf on September 8, 2013.

[4] St.Lawrence University, "Website standards." Accessed from http://www.stlawu.edu/university-communications/website-standards on September 9, 2013.

[5] T. Siegers, "Building Dynamic Websites with Joomla CMS," 2010. Accessed from http://www.slideshare.net/thomasjs/building-dynamic-websites-with-joomla-cms on June 5, 2013.

[6] L. Irby, "Homepage," 2013. Accessed from http://www.2createawebsite.com/ on June 2, 2013.

[7] A. Fenton, "Static vs. Dynamic Websites — What is the Difference?," 2013. Accessed from www.edinteractive.co.uk/article/?id=4 on June 3, 2013.

[8] E. Chapman, "Facebook facts revealed," November 2011. Accessed from http://www.collegiatetimes.com/stories/18784/facebook-facts-revealed on June 11, 2013.

[9] P. Conrath, "Notes taken at St. Olaf Round Table on Academic Support," 2013.

Part IV

Staffing, Hiring, and Training

Staffing and Hiring

© Christine Tinsley,
University of Oklahoma

1 Introduction

Some QMaSCs may have just a few employees, while others may have 50 or more employees, depending on the needs and budgets of each institution. No matter the size of the QMaSC, a good, qualified staff is the most important asset. A larger center may have a director, assistant director, course coordinators, and a tutoring staff, whereas a smaller center may have just a director and a tutoring staff.

2 Hiring an Assistant Director

To begin the process, the director will confer with the human resources department in order to understand all posting and hiring policies. After consulting with that department, the director or a hiring committee will create a job description to be advertised. The university or college's human resource department can be helpful in posting the position to the institution's website and in submitting it to professional journals and organizations. The position might also be posted to the QMaSC's own website and to quantitative reasoning list-servers. It may also be helpful to forward the announcement to departments served by the QMaSC. Once the director or hiring committee has selected applicants who will be interviewed, the director or hiring committee should have a set of questions that they will ask of all applicants being interviewed. There should also be a timeline

Suggested Citation: Christine Tinsley (2016), "Staffing and Hiring", http://dx.doi.org/10.5038/9780977674435.ch13 in G. Coulombe, M. O'Neill, M. Schuckers (Eds.) *A Handbook for Directors of Quantitative and Mathematical Support Centers*, Neck Quill Press, http://scholarcommons.usf.edu/qmasc_handbook.

This material is based upon work supported, in part, by the National Science Foundation under Grant DUE-1255945. Any opinions, findings, and conclusions or recommendations expressed in this material are those of the author(s) and do not necessarily reflect the views of the National Science Foundation

with dates for interviews, for making the first offer, for making a second offer (if necessary), and for notifying other applicants that the position has been filled. Some possible interview questions are:

1. What do you already know about our institution and our students?
2. Tell me about yourself as a tutor, as a supervisor, and as a coworker.
3. How do you handle correcting a staff member?
4. How do you wish to be corrected?
5. How would you handle a difficult student?
6. With respect to instruction or training, what are your strengths?
7. What qualities do you wish to develop through this position?
8. In working with faculty to develop quantitative material, what specialization can you offer?
9. How will this position fit into your career trajectory?
10. Where do you see yourself in five years?
11. Do you have questions for me (us)?

At the end of the interview, the interviewer will let the applicant know when the search will culminate in a hiring decision.

Normally, the director will make a formal offer of the position to the top applicant. (If the offer is not accepted, the offer will go to the second applicant, etc.) After one of the applicants has accepted the offer of the position, the director will follow up with all other applicants thanking them for their applications (and, if interviewed, for their time), letting them know that despite their strong applications, the position has been filled.

In some institutions, the human resources department or the home department provides training in FERPA [1], sexual harassment, and academic misconduct. If the human resources department does not provide this training, the director should provide or arrange training in these areas. Once the employee has finished all training, he or she should sign a letter (see appendix "Acknowledgement of Policies") recognizing that she or he understands the institution's policies concerning these matters.

3 Hiring Graduate Teaching Assistants

Larger institutions might hire graduate teaching assistants to staff the QMaSC. Consult with the human resources department or graduate program to find out what policies need to be followed to employ teaching assistants as tutors or staff in the QMaSC. Whether or not these employees need to be hired, the director still needs to make sure they have completed the sexual harassment, FERPA, academic misconduct, or other training that is required by the institution's human resources department. If the human resources department or graduate school does not provide training in sexual harassment, FERPA, or academic misconduct, the director should provide the employee with training in these areas. Once the employee has finished all training, have the employee sign a letter, (see appendix "Acknowledgement of Policies") recognizing that she or he understands the institution's policies concerning these matters. Depending on the institution and the teaching assistants themselves, teaching assistants may need coaching or training in order to be effective tutors. A separate training program can be designed for them, or they can be included in training activities with the student peer tutors.

4 Hiring Professional Tutors

As with the Assistant Director position, the director begins by consulting the human resources department in order to understand all posting and hiring policies. Most of the discussion about hiring an assistant director can apply to hiring professional tutors. Qualified applicants should have a demonstrated proficiency in math (or the content that they will tutor), strong interpersonal and communication skills (able to work with individual students or with small groups of students), and good employee skills (submitting documents in a timely fashion and responding to email promptly). In addition to tutoring and encouraging students as they learn content, the tutor should teach students good study skills that can help them to become independent learners, so interviewers should seek to find out what the prospective tutor knows about teaching study skills and about different types of learners.

5 Hiring Student Tutors

Many QMaSCs have each potential tutor submit an application, a copy of his or her transcript, and, if he or she has one, a résumé. (See the "Sample Application" in the appendix.) If names of potential tutors come from departments that have vetted the potential tutors, then the students' abilities with the material content have been verified. If a student who has not been vetted through a department applies to become a tutor, he or she should be asked to provide the names of two professors who can verify the student's proficiency with the material. In both cases, it can be helpful to ask a student to provide the name of a faculty or staff member outside the department who can provide a reference regarding reliability and interpersonal skills. Some QMaSCs might ask the student to provide a letter of reference as part of his or her application; others might send an email to the student's references asking for verification independently (see the "Quick Reference Request" in the appendix). The QMaSC might also administer a proficiency exam in case further verification of the student's abilities with math (or other topics) is required. Such an exam can have distinct levels of math in order to determine the applicant's level of ability (see the "Math Proficiency Questions" in the appendix).

After applications and references have been obtained, an interview can be set up. All applicants should be asked the same questions in the interview so that the director can fairly assess their qualifications. As part of the interview, the applicants can be asked to explain a solution to a particular problem, and they can then be asked to play the role of a tutor coaxing a student to arrive at a solution to a problem. Other questions that can be useful are:

1. What is your attitude about mathematics (or other subject that will be tutored)?
2. You are being considered as a tutor for (*course or department*): What was your experience in the class?
3. How do you feel about students who have a hard time understanding the material?
4. What would you say to a student who comes in and talks negatively about a professor?
5. How would you handle a student who just wants you to do their homework?
6. What would you do if you were unable to answer a student's questions?

The interview should inform the applicant about expectations of tutors, and it should include a tour of the QMaSC. Tell the applicants about various trainings that will be required of them if they are hired (FERPA, sexual harassment, academic misconduct, tutor training) and about any documents that will be required (social security number, I-9, W-4, direct deposit). When the QMaSC staff or hiring committee has made decisions about which applicants to hire, the hiring letter should outline the expectations and requirements for tutors. Applicants should be asked to sign or initial the letter to indicate that they understand and accept these requiements and expectations. Additionally, once the various trainings are completed, the QMaSC may wish to have the tutor sign a letter recognizing that she or he understands the institution's and QMaSC's policies.

6 Appendix

Acknowledgement of Policies

Academic Misconduct

"Academic misconduct is defined as any act which improperly affects the evaluation of a student's academic performance or achievement. It specifically includes cheating, plagiarism, fabrication, fraud, destruction of property, and bribery or intimidation, as well as assisting others or attempting to engage in such acts. It is the responsibility of each student to be familiar with the definitions, policies and procedures concerning academic misconduct; and unfamiliarity with the code alters none of a student's rights or responsibilities thereunder." [2]

Sexual Harassment

"Sexual Harassment is unlawful and may subject those who engage in it to University sanctions as well as civil and criminal penalties. Since some members of the university community hold positions of authority that may involve the legitimate exercise of power over others, it is their responsibility to be sensitive to that power. The University is committed to providing an environment of study and work free from sexual harassment and to ensuring the accessibility of appropriate grievance procedures for addressing all complaints regarding sexual harassment." [2]

Confidentiality and Privacy

"Staff members may become aware of personal information concerning students for example grades, personal problems, etc. this information is protected by the Family Educational Rights and Privacy Act (FERPA). Staff members should not discuss this information with other staff members or students. If a student discusses emotional, physical or personal problems you should refer them to the appropriate resources." [2]

I understand and acknowledge the following policies.

______________________________ ____________

Tutor Signature Date

Sample Application (University of Oklahoma)

All applicants need to include: 1) completed application, 2) resume, 3) copy of transcript, and 4) one letter of recommendation from a math professor.

First: ______________________ MI: ______ Last: ____________________________

Local Address: ____

Phone: ______________________ Email Address: ____________________________

Date of Birth: ______________________ ID Number: ______________________

Are you currently enrolled: __________ How many credit hours are you taking: __________

Status/ Class standing: Sophomore Junior Senior Graduate Student

Major: ________________ Overall GPA: __________ GPA in Math Courses: __________

Are you currently employed somewhere else on campus: ________________ If yes, where:

____

____

Based on my performance in class or my previous education, I feel competent to tutor students in the following courses/disciplines: Specific Course(s) / Discipline(s):

____

____

Education

Name and Location. Full Address include country if located outside of the USA	Dates attended	Graduation date and degree earned
High School:		
Undergraduate College:		
Graduate School or Other Education:		

Employment History

Position Held or Title	Dates of Employment	Employer's Name and Location. Full Address. Include country if located outside the USA.

Professional Reference

Make sure to include a letter of recommendation from a professor

Name and Title	Contract Information

Please write a short paragraph discussing what skills you can bring to the position. Include any relevant jobs or past experiences that make you qualified for the position.

____

____

____

____

____

____

______________________ ______________

Applicant Signature Date

Quick Reference Questions

Subject: RSVP requested by Friday: 2-minute reference for *Student Name*

Student Name has listed you as a reference outside of the department for which she has applied to become a tutor. Please help us by answering the two questions:

Does this student possess qualities that would indicate promise as a tutor? (Is she/he approachable, patient, enthusiastic, solid/unflappable, excited about the material, and/or understanding of others' confusion?)

_____ Yes.

_____ No.

_____ I have no basis for evaluating these qualities.

_____ See comments.

Does this student possess qualities that would indicate promise as an employee? (Is she reliable? Is she organized? Is she on time for class, appointments, and meetings? Does she respond quickly and in a timely way to email communication?)

_____ Yes.

_____ No.

_____ I have no basis for evaluating these qualities.

_____ See comments.

Comments:

Thank you for your time!

Math Proficiency Questions

Please go step-by-step, using the space below to explain how you would solve the following problems.

1. Find the center and radius of the circle: $x^2 + y^2 + 8x - 6y - 15 = 0$

2. Use u-substitution to solve the following problem: $\int \sqrt{x} \sin(1 + x^{3/2}) dx$

3. Given: $f(x, y) = ln(x^2 + y^2)$ find f_x, f_y and f_{xy}

Official Use ONLY

Comments and Notes:

7 Bibliography

[1] U.S. Department of Education, Family Policy Compliance Office, "Family Educational Rights and Privacy Act (FERPA)." http://www.ed.gov/policy/gen/guid/fpco/ferpa/index.html, 2013.

[2] The University of Oklahoma, "OU general catalog." http://catalog.ou.edu., 2013.

Training Materials and Methods

© Johnathan Nelson,
University of St. Thomas

1 Introduction

This chapter opens with a discussion of tutoring practices and methods that a director of a QMaSC center would be wise to go over with his or her tutors; it continues with a discussion of different ways that a director could conduct training sessions, and closes with a look at different approaches that different directors take towards training tutors in content knowledge.

2 Training Tutors in Tutoring Practices and Methods: What Should be Covered

It does not take an expert to tell you that knowledge is not sufficient for being a good tutor; one needs to know how to communicate with the individual being tutored in a way that will allow him or her to retain the knowledge, and use it on his or her own. QMaSC directors should and do recognize this when setting up their tutor training programs. According to a survey of 22 QMaSC directors at schools across the country[1], the majority of tutor training that QMaSC's conduct covers tutoring methods and practices, i.e., strategies and techniques used when helping students. There is a good deal of consensus among the directors surveyed regarding what methods and practices are the best.

Suggested Citation: Johnathan Nelson (2016), "Training Materials and Methods", http://dx.doi.org/10.5038/9780977674435.ch14 in G. Coulombe, M. O'Neill, M. Schuckers (Eds.) *A Handbook for Directors of Quantitative and Mathematical Support Centers*, Neck Quill Press, http://scholarcommons.usf.edu/qmasc_handbook.

This material is based upon work supported, in part, by the National Science Foundation under Grant DUE-1255945. Any opinions, findings, and conclusions or recommendations expressed in this material are those of the author(s) and do not necessarily reflect the views of the National Science Foundation

Relational Understanding

Perhaps the single most important principle for tutors to know is that their role is to help students achieve real understanding of the material, as opposed to merely helping students find the right answers. Richard Skemp, author of the 1976 math education classic "Relational Understanding and Instrumental Understanding," says that tutors should help students achieve relational understanding, by which he means "knowing both what to do and why," instead of mere instrumental understanding, which he describes as "rules without reasons."[2] Although QMaSC directors may not use these terms, much of the training that they conduct has the aim of making tutors see the importance of helping students achieve relational understanding.

Patrice Conrath of Bethel University instills this principle in her tutors by telling them that she sees the main job of their center as "working itself out of a job," i.e., helping students to truly understand the material and adopt an approach towards mathematics that will allow them to be better learners in the future. She encourages her tutors to use a "hands behind your back" strategy, in which tutors do as little "showing" as possible. "If we plow through problems," she says "and show them the 'way to do it,' they may think some steps are magic, they might not understand some of the terms, and they may have thought about [how to do a problem] in a completely different, but correct way." She encourages her tutors to build off of the student's own ideas, as this will make it easier for students to integrate new knowledge into their existing framework of understanding.

Rosalie Tepper of Shoreline Community College conveys this same principle to her tutors by telling them that it is not the tutor's role to do the work for the students, but rather to "think of [their role] as asking leading questions and giving hints." This helps students draw on their prior knowledge, which allows them to make connections leading to deeper understanding. Tepper finds that other benefits come from this approach as well: "It keeps them awake and paying attention Also, having the student do as much as possible helps me (when working with a student) gauge their understanding, which in turn allows me to tailor my methods to them." By using questions, she finds it easier to identify the specific part of the problem that she needs to focus on with the student, and what parts of the problem the student already understands. She sees little pedagogical value in just showing students what to do. "Even if (after just showing students what to do) we ask the student 'do you understand what I did?' they might say 'yes' even if it's not true, due to embarrassment at their lack of comprehension, a belief that the tutor can't explain it better, or just a lack of interest," she says. Tepper says that it occasionally happens that a student's prior knowledge will be inadequate to build from, and in these cases the best thing for the tutor to do may involve more "showing" than would normally be recommended. In these instances, she says, it is a good idea to "make up a related example with different numbers, so that the student still has to do his or her homework problem."

Many other QMaSC directors share practices and methods with their tutors that will lead to relational understanding. Marcia Frobish of Grand Valley State University finds it important to train her tutors in questioning skills; she tries to have the tutors help students by "asking them, rather than telling them." Cat McCune of Smith College also tells her tutors to ask questions, not to "teach/tell," and to "keep the pencil in the hand of the tutee." Tiana Bosley of Wayne State University takes a similar approach, saying, "It is important for the tutor to 'not hold the pencil.' He or she needs to ask leading questions to get the tutee involved in thinking and solving the problem. It also helps the tutor assess how much the tutee may know." Maria Belk of Bard College tells her tutors to provide hints and, when these fail, to work through problems similar to the one the student is asking about, rather than simply giving the answer or doing the problem for the student. Thomas Dick of Oregon State University encourages his tutors to have the students "think aloud" about their approach to problems.

Interpersonal Skills and Professionalism

Many QMaSC directors also find it important to train their tutors in interpersonal skills, i.e., forming a friendly-but-professional relationship with the students they help. According to the survey, tutors should be trained in maintaining patience when students get frustrated or belligerent, developing a relationship with the tutee that makes them feel that its safe to ask questions, and in being generally upbeat and positive; in the words of Cat McCune of Smith College, tutors should be "aggressively enthusiastic" when greeting new students.

One common situation that tests tutors' professionalism is when students complain about their professors. It is important that tutors not join in, which would turn the tutoring session into a "professor bashing" session. Patrice Conrath of Bethel University tells her tutors to listen to the comments, but not join in. It is okay to acknowledge the students frustration, she tells them, but then to move on. She encourages her tutors to say things like "I'm sorry you are frustrated, I know it is hard, but let's attack the material." She wants her tutors to have positive attitudes, showing students "the excitement of doing math and the perseverance to plow through when it's tough."

FERPA and Confidentiality

Tutors should definitely be made aware of the Family Education Rights and Privacy Act (FERPA) [3] , which pertains to them as employees of the educational institution. FERPA prohibits school employees from divulging information to anyone other than the student about that student's academic records (including appointments for tutoring sessions or sign-ins to the QMaSC) other than for strictly professional reasons [2]. In general, the best policy would be that tutoring sessions are confidential and, while tutors can discuss experiences with a student for the purpose of improving the help that they give him or her, they need to be careful that they are not gossiping, or talking about particular students that they tutor, with friends who are not tutors.

Other Training Topics to Consider

QMaSC directors mention many other methods and practices that they cover in tutor training, including being aware of what course students are in so that course-appropriate approaches to each problem can be provided, how to work with more than one student at a time, what tutors should do when they get stuck, how to work with students from different cultures or who are English language learners, how to use answer keys properly, how to work with different learning styles, how to avoid being monopolized by especially needy students, strategies for addressing math anxiety, and making sure that help is authorized for the assignment that the student is asking about. Every QMaSC director should consider the approach that they want their tutors to take in dealing with these issues, but each director's preferred approach will likely vary, as differences in the way the QMaSC is set up, institutional or departmental curriculum, testing policies, demographics, and student support and disability services will need to be taken into consideration.

3 Training Tutors in Tutoring Practices and Methods: How to Conduct Training Sessions

Most of the QMaSC directors surveyed conduct training on tutoring practices and methods in an employee orientation session. The directors reported that their sessions lasted between 1 and 6 hours. In addition to covering basic employee information (expectations, how to find a sub, recording hours, etc.) these sessions also can include mock tutoring sessions, where tutors take turns acting as a student or as a tutor and offer one another feedback on how they did, or a scripted, staged session in which tutors (or the director) act out the "dos and don'ts" of tutoring. Many directors also have handouts that they give their tutors for reading and discussion; these handouts can include materials that the director has prepared covering tutoring practices and methods, as well as articles on teaching and learning that can be read and discussed.

Most QMaSC directors also have ongoing training for tutors. A common approach is to have meetings at the beginning of each semester; other directors conduct training more frequently. Most directors require tutors to attend training sessions (especially new tutors), but are generally pretty flexible when tutors have conflicts. Some conduct make-up sessions for tutors who were unable to make the original meeting; others put training materials online or send meeting notes to tutors who were unable to attend.

Rosalie Tepper of Shoreline Community College says that she sees the interview process as the beginning of training. During interviews with prospective tutors, she explains her tutoring philosophy (described in the previous section) with students, and lets them try to put it into action in a mock-tutoring session. She also has ongoing training to reinforce the practices and methods that she recommends to her tutors.

At the University of St. Thomas, "Effective Math Tutoring Tips," written by George Dennis of Lower Columbia College, is used to introduce new tutors to effective practices and methods [4]. Dennis provides a number of tips useful in helping tutors adopt an approach that will encourage students to truly learn the material and become better independent learners. All of the tutors, new and returning, read and discuss Dennis' tips in the beginning-of-the-school-year meeting; this is especially effective because veteran tutors point out which of the tips they find particularly useful, which validates the tips in the minds of the new tutors. Skemp's "Relational Understanding and Instrumental Understanding" is also shared with the tutors, many of whom have likely seen the value of relational understanding in their own studies, but reading this article can solidify its importance in their minds.

Patrice Conrath of Bethel University says she likes to include time for tutor discussion led by her veteran tutors during training sessions. Conrath has a great strategy for making these discussions as fruitful as possible:

> To solicit some of these lead tutor ideas, I will often put up a Google doc about a week before our monthly lab training meetings, and ask the lead tutors to write about current issues in the labs. They interact on this Google doc, brainstorming together, and then I divide up and organize issues and have different tutors lead different discussions at the meeting (usually the ones they brought up, or had good ideas about). I will lead the meeting, but they do most of the talking. It really goes quite well, and other tutors actively engage in the discussions and will feel empowered to contribute as well in this format. I have often wished I had brought a camera to record these fruitful discussions and may do so in the future.

Conrath says that her veteran tutors often have seen for themselves the value of the practices and methods that she advocates, and it can be good for the newer tutors to hear their peers reinforce her ideas. Also, being students themselves, the tutors can provide insight into professor expectations and student culture that will help their peers.

There is also information that a director may find helpful to use in conducting training sessions put out by the College Reading and Learning Association (CRLA) and by the National Tutoring Association. The CRLA guide provides a comprehensive guide to general tutor training topics and also a section specific to mathematics [4]; the National Tutoring Association has a code of ethics and information, as well as information on certification, workshops, and conferences [5].

4 Training Tutors in Mathematical and Quantitative Content

Some QMaSC directors cover mathematical or quantitative content in their training sessions, but many do not. Many QMaSC directors say that, since the tutors that they hire have taken several math classes, have earned good grades in these classes, and have strong recommendations from faculty members, they assume that the tutors will have no problems with the content. Others say that they would perhaps like to train tutors in content, but it is unrealistic to do so because there would be too much material to cover. Centers that focus more tutoring for quantitative courses may find it difficult to cover the wide range of disciplines and courses that they support.

Furthermore, the set-up of each QMaSC can affect the need for content training. For example, Bethel University's QMaSC has large numbers of tutors working at the same time (although it is open for fewer hours than many other institutions' QMaSCs), so even if some of the tutors cannot provide help on a given topic, there is usually at least one tutor who can.

QMaSC directors that do think it important to train tutors in mathematical content believe that tutors especially benefit from training in topics from basic math or pre-calculus and trigonometry, application-type questions from various courses, and topics from mid-level mathematics courses. Tiana Bosley of Wayne State University says that she finds it important to train tutors in content from lower level math courses, as often a good deal of time has passed since the tutors studied some of this content themselves. Darry Andrews of Ohio State University finds that tutors often are "rusty" in explaining how to set up word problems, and so would benefit greatly from training that would cover applications. Dennis Lewandowski of Michigan Technological University notes that his tutors do not always feel comfortable with topics from Calculus 3, Differential Equations, or Linear Algebra; his content training is driven by their suggestions, and is conducted in his meetings with what he calls "open discussion," i.e., having the tutors that are confident with the material share their knowledge with those that are not.

At the University of St. Thomas, tutor training on mathematical content is held, and it is very beneficial. Although the tutors that work in the St. Thomas' QMaSC have earned high grades in their classes, and have strong recommendations from math faculty, their knowledge of many topics could use some shoring up. Many students that come into our QMaSC seeking help are in Finite Mathematics, an applications-based lower level math course. Topics from this course include probability and financial math, which the tutors often do not study until their senior year, if at all. Also, there are concepts in college algebra and trigonometry (such as the rational zero test for polynomials and trigonometric identities, to name a couple) that undergraduate students do not often use after initially learning them (which many tutors did in high school). Finally, there are difficult topics from Calculus II and Multivariable Calculus, such as sequences and series or vector calculus, that even strong students can struggle to understand, and so it is beneficial for the tutors to spend some time reviewing these topics as well.

During the 2012-2013 school year, a new approach to tutor training on mathematical content was tried at St. Thomas; topics, in which many tutors were relatively weak, were chosen and associated worksheets designed and distributed to help the tutors review these topics. Referencing course schedules provided by professors, training sessions for each topic were scheduled on a date when professors were beginning to cover that topic in their courses. During the training sessions, the tutors worked through the material, discussing with one another and getting help when needed. Attendance at the training sessions was optional, tutors who couldn't make training could pick up the worksheet for study on their own. However, tutors were given brief tests on each topic covered in training, and each tutor was expected to pass at least 5 tests throughout the year (a total of 14 tests were given; tutors could pick and choose which tests to prepare for and take). It was pleasing to see that many of the tutors chose to be tested on topics in which they knew that they had previously struggled, and that most tutors had no trouble in passing the minimum number of tests, and some tutors even took well more than the minimum number (the tutors could record hours spent at training sessions or taking the tutor tests, and also were promised an extra raise to tutors who passed at least 10 tests throughout the year). However, it was somewhat disappointing to find that some tutors did not feel like they could put in the time to prepare for the tests; the expectation that tutors pass at least 5 tests ended up being one enforced with some "flexibility." Overall, though, this program seemed to be helpful in making the tutors better able to help students.

5 Conclusions

There is much variety in how QMaSC directors conduct tutor training, and in what they cover. Much of this variation arises naturally from differences between institutions.

It seems, however, that many of the methods and practices advocated by QMaSC directors are oriented towards helping students achieve what Skemp would call "relational understanding," as opposed to "instrumental understanding." Also, QMaSC directors find it important to train tutors in how to interact with students, professional behavior, FERPA, and many other areas, several of which will require institution-specific approaches. Most directors accomplish this training through a combination of orientation sessions for new tutors and ongoing training sessions for continuing tutors, and find mock tutoring sessions, handouts outlining important strategies, and tutor-led discussions to be effective ways of conveying best methods and practices to the tutors.

When setting up a tutor training program, a QMaSC director should also consider whether or not training should be conducted on mathematical content. Factors to be considered include how well the tutors' knowledge will line up with the topics that they will be asked about and the amount of time that can reasonably be spent on training the tutors in mathematical content. The extent of content training can vary greatly, from ad hoc tutor-led discussions of the needed review material to systematic content training with worksheets and tests.

6 Bibliography

[1] J. Nelson, "QMaSC Director Survey Re: Training," survey, 2013.

[2] R. Skemp, "Relational understanding and instrumental understanding," *Mathematics Teaching in the Middle School*, vol. 12, pp. 88–95, September 2006. Reprinted from *Mathematics Teaching*, December 1976.

[3] U.S. Department of Education, Family Policy Compliance Office, "Family Educational Rights and Privacy Act (FERPA)." `http://www.ed.gov/policy/gen/guid/fpco/ferpa/index.html`, 2013.

[4] G. Dennis, "Tutor training handbook." `http://www.webpages.uidaho.edu/bestpractices/peer_train_math.html`, 2013.

[5] K. Agee and R. Hodges, *Handbook for Training Peer Tutors and Mentors*. Independence, KY: Cengage, 2012.

7 Acknowledgements

Thank you to the numerous directors of QMaSCs throughout the country who participated in my survey, especially those who shared more of their thoughts in follow-up emails.

Thank you to all of the participants in this project, whose feedback has been of immense value in bringing this section to its final edition, especially Mary O'Neill, who was specifically charged with editing this section.

Thank you also to the Mathematics Department of the University of St. Thomas for supporting my participation in this project.

Practice and Mentoring

© Cymra Haskell,
University of Southern California

1 Introduction

Explaining is a basic human activity, in which we all engage to a greater or lesser extent. How many of us have given directions to a newcomer in town? How many of us have helped someone with their homework or explained to a friend how we cook one of our favorite recipes? In this chapter we will explore the specific art and practice of explaining quantitative and mathematical material and discuss how to encourage tutors in a QMaSC to develop and refine their craft.

Explaining seems like a natural activity, so some might assume that competent practitioners of an art will be competent tutors of that art. However, think of a time when you were given directions but were still unable to find your way. The problem may not have been that the person you asked didn't know how to get to your desired destination, but they were not able to communicate that information to you. Perhaps they gave you too little information, so you missed a turn or got lost. Alternatively, they may have given you too much information, so you forgot some of it or simply couldn't see the big picture. If something as simple as giving someone directions can require knowledge and skills beyond that of knowing how to get to the desired destination, it is not hard to understand the potentially large disconnect between being a competent practitioner and a competent tutor of complex quantitative or mathematical material. Section 2 will examine what good tutoring of quantitative and mathematical material looks like and the skills and knowledge a person needs to be an effective and masterful tutor.

Suggested Citation: Cymra Haskell (2016), "Practice and Mentoring", `http://dx.doi.org/10.5038/9780977674435.ch15` in G. Coulombe, M. O'Neill, M. Schuckers (Eds.) *A Handbook for Directors of Quantitative and Mathematical Support Centers*, Neck Quill Press, `http://scholarcommons.usf.edu/qmasc_handbook`.

This material is based upon work supported, in part, by the National Science Foundation under Grant DUE-1255945. Any opinions, findings, and conclusions or recommendations expressed in this material are those of the author(s) and do not necessarily reflect the views of the National Science Foundation

The art and practice of tutoring involves both strong interpersonal skills as well as quantitative or mathematical habits of mind and deep content knowledge. Although a QMaSC is not a school of education, there are things it can do to enhance the effectiveness of its tutors. Section 3 will explore some of these and, in particular, will discuss mechanisms for mentoring tutors and how mentoring can accelerate the progress of tutors toward becoming masterful tutors. By providing its tutors with the framework, time, and space to refine their craft, a QMaSC is actually providing them with a very rich opportunity to develop both personally and professionally. In particular, tutors will deepen their content knowledge by asking questions they might never otherwise have thought to ask and by exploring connections between different areas in their subject they might never otherwise have explored.

2 The Art and Practice of Tutoring

Tutoring quantitative and mathematical material is an incredibly complex activity that draws upon both interpersonal skills as well knowledge, skills, and attitudes of mind specific to the subject. In this section we attempt to identify more precisely what these entail. The act of articulating them by necessity considers them distinct, however, each skill reinforces and draws upon the others, so in reality they do not fall into neat packages.

It is unrealistic to expect every tutor to possess all the knowledge and skills listed here. Instead, we attempt to articulate the knowledge and skills of masterful tutors. In the next section we will discuss mechanisms for mentoring tutors to increase their knowledge and skills and accelerate their progress toward becoming a masterful tutor.

Interpersonal Skills

To learn new material students need to venture outside of their comfort zone in order to broaden it. They can only do this if they feel secure and confident. It is the responsibility of the tutor to create an environment conducive to this. This involves a delicate juggling act between providing encouragement and validation while simultaneously addressing and correcting errors and misinterpretations. It also requires sensitivity and the ability to read body language in order to assess how the student is feeling. One mechanism tutors may use to create this environment is interacting with students at other levels; for example, they may discuss their hobbies or hopes and aspirations.

masterful tutors solicit input from students and listen carefully in return in order to assess students' understanding. This kind of listening is a little different from the kind of listening we do when chatting with our friends. Accurately articulating thoughts and understanding is very difficult at the best of times; it is all the more difficult for a student who is learning a new subject. As a consequence, students often express themselves imprecisely. masterful tutors listen empathically, expecting to find correct understanding but also being very careful not to let incorrect understanding go unnoticed. This is another delicate balancing act involving a host of skills, including familiarity with cultural norms and expressions as well as familiarity with common misunderstandings and misinterpretations of the material. One mechanism tutors may use to ensure they are assessing a student's understanding correctly is to rephrase what they have understood and verify with the student this was their intention.

masterful tutors are able to juggle the needs of students with the demands of the discipline. Students, learning new material for the first time, will often say things imprecisely or use incorrect mathematical notation. Constantly correcting their language and notation can undermine their confidence. A tutor will balance the needs of the discipline as regards attention to detail with the needs of maintaining student confidence. One mechanism for doing this is to let imprecisions ride while a general discussion is happening but then insist on correct notation as the student gains understanding and confidence.

masterful tutors do not let themselves be drawn into competition with students. Although this may sound almost ridiculous to have to articulate, it can occur. It is likely to happen when a tutor is explaining something to two or more students, one of which already has a relatively firm grasp of the content. This student may interrupt and try to guide the conversation to use an approach that makes sense to them but is different from the tutor's approach. Some tutors may feel the need to demonstrate their understanding is as great as this student's understanding and will react to this by skipping over the explanation, only dwelling on the technical points, or diminishing this student's approach or contribution. The effect is the other student has been abandoned and the confidence of the student that had a firm grasp of the material may have been undermined. Instead, masterful tutors do not feel any need to compare their understanding of the material with students' understanding and have the self-confidence to recognize and relish occasions when a student's understanding surpasses their own. These tutors remember that their obligation is to ensure all students understand the material and all students are validated in their own thinking. They will carefully consider the approach suggested by the student. If they think it is a better approach than their own, they will adopt it. If they think it is problematic, inferior to their own, or is an approach that does not permit them to address an issue that needs addressing, they will still validate the student's contribution, but continue with their explanation and possibly return to the student's approach when they are done. Of course, being able to do this requires the tutor to have a solid enough command of the subject to effectively evaluate alternative approaches and their implications.

Masterful tutors do not talk too much. It is a common complaint of students that they are able to follow the work of a tutor or teacher, but they are not able to do problems by themselves. To address this issue, masterful tutors make sure students drive discussions as much as possible. Some mechanisms tutors use to ensure they are doing this is to have the student be the one holding the pencil or to simply stop themselves, even in midstream, and consciously ask the student for input. To be effective in this, tutors have to be comfortable enough themselves with the material so, when a student takes an unexpected, but not incorrect, approach, they are able to assess the situation and allow the student to continue. They also need to reflect on how and why they do problems the way they do and be able to assess when other approaches are mathematically sound and will lead to correct solutions.

Masterful tutors are able to motivate students and engage them in their learning. They are able to generate excitement in the student regarding the material and a desire to perform well. One mechanism they may use to do this is to show excitement themselves or to talk about how the subject has been relevant to them, be it in 'real' life or in their other classes.

Knowledge, Skills and Attitudes of Mind Specific to the Subject

Masterful tutors have a sound knowledge of the subject they are tutoring. They are able to do the exercises required of the students and have performed or could perform well on a test on the content. However, competence in a subject as measured by the ability to excel in a test on the material does not constitute sufficient knowledge for masterful tutoring. In addition to competency, masterful tutors have both an attitude of mind similar to researchers and a profound understanding of the specific content they are tutoring. In the next few paragraphs we attempt to articulate more precisely what this attitude of mind and understanding encompasses.

Quantitative subjects tend to be very precise subjects in which every idea builds on previous ideas. Students do not usually learn in such a linear way; they have jumps and holes in their understanding of the big picture, and they go back later and fill in the holes. When explaining material, masterful tutors accommodate human learning styles by painting the big picture but also accommodate the needs of the discipline by carefully and accurately filling in the details. Most tutors who perform well on tests can do the latter but may have trouble with the former. Masterful tutors have done more than develop excellent problem-solving skills in the subject. They have reflected on the nature of the processes and procedures used, why they are valid, how they relate to other procedures, and why they are helpful in the situation at hand. They can articulate the larger goal and combine all the little steps into groups of steps, each of which has an intermediary goal. In this way they are able to help the student see the bigger picture.

Masterful tutors can solve a single problem in multiple ways and are cognizant of the principles on which these different ways are based. They understand the advantages and disadvantages of the different methods and when those methods are applicable. They have open minds, realize there are doubtless other methods of solution they have not yet entertained, and they are willing and able to assess the advantages and disadvantages of these other methods.

Masterful tutors are familiar with common misconceptions held by students who are learning the subject for the first time and with common errors made by students. Moreover, they understand the sources of these misconceptions and errors. In particular, when the source of misunderstanding is an unconscious comparison of the new practice with another, previously learned practice, the tutor is able to identify the source, draw it to the attention of the student, and enter into a dialogue with the student comparing the two practices.

Masterful tutors are able to analyze the conceptual steps needed to master a particular quantitative practice and can identify the concept with which a student is having difficulty. Moreover, they can formulate examples and exercises on the spur of the moment that illustrate the concept and provide students with a mechanism for internalizing it and putting it into practice.

Masterful tutors are familiar with different models and metaphors used for understanding a mathematical practice. They are able to identify the metaphor to which the student is relating the practice. They are able to come up with examples that employ that metaphor and, if appropriate, are able to enter into a discussion with the student about other metaphors that also capture the practice.

Masterful tutors understand how the particular class relates to other areas of the discipline or to other disciplines. Not only does this give them the tools to generate excitement for the subject and motivate the questions being asked; but, by comparing a particular mathematical practice with a practice in another field of mathematics or another discipline, they can solidify students' understandings of that practice.

To aid in understanding the differences between competency in a mathematical subject and the more profound understanding needed to be a masterful tutor of that subject, we draw on an analogy inspired by Richard Skemp [1]. Consider the city of Los Angeles. This is a large city, situated in an even larger megalopolis covering hundreds of square miles. Some of the cities in the megalopolis (like Beverly Hills and Santa Monica) have completely merged with each other and a driver cruising in Los Angeles could pass from one to the other without even being aware of it. Other cities (like Temecula or Oceanside) are clearly separate from Los Angeles though there are many roads that connect these cities to Los Angeles and to each other. The megalopolis is analogous to the whole field of mathematics and Los Angeles is analogous to the subject area of mathematics being tutored. Students learning the subject for the first time are like tourists visiting Los Angeles; this may or may not be their first time visiting the city, so the city may be totally new to them or they may be somewhat familiar with some parts of it. Tutors who have mastered the subject as measured by their ability to excel on a typical test on the material but do not yet have the profound knowledge of a masterful tutor are like typical residents of Los Angeles. They are familiar with their neighborhood and have learned the quickest routes to get from home to work and from home to their favorite haunts in the city. However, when they venture out of these familiar stomping grounds they can get lost and they may not even know the most direct route from one of their favorite haunts to another without going via home. They do not have the big picture of the city and they are not aware of how the city is laid out and how the different parts of the city are interconnected. masterful tutors are like taxi drivers in the city. They are very familiar with the whole city and, given any two points in the city they can find many routes from one to the other; to a tourist wanting to sample the delights of the city they will give one set of directions and to the business man trying to get there in the most efficient way possible they will give another. They can probably also find their way from Los Angeles to one of the neighboring cities like Temecula, though they won't have the same familiarity with Temecula as they do with Los Angeles.

In Appendix B, there hypothetical tutoring sessions are presented and discussed. The three sessions differ primarily in the tutors' knowledge, skills and attitudes of mind specific to the subject and illustrate how important these are for effective tutoring.

3 Mentoring Tutors in Order to Improve their Art

In whatever way your QMaSC decides to train and mentor tutors, it is important to remember and practice all the principles of teaching you are seeking to elicit from the tutors. In particular, you should endeavor to build their self-confidence, not break it, you should make them feel you are their ally and not their judge, you should encourage them to critically evaluate their own effectiveness as tutors, and you should empower them to own their learning and try out new techniques.

We have identified three different (but interrelated) activities that will accelerate the progress of tutors toward becoming masterful tutors. We recommend that QMaSCs find ways to ensure their tutors are engaging in these three activities.

Identifying the Knowledge, Skills, Attitudes of Mind, and Behavioral Traits of a Masterful Tutor

Unlike mathematics in which every step is carefully constructed and can be pondered for several minutes, hours, days, or even years, tutoring happens in real time. When we tutor we make thousands of largely unconscious decisions every minute. Will we speak or remain silent? Will we smile or frown? What timbre of voice will we use? Will we take hold of the pencil ourselves or insist that the student write? Will we interrupt to address an issue or let the student continue? Which misunderstanding will we address first? Will we insist on proper notation or let it ride? Will we address an issue at a purely procedural level or will we embark on a lengthier conceptual discussion of it? In order to adopt the largely unconscious behavioral traits of a masterful tutor it can help enormously to articulate exactly what it is that constitutes these traits. Since these traits form a complex set of contextual behaviors, articulating them is a far from easy task and is something that should be an ongoing process. We can help tutors in this process by directing them to literature on the subject as well as providing the framework, time, and space for in-depth discussions. Some examples of literature on the subject include 'Knowing and Teaching Elementary Mathematics,' by Liping Ma [2], 'Relational Understanding and Instrumental Understanding,' by Richard Skemp [1], 'Measuring Teacher Quality in Practice,' by Deborah Ball and Heather Hill [3], a report put out by the Conference Board of the Mathematical Sciences, 'The Mathematical Education of Teachers II' [4], and Appendix B of this chapter. Periodic assignments of such readings followed by group meetings of the tutors for discussion can be valuable for advancing tutors toward becoming masterful tutors.

Watching, Analyzing, and Reflecting on Tutoring Sessions

Articulating the traits of a masterful tutor is one thing, internalizing and adopting those traits is another. Watching, analyzing, and reflecting on tutoring sessions can accelerate the adoption of these traits. The challenge here is to provide tutors with these sessions. Three ways to do this are given below.

All tutors have experienced their own tutoring sessions, so they should be encouraged to reflect critically on these sessions. To this end a QMaSC might consider requiring its tutors to keep a journal cataloging their tutoring sessions including reflections on the sessions. To stimulate deeper reflections, the tutors could be provided with prompts and there could be regular group meetings in which the tutors discuss each other's experiences. We give two examples of such prompts below.

Reflective Prompt 1: Choose one of your tutoring sessions in the QMaSC and answer the following questions.

a) Describe the student's problem.

b) What are the potential misunderstandings that a similar student might have when trying to solve such a problem?

c) What questions did you ask the student in order to assess their level of understanding and any misunderstandings they may have? What were their responses? Did you anticipate most of their responses or were you surprised by any of them? What were the main things that prevented the student from being able to solve the problem independently?

d) How did you address these issues in your session with the student?

e) What other problems did you develop in order to cement the student's understanding of the concepts that you worked on with him or her?

f) In hindsight, can you think of any ways you may have improved this session?

Reflective Prompt 2: Choose one of your tutoring sessions in the QMaSC in which your initial approach to solving the problem was not the approach you finally settled on with the student. Describe the student's problem. Explain your initial idea of how to solve the problem. Explain how you ended up solving the problem with the student. Describe why and how you ended up doing the problem in a different way with the student. Did you also explain your way to the student? Why or why not?

A second way to provide tutors with tutoring sessions to watch, analyze and discuss is to have tutors observe each others' sessions. This can be very valuable though its success depends significantly on the way it is implemented. The problem is three-fold. If the observer is not a masterful tutor, she may not notice missed opportunities or imagine alternative approaches the tutor could have taken. Secondly, the observer is often not critical enough for fear of hurting the other tutor's feelings. Thirdly, since tutoring involves a huge number of split-second decisions that can be based on the tutor's highly nuanced assessments of the student's level of understanding, the observer may forget the particular criticisms she had and, even if she does remember, the observer and/or the tutor may have forgotten the exact contexts in which they arose. This method works best when the observer is herself a masterful tutor, the tutor trusts the observer and can constructively handle criticism from her, and no time elapses between the session and the discussion of the session.

A third way to provide tutors with tutoring sessions is to use videos of sessions. There are many advantages to watching, analyzing, and discussing a tutoring session on video as compared to one that has just been observed. One big advantage is that the viewers do not need to rely on their memories when discussing particular events in a video; instead, they can zoom in on those few minutes. A second advantage is that you can ensure you are watching a session that provides a rich source of material for discussion. If the video does not involve tutors in your QMaSC, a third advantage is that tutors watching the video won't feel compelled to mitigate their criticism in order not to hurt the tutor's feelings. However, using videos requires the QMaSC to have a rich source of video material. A QMaSC can explore the possibilities of making videos of sessions in their own QMaSC; however, because of issues relating to student confidentiality, this may not be possible at all or may require the students involved to sign permission forms. Commercial or online sources of videos are mostly of classrooms rather than tutoring sessions but these can still be highly valuable for discussing many of the issues that arise in tutoring. One commercial source is the book and DVD by Jo Boaler and Cathy Humphreys 'Connecting Mathematical Ideas' [5]. The DVD contains videos of Cathy Humphreys teaching in a middle school classroom and the book has written commentary by both her and the mathematics educator, Jo Boaler. On the Max Ray Blog [6] there is a collection of links to videos of people doing and teaching mathematics.

Engaging in Profound, Professionally Oriented Explorations of the Subject

As we have seen, the profound, professionally oriented understanding of the master tutor goes beyond that of a person who can excel in a test on the subject. Furthermore, it is unrealistic to expect tutors who are newly hired in the QMaSC to have this kind of understanding. If tutors do not obtain this understanding in their classes, when, where, and how will they obtain it?

In her comparison of elementary school teachers in China and the United States [2], Liping Ma found that the Chinese teachers in her study who had profound understanding had obtained it primarily on the job through the study of teaching materials, mathematical and pedagogical discussions with their colleagues, and deep reflection on the material they were teaching and the students' responses to their teaching. If a QMaSC wants its tutors to gain this understanding, it is important that it provide them with the time, space, and framework for engaging in such study, discussions with their colleagues, and reflections. To that end we recommend it organize professional development workshops for its tutors. A good format for such a workshop is to divide the tutors into small groups (say, three or four tutors each) and give each group a mathematical question to ponder and discuss. The question could be the same for each group or different for the different groups. After a period of time the groups reconvene into one large group, each group presents their thoughts and findings to everyone else, and a discussion follows. (Chart paper can be useful for this.)

Appendix C provides examples illustrating what such workshops might look like. The modules in this appendix were developed by the author of this chapter for a class in the Department of Mathematics at the University of Southern California called 'The Foundations of Mathematics and the Acquisition of Mathematical Knowledge.' In this class K-12 mathematics is used as a conduit for a deeper exploration of mathematics itself and for an exploration of how we acquire and learn mathematics. Most of the activities in this class are suitable for professional development workshops for tutors in mathematics. Some of the modules you'll find in the appendix are based on elementary mathematics and might not, at first, seem relevant to math tutors in a QMaSC situated in a college. However, they have been included for a variety of reasons. First of all, these workshops should be accessible to all tutors in the QMaSC, so they could be done in a group meeting of all the tutors together. Secondly, they certainly do not preclude very rich mathematical discussions and they provide a good starting point for all tutors; some of the issues that arise in more advanced mathematics stem from these issues in elementary mathematics and starting with examples in more advanced mathematics may be too intimidating for some tutors. Thirdly, for QMaSCs that service courses outside of mathematics departments, the issues raised in these workshops might indeed be relevant to the material they are tutoring. Finally, since elementary mathematics will be very familiar to all the tutors in your QMaSC, providing professional development in this area will serve the purpose of highlighting the kind of profound and professionally-oriented knowledge for which tutors should be aiming. This will encourage them to look for that kind of knowledge in the more advanced material they are tutoring. There are also other modules in the appendix that involve college level mathematics.

For many years, mathematics educators at the University of Michigan have been studying the mathematical knowledge needed for teaching mathematics at the elementary and middle school levels. Their work is described in [7]. More information can be found at the website of the Study of Instructional Improvement [8] or the website of the Learning Mathematics for Teaching project [9]. They have released a series of questions that are posted on these websites that could be used as the framework for one or more professional development workshops. Although the mathematical material in many of these questions is probably less advanced than the subject matter being dealt with in a QMaSC that is housed in a college, as we noted in the paragraph above, having professional development workshops on such material can be very valuable and we recommend their use. In particular, many of the questions concerned with the appropriateness of examples or whether examples provide illustrations of particular metaphors, can provide good food for thought and stimulate tutors to start thinking about the more advanced mathematics they are tutoring in these terms.

4 Summary

Tutoring in a QMaSC is a highly sophisticated activity that involves well-developed interpersonal skills as well as a profound understanding of the subject matter. As such it provides rich growth and learning opportunities, and tutors will undoubtedly develop both personally and professionally through their employment in a QMaSC. In this chapter we have attempted to articulate some of the skills and knowledge needed to be a masterful tutor and have suggested some steps QMaSCs can take in order to accelerate the progress of their tutors toward becoming masterful tutors.

5 Bibliography

[1] R. Skemp, "Relational understanding and instrumental understanding," *Mathematics Teaching in the Middle School*, vol. 12, pp. 88–95, September 2006. Reprinted from *Mathematics Teaching*, December 1976.

[2] L. Ma, *Knowing and Teaching Elementary Mathematics: Teachers' Understanding of Fundamental Mathematics in China and the United States*. New York: Routledge, 2010.

[3] D. L. Ball and H. C. Hill, "Measuring teacher quality in practice,"

[4] Conference Board of Mathematical Sciences, "Statistical Abstract of Undergraduate Programs in the Mathematical Sciences in the United States," 2013.

[5] J. Boaler and C. Humphreys, *Connecting Mathematical Ideas: Middle School Video Cases to Support Teaching and Learning*. Portsmouth, NH: Heinemann, 2005.

[6] M. Ray, "Max ray blog." http://mathforum.org/blogs/max/.

[7] H. C. Hill, S. G. Schilling, and D. L. Ball, "Developing measures of teachers' mathematics knowledge for teaching," *Elementary School Journal*, vol. 105, no. 1.

[8] "Study of Instructional Improvement (SII)." http://www.sii.soe.umich.edu.

[9] "Learning Mathematics for Teaching (LMT) Project." http://www.sitemaker.umich.edu/lmt.

6 Appendix A: The QMaSC as a Community

In this chapter we discussed the art of tutoring and saw that by providing tutors with the framework, time, and space to refine their craft, a QMaSC not only improves the tutoring services it provides but also provides the tutors with a valuable opportunity to develop personally and professionally. This, of course, benefits the institution, its students, and, most especially, the tutors themselves.

Some QMaSCs may hire students in administrative roles as well as, or instead of, as tutors. Although employment in this capacity doesn't provide students with the same learning opportunities as tutoring, it still has advantages both for the students and for the institution or department in which the QMaSC is housed. In particular, it provides students with the opportunity to do a job that is meaningful to them (rather than, for example, wash dishes at the cafeteria), it provides them with an opportunity to demonstrate responsibility, and it is a valuable work experience they can put on their resume. Students working in a QMaSC will feel they are contributing to the mission of the institution (which, indeed, they are) and, as a result, will be more invested in the institution and their own degree. This will probably result in greater retention and recruitment of students. Most importantly, through their employment, the students may get to know the faculty and other students (including graduate students, should the institution have a graduate school). These connections can be invaluable to them. Not only will they draw upon them when making important life decisions, including what courses they should take, whether or not they should apply to graduate school (for undergraduates), and to what kinds of jobs they should apply, but through them they will learn what it means to be a mathematician or other academic professional. This can help them imagine themselves in one of these roles and can provide motivation in their classes. Of course, it also benefits the institution when students are happy, motivated, and successful.

Notice that essentially all of these benefits arise out of a strong sense of community within the QMaSC and the department in which it is housed, and the best way for the QMaSC to enhance these benefits is to enhance the sense of community. The easiest way to do this is to have social gatherings with food! The power of these gatherings cannot be overemphasized. They could be gatherings for just the employees of the QMaSC, or they could include faculty and/or graduate students in the departments served by the QMaSC. Employees should also be encouraged to suggest improvements in how the QMaSC is run, and regular group meetings of the employees to discuss the running of the QMaSC can be very valuable both for ensuring the employees have a common vision and for soliciting feedback from the people involved in the day-to-day affairs of the QMaSC. It is good to blur the lines between the responsibilities of different types of employees; however, this has to be done with care. It can be problematic if an employee who is hired essentially as an administrator is not qualified to act as a tutor takes on the role of tutor. Also be careful not to oblige employees to take on responsibilities that are not in their job description and might make them feel overwhelmed. Soliciting employees' help and allowing them to have a say in their responsibilities is a good idea so all employees have a sense of being valued for any contribution they can make to the QMaSC. If you do not package your employees in a box you are more likely to access the personality traits, unique to them, that will benefit the QMaSC, while also giving them the sense of being valued.

7 Appendix B: Examples Showing Levels of Tutor Understanding

To illustrate some of the knowledge, skills and attitudes of mind of a masterful tutor, we'll consider a scenario in which a student walks into a QMaSC with a question and we compare the hypothetical responses of different tutors. Since many readers of this chapter may not be mathematicians we have deliberately chosen an example involving elementary mathematics to which essentially all readers will be able to relate. This serves the added purpose of highlighting the sophisticated mathematical thinking in which masterful tutors engage even when tutoring elementary material. We will look at examples involving more advanced mathematics in Section 3.

Example: To set the stage, imagine the QMaSC situated at a high school. A new student at the school and recent immigrant to the country has come to the QMaSC with a question about how to perform subtraction. He is attempting to evaluate $43 - 25$. His work is shown in the picture below.

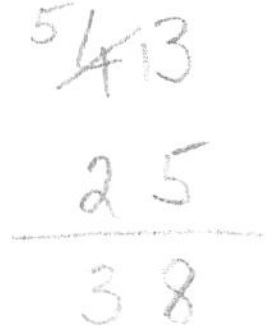

The student knows he has the wrong answer because when he adds 38 and 25 he gets 63 instead of 43, but he doesn't know what he did incorrectly.

Here are the hypothetical responses of three different tutors in the QMaSC.

Tutor 1

Tutor:	Can you draw a picture to illustrate what you did?
Student:	What do you mean?
Tutor:	Well, can you make up a story of when you might want to do this subtraction problem?
Student:	How about, I have 43 candies and I give 25 of them to my friend. How many do I have left?
Tutor:	Excellent! Can you draw me a picture now?
Student:	You mean of the candies?
Tutor:	Yes.
Student:	Okay.

Student dra
candies cros
He loses cou
himself.

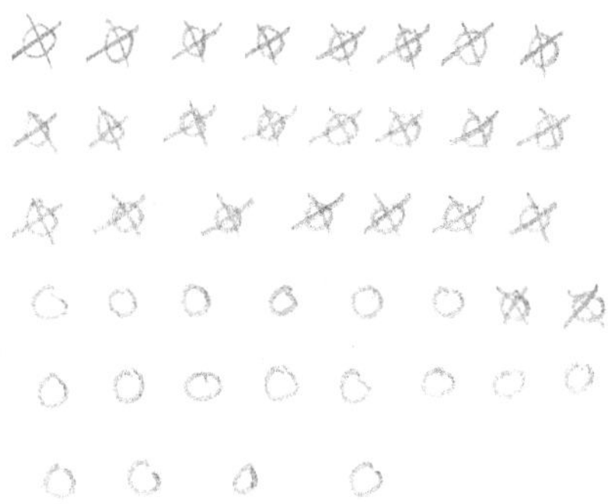

Tutor:	So how many do you have left?
Student:	18!
Tutor:	Excellent!

The session ends.

Tutor 2

Tutor:	Oh, you're just making a small mistake; you need to take 1 away from the 4 instead of adding it.

Tutor points at the student's work.

Student:	You mear

Student works the pr

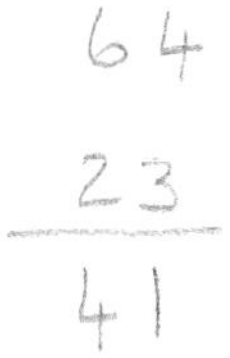

Student:	That would give me 18. Yes, that's right because $18 + 25 = 43$! Hmm, but I always thought we added.
Tutor:	No, you should take 1 away. Let's do another problem to reinforce the idea. How about $64 - 37$?

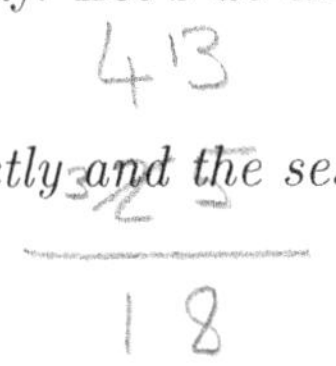

The student works the problem correctly and the session ends.

Tutor 3

Tutor:	Let's see. It is great that you noticed it was incorrect! Can you explain to me what you did?

Student: Well, I want to take 5 from 3 but 5 is bigger than 3 so I can't. So I go to the ten's place. Since I'm giving a ten to the 3, I need to give it to the 4 as well. That gives me 5 here and 13 over here. Now 5 from 13 is 8 and 2 from 5 is 3.

Tutor: Good. You seem to have a really good understanding of what you're doing but you're making a mistake. Let's see if I can't help you find your mistake. As you noticed, the number forty-three is 4 tens and 3 ones. So, can you draw me a picture of 4 tens and 3 ones?

Student: You mean like rods?

Tutor: Yes, exactly.

Student draw.

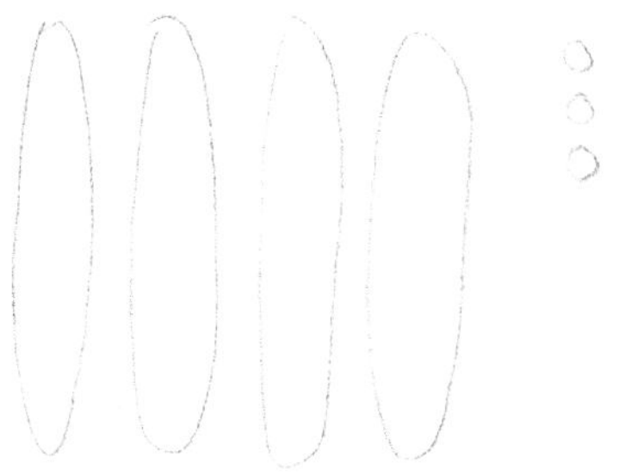

Tutor: Let's put some lines in the rods so that we can see that they are each composed of ten ones. Try to do it carefully so that all the ones look the same size.

Student modi

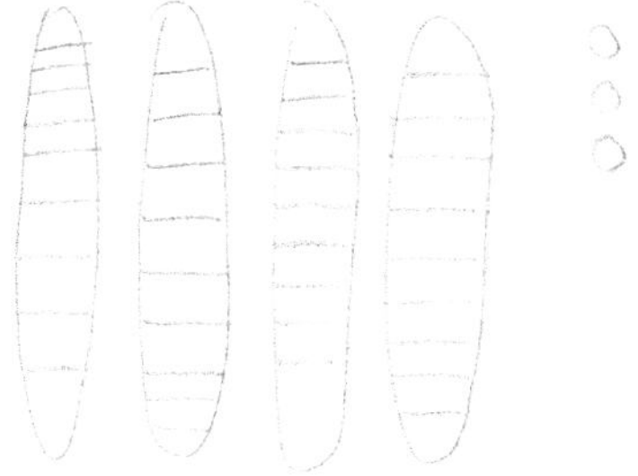

Tutor: Okay, that'll work. Just remember that all the little pieces are the same size as each other. Now, we want to take away twenty-five which is 2 tens and 5 ones. How can we do that?

Student crosses out 2 tens and 5 ones as shown below.

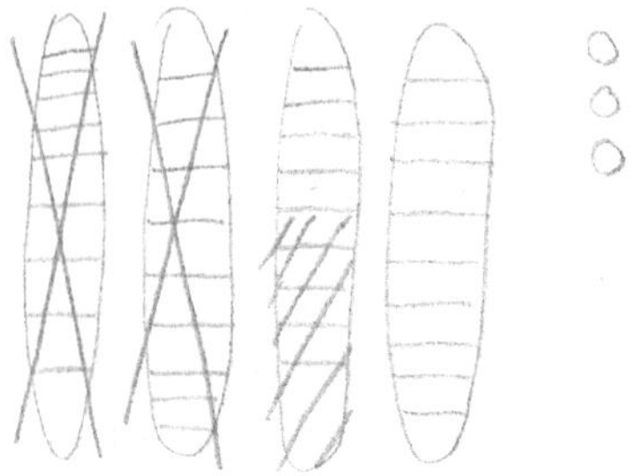

Student: I have 18 left!

Tutor: All right. Good. We've got the correct answer. But we want to understand what's going on when we do the problem on paper so that we can find your mistake. So, let's try to copy the work you did on paper in our picture. Let's start again with our picture of 4 tens and 3 ones. Can you draw it again? Since we're doing it again anyway, try to draw it a little more carefully. Here.

Student a

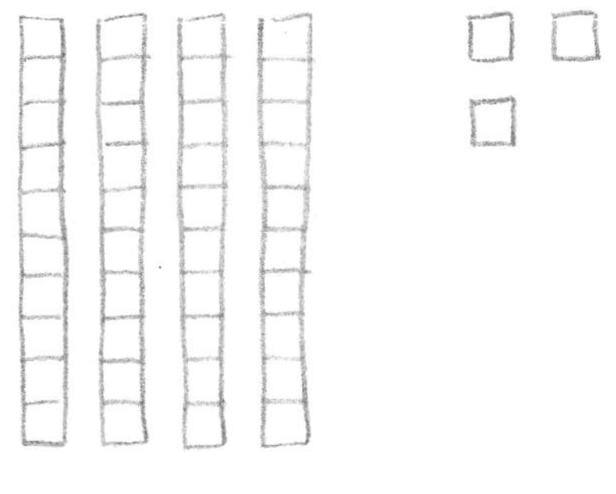

Tutor: Now, what's the first thing you did when you tried to do this problem on paper?

Tutor points to the student's calculation on paper.

Student: I tried to take 5 from 3.

Tutor: That's right. Notice, that means that you're trying to take 5 ones from these 3 ones here.

Tutor points at the 3 single ones in the picture.

Tutor: Actually, let's label our picture so that we can really see how it relates to what we're doing on paper.

Tutor labels the picture as shown below.

Cymra Haske

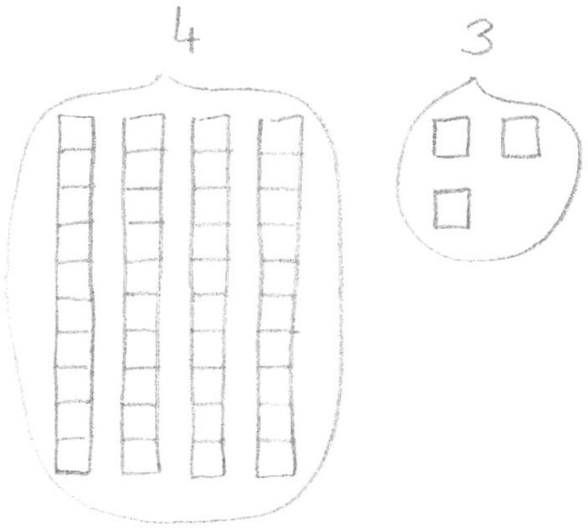

Tutor: Now we can see that it is labeled as 4 tens and 3 ones which is how it looks on paper. Okay since we can't take 5 ones from 3 ones, where are the extra

Student

Tutor:

Student

Tutor:

Tutor is

Tutor:

Student

The stu

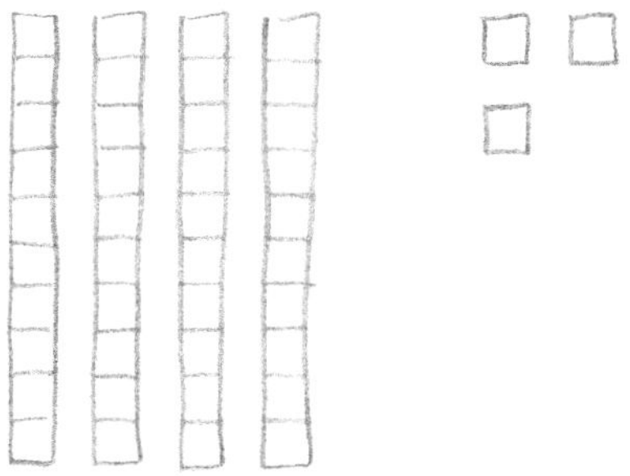

Tutor: Great! So how many tens and how many ones do we have now?

Student: We have 3 tens and 13 ones.

Tutor: Can you label that on the picture the way we did before?

Student: Like this?

Student labels the picture as shown below.

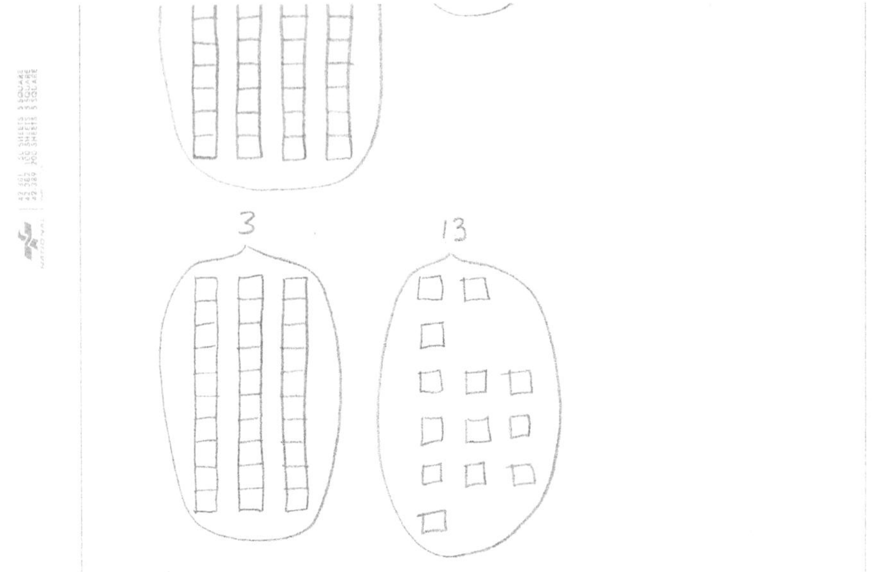

Tutor: Awesome! All right. Remember what we're doing. We're taking 2 tens and 5 ones away from these 3 tens and 13 ones. Let's compare that with what you wrote on your paper.

Tutor points to the student's original work.

Student: What do you mean?

Tutor: Well, we started with 4 tens and 3 ones. That's over here in the picture.

Tutor points to the 4 tens and 3 ones in the calculation on paper and in the picture.

Tutor: We wanted to take away 2 tens and 5 ones. On paper we started by trying to take 5 from 3.

Tutor points to the calculation on paper.

Tutor: That corresponded to trying to take 5 ones from the 3 ones here.

Tutor points to the 3 ones in the picture.

Tutor: We couldn't do that so we went over to the long rods and turned one of those long rods into ten ones.

Tutor is pointing at the picture.

Tutor: So over here on the paper, we crossed out the 4 and put a 1 in front of the 3 to indicate that we now had thirteen ones.

Tutor points to the calculation.

Tutor: But look at your picture. When you broke up that long rod into 10 pieces how many long rods did you have left?

Tutor is pointing at the picture.

Student: There are 3 left.

Tutor: Now look at what you wrote. How many tens did you say that you had left?

Tutor is pointing at the calculation.

Student: Oh, I see. Over here I have 3 long rods but over here I wrote 5.

Tutor: Yes. Can you correct your work now?

Student: So I should put 3 here instead of 5?

Tutor: Exactly! Can you finish it?

Student: Yes. Over here I take 5 ones from the 13 ones leaving 8 ones, and 2 tens from 3 tens leaving 1 ten. That's 18!

Tutor: Shall we try another example?

Student: Okay.

They do another example drawing both the picture and writing the algorithm down on paper. The student realizes that they should subtract 1 when they 'borrow' from the ten's place instead of adding one.

Student: But I'm confused. I've always added 1 to the ten's place.

Tutor: Really?

Student: Yes, I'm sure I always added 1 and I never seemed to have a problem before.

Tutor: Hmm. Were you using a different method? You're new here, aren't you? Did you

Student: Yeah, I j

Tutor: Can you

Student: Yes, I th

The student shows t

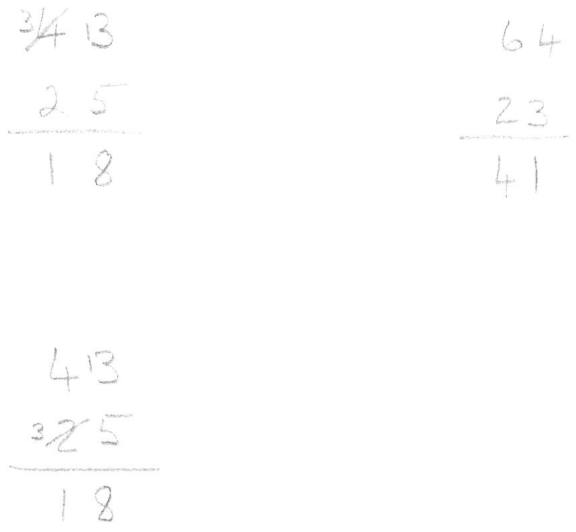

Student: We always added 1. I started adding to the top number when I came here, because that's what I saw people doing here. But it gives a different answer.

Tutor: Oh, that's interesting. Let's see if we can't work out what you're doing by drawing some pictures. We have the 4 tens and 3 ones that we had before. This time, since we're changing what we're taking away, we'd better draw a picture of what we're taking away too.

Tutor draws the picture below.

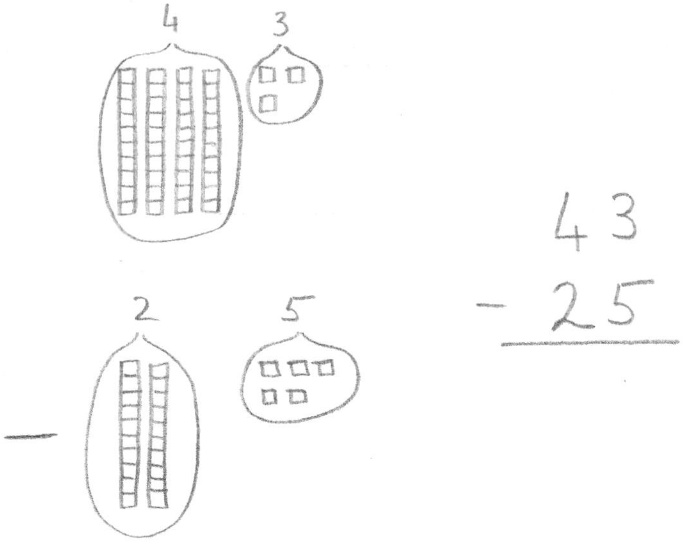

Tutor:

Tutor modif

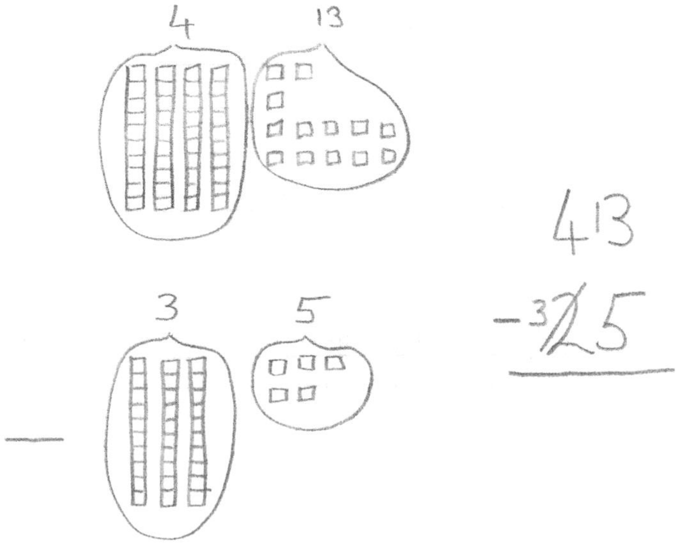

Tutor: Oh, that's cool! I see what we're doing. Do you understand how this works? Why can I add 1 ten over here and 10 ones over here?

Tutor is pointing at the picture.

Student:	Oh yes! If we add ten to both terms then the difference is still the same! That's cool.
Tutor:	Right. We've changed the problem to a new problem that has the same answer! The new problem is to subtract 3 tens and 5 ones from 4 tens and 13 ones.
Student:	Cool.
Tutor:	Let's try to do another problem this way. Can you think of another problem for us to do?
Student:	How about $83 - 57$?

Student and tutor work this problem using this other method, drawing pictures as they go along. The session ends.

Let's take a closer look at the three approaches taken by the different tutors to this situation. All three tutors had pretty good people skills and in no way made the student feel uncomfortable or stupid. All three tutors involved the student in the solution; they didn't just show the student how to do the problem but made sure the student was working the problem himself. All three tutors knew how to do the problem. (Although this is not obvious from the dialogue in the case of Tutor 1, we will assume it was the case in this hypothetical situation.) In summary, the tutors were similar in terms of their interpersonal skills and their ability to perform well on a test on the subject matter. Where the tutors differed was in the mathematical approaches they took to the problem and the degrees to which they engaged the student in a mathematical discussion.

Tutor 1 had excellent intentions in getting the student to think about the problem concretely. With mathematically less sophisticated students than the student in this example, this may have been an appropriate starting point and it probably wouldn't have been harmful as a starting point even with this student. What is problematic is that the tutor stopped there. The picture that was drawn and the method of calculation bore almost no relationship to how the calculation is represented and performed on paper. Indeed, the mathematical technique employed was that of counting. Although it is important for students to realize that counting is a valid approach to solving subtraction problems and is the theoretical underpinning of the subtraction algorithm, it is impractical to perform subtraction problems by counting when the numbers are large and that is the reason we use the subtraction algorithm. Indeed, this was highlighted by the fact that the student made multiple errors when trying to perform the calculation this way the first time. By stopping where he did, this tutor may have sent a dangerous message by suggesting that counting is the appropriate approach to take towards subtraction problems of this type. Moreover, the tutor missed an important cue; the fact that the student knew how to check his answer and actually did so is a strong indication that the student was more mathematically adept than one might first suppose in a student having trouble with the subtraction algorithm. This meant that the tutor's approach was probably inappropriate for such a student. In the final analysis, although the student left with the correct answer to that particular problem, he gained essentially nothing from the interaction; he learned no tools for tackling subtraction problems in general and gained no understanding of the mathematics underlying the algorithm. In fact, the interaction may have been unintentionally harmful; by not attempting to access the student's mathematical thinking, this tutor may have sent a message that such thinking will only lead to confusion and it is better to do problems in a simplistic way.

Tutor 2 had a completely procedural approach to the problem. He identified the student's mistake and was effective at correcting the mistake without making the student feel stupid. He was adept at teaching the algorithm procedurally, as indicated by his ability to create an appropriate problem to solidify the student's mastery of the procedure, and the student probably left the QMaSC knowing how to perform the algorithm. However, the student learned little or no mathematics. Moreover, what drew the student to the QMaSC in the first place was confusion arising from having seen two different approaches to the algorithm, and this was never addressed. Like Tutor 1, this tutor missed the cue indicating the mathematical sophistication of the student. Moreover, the student gave him another cue that he also missed; when the student said, "I always thought we added," the tutor jumped to the conclusion that the student was remembering the standard algorithm used in this country incorrectly. In fact, the student had learned a different algorithm in which you do add instead of subtract; his problem arose when he came to this country, saw the algorithm used here, and mixed it up with what he had learned before. By not catching this cue, Tutor 2 missed the opportunity to validate the student's mathematical thinking and engage in a mathematical discussion with the student on the nature of the two algorithms. He also missed the opportunity to deepen his own mathematical thinking and understanding of the subtraction algorithm.

We can only speculate about why the first two tutors missed these important cues (especially since the example is purely hypothetical!). It was probably a combination of reasons. One reason may be that these tutors felt the need to be authority figures in the room and, as a result, were not mentally prepared to learn from the student. Another reason may be they did not think mathematical flexibility is important; once they had learned one way to solve a problem they didn't think it interesting or instructive to explore other ways. Yet another reason may be they were not very confident in their own mathematical abilities and were not accustomed to engaging in mathematical exploration of a problem; although they were good at learning procedural techniques they had not developed mathematically to the level where they were able to analyze and explore those techniques.

It is interesting to observe that although Tutor 1's approach in its use of manipulatives (in the form of pictures) was more modern than Tutor 2's, in this case he was probably less effective than Tutor 2, since his choice of manipulatives didn't relate to the way the problem is performed on paper. At least with Tutor 2 the student left the QMaSC able to perform the subtraction algorithm correctly. That was not the case with Tutor 1.

With Tutor 3 we see masterful tutoring taking place and it is immediately recognizable as such. He started by congratulating the student on noticing his error thereby reinforcing the student's sense of his own mathematical competence. Before proceeding with any kind of explanation, he asked the student to explain his method of approach and listened carefully to the student's answer. This was instrumental in assessing the mathematical maturity of the student and his level of understanding of the subtraction algorithm. The student's explanation revealed he had a relatively sophisticated understanding of place value and he knew place value played a role in the calculation. The tutor hooked onto this immediately by asking the student to draw a picture of the number forty-three that illustrated how the number appears on paper. They then worked through the algorithm reinforcing each step with the picture, so that the pictures illustrated and motivated the algorithm. The tutor knew when to enforce precision and when not to do so; the first time the student drew 4 tens and 3 ones, the tutor did not insist that all the ones be the same size. The second time though, when the student had gained confidence, the tutor suggested they pay more attention and make it more precise. After completing the problem the tutor realized that, although the student had understood the error he had made in that particular problem, that didn't necessarily mean the student could recognize the algorithm he had in his head was incorrect. However, the tutor didn't feel compelled to articulate this himself for the student. Instead they worked through another problem together with the pictures and on paper, so the student saw again the mistake he was making when doing the problem on paper. Doing this permitted the student to identify the cause of his mistake himself. It also made space for the student to articulate the source of his confusion; namely that he had modified his method to account for what he saw people doing in this country. By this time the tutor knew the student was quite sophisticated mathematically, so rather than assume the student was incorrectly remembering the subtraction algorithm, he drew on his knowledge that this student was a recent immigrant and questioned him to find out what he might have been doing in the past. This led to the incredibly rich discussion and discovery on part of both the tutor and the student of this alternative way to do subtraction. The student left this encounter not only competent at performing the subtraction algorithm correctly, but with an increased sense of his own mathematical competence, a richer sense of the power of analytical and mathematical thinking, and a much deeper understanding of place value and the role it plays in subtraction.

There was a potentially rich discussion in which Tutor 3 did not engage with the student, be it consciously or unconsciously. The student said, "since I'm giving a ten to the 3, I need to give it to the 4 as well." What was the source of this misconception? Since we now know the student was accustomed to a different algorithm in which you give a ten to both the minuend and the subtrahend, this misconception probably originated in the explanation his teacher in his home country gave for the algorithm. However, this student is a high school student probably studying algebra, so this misconception was no doubt reinforced by the idea that whatever you do to one side of an equation you should do to the other. There could have been a whole discussion of why you need to do this when you are working with an equation but you do not need to do this in this setting. Following the student's explanation of the algorithm, the tutor needed to make a choice; embark on this discussion or investigate the algorithm directly. The tutor chose the latter and that was probably the best choice. Without a good understanding of the algorithm, the student would probably have had difficulty understanding why you perform different actions to the number in the ten's place and the number in the one's place, so a discussion of this misconception at that point in time would probably have been confusing and unproductive. Unfortunately, after the student understood the algorithm, it is unlikely he would have had the patience to confront his former misconception and the source of that misconception. So, that was a discussion they didn't get to have. Tutors have to make these kinds of decisions all the time.

8 Appendix C: Examples of Professional Development Workshops for Tutors

This appendix contains examples of professional development workshops for tutors. They are modules developed by the author of this chapter for a class taught in the Department of Mathematics at the University of Southern California called 'The Foundations of Mathematics and the Acquisition of Mathematical Knowledge.' Engagement in such workshops will increase tutors' profound, professionally oriented understanding of the subjects in which they tutor and accelerate their progress toward becoming masterful tutors. QMaSCs should adapt the workshops to fit the mathematical sophistication of their tutors and their time constraints.

Models for Multiplication and Division of Positive Numbers

Materials Needed: Ample supplies of chart paper, markers, and whatever is needed to attach the chart paper to the wall of the room for students to display their work. Moderator should be equipped with plenty of examples of story problems both using fractions and not using fractions that illustrate each interpretation of multiplication and division for distribution to the groups for item (11). These stories should contain multiple examples of each interpretation of multiplication and of each interpretation of division. They should also contain many examples involving multiplication and division by fractions. They could be written on individual cards or could be put together on a sheet of paper to be distributed to the groups.

Mathematics covered: We use models and metaphors to understand the arithmetic operations we do with numbers and, in applications, to know when we should use these operations. The primary model we use to understand addition of positive numbers is *combine* and the corresponding model for subtraction is *take away*. Notice these models break down when we expand our number system to include negative numbers. This module is about the models we use for multiplication and division. There are two primary models for multiplication, in one of which the two products play very different roles ($a \times b$ means a copies of b things) and in the other of which they play the same role (the area of a rectangle whose one side has length a and other has length b). The consequence of the second is that multiplication is commutative; when combined with the first this means there are actually *three* ways to interpret a specific expression $a \times b$. The three ways are distinguished by the *units* of the numbers involved. This in turn means there are three models for division.

Multiplication: The product $c = a \times b$ has three different interpretations.

1. a) It is the total size of a objects each of which has size b. In this case a is unitless, b can have units, and c has the same units as b or, more generally, a has units, the units of b are some (other) units divided by the units of a, and the units of c are those other units.

   b) It is the total size of b objects each of which has size a. In this case a has units, b is unitless, and c has the same units as a or, more generally, b has some units, the units of a are some (other) units divided by the units of b, and the units of c are those other units.

2. It is the area of a region that is a units in one direction and b units in the other direction. In this case a and b have identical units and the units of c are those units squared.

Notice how a and b play very different roles in (1)(a) and (1)(b).

Division: The quotient $a = c \div b$ has three different interpretations.

1. It is the the number of 'pieces' when something of size c is divided into pieces each of which has size b. In this case c has whatever units it has, the units of b are some (other) units divided by the units of c, and the units of a are those other units. This is called the *measurement model of division.* It is the 'inverse' of (1)(a) of multiplication.

2. It is the size of each piece when an object of size c is divided into b pieces. In this case, c and b have whatever units they have and the units of a are the units of c divided by the units of b. This is called the *partitive model of division.* It is the 'inverse' of model (1)(b) of multiplication.

3. It is the length of one side of a rectangle when the area is c and the width is b. In this case a and b have the same units and the units of c are the square of these units. This is called the *area model of division.*

Adults who have successfully mastered multiplication and division often do not consciously distinguish between the different meanings of $a \times b$ and $c \div b$ given above. Also, they may not distinguish between multiplying by, say, 1/2 and dividing by 2. However, to children learning this material for the first time, the different interpretations of these expressions can be very different and they may have mastered one interpretation but not another.

Notice that the interpretations (1)(a) and (1)(b) for multiplication are the same; the only difference is which number plays which role; one number measures the *size* of each object and the other number measures *how much or how many* of the objects we have. In interpretation (2), the two numbers are playing equal roles. Here are examples analyzing the interpretations used in different multiplicative expressions.

Story Problem: Chengxi has 6 bags of cookies each of which contains 4 cookies. How many cookies does he have total?

Discussion: The answer is 6×4. This uses interpretation (1); the 6 is a measure of how many bags we have, the 4 is the size of each bag and the product is the size of the total. In terms of units the 6 is measured in bags, the 4 is measured cookies per bag, and the product is measured in cookies.

Story Problem: Megan runs at a speed of 12 miles per hour. How far can she run in half-an-hour?
Discussion: The answer is $(1/2) \times 12$. This uses interpretation (1); the half measures how much time passes, the 12 measures the 'size' of one unit of time measured in miles, and the product is the size of the resulting combination in miles. In terms of units, the half is measured in hours, the 12 is measured in miles per hour, and the product is measured in miles.

Story Problem: A farmer has a field that is 40 meters long by 30 meters wide. What is the area of her field?
Discussion: The answer is 40×30. This uses interpretation (2); each number has units of meters and the product has units of square meters.

Here are examples analyzing the interpretations used in different expressions involving a division.

Story Problem: Sylvie has 24 cookies total that she packs in bags with 4 cookies per bag. How many bags of cookies does she have?
Discussion: The answer is $24 \div 4$. This is the measurement model of division (1); the answer is the number of collections obtained when 24 cookies are divided into sets each of which has size 4 cookies. In terms of units, the 24 is measured in cookies, the 4 is measured in cookies per bag, and the quotient is measured in bags.

Story Problem: If Lisbeth strolls at a rate of half-a-mile per hour, how long does it take her to go one third of a mile?
Discussion: The answer is $(1/3) \div (1/2)$. This is the measurement model of division (1); it tells us how many pieces of size 1/2 mile lie in a piece of size 1/3 mile. In terms of units, the 1/3 is measured in miles, the 1/2 is measured in miles per hour, and the quotient is measured in hours.

Story Problem: Aditya has 24 cookies total that he evenly distributes into 6 bags. How many cookies are in each bag?
Discussion: The answer is $24 \div 6$. This is the partitive model of division (2); the answer is the size of each bag when 24 cookies are divided into 6 bags. In terms of units, the 24 is measured in cookies, the 6 is measured in bags and the quotient is measured in cookies per bag.

Story Problem: A rectangular cake has one side that is 9 inches long and has an area of 99 square inches. How long is the other side?
Discussion: The answer is $99 \div 9$. This is model (3); the answer is the length of one side of a rectangle whose area is 99 and whose width is 9

When students have completed this module they should be cognizant of the two models that are used for multiplication and the three models that are used for division and be able to identify when each model is being used. Moreover, given a calculation and a model for the calculation, they should also be able to come up with a story in which that calculation would be performed based on that model. Furthermore, they should be able to identify the difference between a problem involving division by an integer and a problem involving multiplication by its reciprocal.

Accessibility: This module should be accessible to any student that has completed high school mathematics.

1. The moderator asks the students to write down a short story problem whose solution is the product 6×4. Each student should write their story on chart paper.
2. The students share their stories. If one of the categories is not represented, the moderator should supply a story problem in that role. Also, if not already included in the students' stories, the moderator should present stories where both the numbers 6 and 4 clearly have units (as in the example of Megan above), since these examples are a little harder to interpret. The moderator categorizes the stories into three categories: those that have 6 groups of size 4, those that have 4 groups of size 6, and those in which 4 and 6 play similar roles. The moderator should not explain the rationale behind the categories; to divine it will be the next task assigned the students.
3. The moderator divides the students into groups. The groups are asked to identify what is similar about the stories in each of the categories and what is different about the stories in different categories. For each story they should also work out the *units* of each of the numbers 6, 4, and the product 24. Have them display the units in a chart as shown below. They should say what's similar about the units in stories that are in the same category and what's different about the units in stories that are in different categories.

Story	units of 6	units of 4	units of 24
Halloween	bags	candies per bag	candies
⋮	⋮	⋮	⋮

4. The class reconvenes and the groups report their discussions back to the whole class. In the discussion that follows it should be made explicit that there are three models for the multiplication $a \times b$ as identified above. These models should be understood intuitively as the size of a copies of b things, the size of b copies of a things and the area model. Students should also understand that these three ways are characterized by the *units* assigned to the

numbers 6, 4, and 24; in the first model the 6 has simple units, the units of 4 are some (other) units *per* units of 6, and the units of 24 are these other units etc.

5. Students go back to their groups charged with creating 3 stories for the multiplication 15×7 (one of each type) and 3 stories for the multiplication $20 \times (1/4)$ (one of each type). When the moderator visits the different groups she should check that their stories for $20 \times (1/4)$ are really for that and not for $20 \div 4$ instead (the result is the same, of course, but the concept is different).

6. Class reconvenes and students share their stories. In the discussion that follows, the difference between $20 \times (1/4)$ and $20 \div 4$ should be discussed. The stories of the different groups should be pooled and displayed in categories according to how the different numbers are used.

7. Students are asked to write down a story in which you'd want to do the calculation $24 \div 6$. Students write their story on chart paper.

8. Students share their stories. The stories should be displayed on the wall where everyone can see them. If any of the three models of division are not represented, the moderator should provide a story with that model. Also, if not already included in the students' stories, the moderator should include examples in which both numbers 24 and 6 clearly have units, as in the example of Lisbeth above.

9. In groups, students are asked to chart the units of each number 24, 6, and 4, in each story (just as they did for the multiplication stories) and to categorize all the stories according to the model they illustrate of division (similar to the way the moderator had categorized the multiplication stories). They shouldn't be told *how* to categorize the stories; that is for them to work out.

10. Class reconvenes and groups report back. Hopefully they will have found the three categories based on an analysis of the units involved. In the discussion that follows the moderator should try to make them articulate that one category consists of all the stories in which the quotient $24 \div 6$ is the number of groups when 24 things are divided into groups each of size 6 and another category consists of all the stories in which the quotient is the size of each group when 24 things are divided into 6 groups. There should be a discussion of why the number of things has plain units but the size of each group has units of one thing *per* units of another.

11. Students divide into groups. Each group is given a collection of, say, 10 story problems whose solution is given by multiplication or division of numbers. In the case of multiplication, students are asked to identify if the model being used is (1) or (2) and if it is (1) which number has units and which is indicating how *many* or how *much* of that thing is being combined. In the case of division, they should identify which of the three models is being used and the units of all the numbers involved. Here are some examples of stories that could

be used. (The stories listed here are in order; they should be mixed up when presented to the students.)

- *Story:* If one scoop of ice cream has 260 calories, how many calories do 5 scoops of ice cream have? (*Answer*: 5×260. Category (1); 5 has units of scoops and 260 has units of calories per scoop.)
- *Story:* If a recipe calls for 3 cups of flour and you want to make half that recipe, how much flour should you use? (*Answer:* $(1/2) \times 3$. Category (1); (1/2) has units of recipes and 3 has units of cups of flour per recipe.)
- *Story:* If lumber costs $1.50 per board foot, how much does it cost to buy 30 board feet? (*Answer*: 30×1.5. Category (1); 30 has units of board feet and 1.5 has units of dollars per board foot.)
- *Story:* If Sally runs at 12 miles per hour, how far can she run in half-an-hour? (*Answer*: $12 \times (1/2)$. Category (1); (1/2) has units of hours and 12 has units of miles per hour.)
- *Story:* If a rectangular piece of paper has a width of 5 inches and a length of 8 inches, how much area does it cover? (*Answer*: 8×5. Category (2); both 8 and 5 are measured in inches.)
- *Story:* If you have 14 lb of turkey and each serving is 1/4 of a pound, how many servings of turkey do you have? (*Answer:* $14 \div (1/4)$. Category (1).)
- *Story:* Benny has 2 cups of sugar and his recipe for cookies calls for one-and-a-half cups of sugar. How many recipes of cookies can he make with the sugar he has? (*Answer:* $2 \div (3/2)$. Category (1).)
- *Story:* If you divide 15 candies between 5 friends, how many candies will each friend get? (*Answer:* $15 \div 5$. Category (2).)
- *Story:* If Sally walks one third of a mile in half-an-hour, how fast is she walking? (*Answer*: $(1/3) \div (1/2)$. Category (2).)
- *Story:* If a farmer has a rectangular field that is 1200 square feet and one side has a length of 30 feet, how long is the other side? (*Answer:* $1200 \div 30$. Category (3).)

Stories should include many examples with multiplication and division by fractions, since it is easy for students to get confused and think they are, for example, dividing by 2 when actually they are multiplying by 1/2 (the result, of course, is the same, but the concept is different). They may also confuse division by 1/2 with multiplication by 1/2.

12. Class reconvenes and groups report back. The discussion should include the difference between story problems whose solution is 12×2 and story problems whose solution is $12 \div (1/2)$. Many students won't distinguish between these two though they are quite different for children

learning the subject for the first time and it is an important and non-trivial result that $12 \div (1/2) = 12 \times 2$.

13. Students go back to their groups. Each group is asked to create three story problems whose solution is the division $36 \div 4$ (one for each model) and three story problems whose solution is the division $12 \div (1/2)$ (also one for each model). Tell the students to be careful in the second case; it is easy to come up with stories that illustrate, say, $12 \div 2$ or 12×2 instead of $12 \div (1/2)$.

14. Class reconvenes and groups share their stories. Discussion follows.

Area

Materials needed:

- Ample supplies of chart paper, markers, and whatever is needed to attach the chart paper to the wall of the room for students to display their work.
- A blackboard in the room could be useful but is not absolutely essential.
- Enough rulers for every group to have at least one.
- Pictures of the two triangles in item (14) to distribute to the groups (one per group).
- Copies of the triangle in item (16) to distribute to the different groups (one per group).
- Sheets of graph paper with shapes drawn on them for item (18); on each sheet the shapes should include a regular pentagon, a regular hexagon, an irregular convex polygon, a non-convex polygon, and a shape with curved edges. Each group gets one sheet of shapes.
- Extra graph paper with smaller squares than the squares used for the sheets of shapes would be desirable but not absolutely essential. (This is so the students can use the smaller squares to get a better estimate of the area of the shape with curved edges.)

Mathematics covered: This module is about area. Area is a quantity we associate with regions in the plane and surface area is a quantity we associate with regions in three-dimensional space. The area of a region is somehow a measure of how much two-dimensional 'stuff' is needed to cover that region. When we *measure* area we assign a *number* to this rather intangible quality of a shape. There are many subtle issues involved with how we make this assignment and a full investigation of this topic belongs to the fields of differential geometry and measure theory which are usually graduate level classes in mathematics.

In this module and at this level, the goal is to get students to gain a better understanding of the familiar area formulae that are used in K-12 mathematics; namely the areas of a rectangle, triangle, and circle, and the surface areas of a cylinder, cone, and sphere. In particular, with only three conditions that describe how we want addition of numbers to correspond to the intangible quality that we call area, and a choice of unit, we obtain these familiar formulae. The three conditions are:

- The area of a region is non-negative.
- The area of a region is a property of its shape only and not its location in space. In other words, if you have a shape that is located in two or three dimensional space and you move it (i.e. you perform a Euclidean transformation) then its area will remain unchanged.
- The area of the union of two disjoint regions should be equal to the sum of their individual areas. The common metaphor of addition of numbers is that it should be what you get when

you *combine* things, so we get this condition by insisting that metaphor continue to hold in the case of areas.

The usual choice of unit is that a square whose side has length 1 and whose width has length 1 has an area of 1.

After completing this module students should have a better understanding of what area is and how we use numbers to measure areas. They should understand why the standard formula for the area of a rectangle is length times width, why the area of a triangle is equal to one half its base times its height, and should understand how to calculate the area of any polygonal region by dividing it up into triangles. They should have the concept of finding the areas of regions with curved edges by approximating them by polygons and should have an intuitive understanding of why the area of a circle is equal to πr^2 based on the fact that the ratio of the diameter to the circumference is π. They should be able to derive the formulae for the surface area of a cone and of a cylinder. They will see that the concept of the surface area of curved surfaces in three dimensional space is different from that of area in the plane, will have explored its meaning and will have seen a derivation of the formula for the surface area of a sphere that does not use calculus. They should also have gained some consciousness of what we mean by a feature of an object and how numbers are abstract objects that we use to measure the feature of an object.

Accessibility: This module should be accessible to all students that have successfully completed high school mathematics and will be relevant to all undergraduate students, math majors or not. However, this module is very challenging because it requires students to question the meaning of area and many students may have no idea how to start thinking about this. For this reason, this module shouldn't be done too early; students will need to have developed a thirst and delight for mathematical inquiry otherwise they are likely to shut down.

Note: This is a long module that might best be done over (at least) two days. A recommended breaking place is given.

1. The moderator divides the students into groups and charges them with discussing, 'What do we mean by the area of an object? ' The moderator should go around to the different groups to ensure they are thinking constructively about the question. In particular, many groups will probably define area in terms of itself, saying things such as, 'the area of an object is the number of square feet that it takes up.' The moderator should bring this to their attention. For example, the moderator might ask, 'What is a square foot?' If they say that it is a square whose length and width are both equal to 1 then the next question should be, "What do we mean by the area of some non-rectangular object?" (The point here is that you can't fill the object up with squares of length 1 and side 1, instead you have to chop things up into smaller pieces.)

2. The class reconvenes and the groups report their discussions back to the whole class. In the class discussion that follows the moderator should bring their attention to any *properties* they said area should have. For example, they might somehow articulate that if a shape is moved from one location to another its area shouldn't change, or they might say the area of a stick-like object should be equal to zero, or areas should be positive, or if two shapes butt up next to each other, then the area of the combined shape should be the sum of the two areas.

3. After the discussion, have students go back to groups. They should be charged with discussing *why* the area of a rectangle is equal to its length times its width. For example, why is not the area of a rectangle equal to any of the following? Or, maybe we *could* use one or more of these as our notion of the area of a rectangle? Why or why not?

   a) the length of the diagonal

   b) the length of the perimeter

   c) the length of the diagonal multiplied by the width of the rectangle

   d) twice the base times width

   e) the base times twice the width

   f) the base times the width squared

   The moderator should circulate among the groups. If a group is ruling the first two out based on units, then prod the group to think why the units of area should be length squared. Hopefully the groups will come up with things similar to the following.

   a) A long, skinny rectangle doesn't have much area compared to a square but its diagonal is much longer, so that's why the diagonal doesn't work.

   b) If we rotate a rectangle its area doesn't change but its base times the width squared will change because its base and width will be different.

4. Class reconvenes and the groups report back to the whole class. In the discussion that follows, if a group says, for example, that one rectangle has a larger area than another but its diagonal is shorter than the other's, then the moderator should ask them how they know that the area of the one is bigger than the area of the other. If they are using properties of area in their explanations, then the moderator should highlight the fact they are doing this and the property that they are using.

5. The moderator should now talk a little about how we use numbers to measure *features* of objects. For example, we use numbers to count how many objects there are in a collection (the objects are the different collections and the feature is how many things the collection contains), we use numbers to measure lengths of line segments (the objects are line segments and the feature is how far the segment extends), and we use numbers to measure the weight

of an object (the objects here are physical objects and the feature is how heavy the object is). Students can come up with other examples, no doubt. Area is a feature of a region in the plane (the objects are regions in the plane and how much 'stuff' the region occupies is the feature). The discussion the class has been engaged in is, how should we assign a number to the amount of stuff there is? This is a little more subtle than how we do it for lengths, so, first the class will explore how we do this for lengths.

6. The students go back to their groups charged with discussing how we decide whether a line segment has length 10 or length 200 or length 127.3?

7. The class reconvenes and groups report back. Hopefully they will say things such as, "We choose what 1 unit is and then, when we want to measure something else, we take that unit length and move it to see how many of those fit into our length." Moderator can ask questions such as, "What if more than 4 fit in but less than 5?" Hopefully they will respond with something to the effect, "We can chop our unit up into little pieces that are all the same length." Moderator should ask how they know they all have the same length. Hopefully the response will be if you put them all next to each other you can see they have the same length. The details of the discussion are not too important, this is just an illustration of the kinds of discussions that should be going on at this point.

8. The moderator summarizes. To associate a number with the length of a line segment we need to do the following.

   - We choose a line segment whose length will be 1. This is our choice of *unit*. When we measure in centimeters this is different from when we measure in feet.
   - The lengths of all line segments should be *non-negative*. (Length 0 is okay since dots have length 0.)
   - If a line segment is moved from one place to another then its length doesn't change.
   - If we butt two line segments up next to each other then the length of the resulting line segment should be the sum of their lengths.

   These properties are enough to determine the length of *any* line segment. There could be a discussion here of how, in general, when we associate a number to a feature we want to do that in such a way that *combining objects* should correspond to *addition of numbers*.

9. Now it is time for the class to return to the method of assigning areas to regions in the plane. How do we do this? The class should understand that area is a *numerical* assignment to a *shape* in the plane that has the following properties

   - We choose a region whose area will be 1.
   - The area of any region should be non-negative.

- When a region is moved from one place to another its area doesn't change.
- The area of the union of two disjoint regions should be equal to the sum of their two areas.

10. The students go back to their groups to discuss, based on this, what possible expressions we could come up with that would capture the area of a rectangle in terms of the dimensions of the rectangle. In particular, recall the expressions suggested above. Would any of these work as the area of a rectangle?

11. The class reconvenes and the groups report back. From the discussion that follows it should come out, given these properties of area, the area of a rectangle *must be* equal to a constant times its base times its width. The constant corresponds to the choice of unit. If a square of length 1 and width 1 is said to have 1 unit of area then the constant is equal to 1.

12. The students go back to their groups charged with *proving* that the area of a triangle is equal to one half its length times its height.

13. The groups report back. Hopefully most groups will be able to come up with a relatively convincing argument probably based on taking two congruent triangles, chopping them up, and reconstructing the pieces to form a rectangle. However, it is likely their arguments may not work for some triangles. The discussion that follows should focus on what we mean by the base of the triangle and what we mean by its height. Could *any* side be the base or did the proofs assume that one particular side was the base (namely the side that connected the two angles that were both less than 90°). If we're assuming the base is the side that connects these two angles (which they almost certainly all were doing) does every triangle always have such a side? (The answer is yes, because the sum of angles of a triangles is 180°: if there is at most one angle whose measure is less than 90° then there must be two angles with measure greater than 90° which would contradict the fact that the sum is 180°.)

This may be an appropriate place to break for the day.

14. The students return to their groups charged with discussing why the area of a triangle is one half the base times the height, even when the base is taken to be a side that is adjacent to an angle that is greater than 90°. The moderator should give them a triangle to work with as shown below. Notice that the triangle is chosen so that when you double it, the parallelogram that is formed has to be chopped up into more than two pieces in order to form a rectangle whose width is the base and whose height is the height of the triangle.

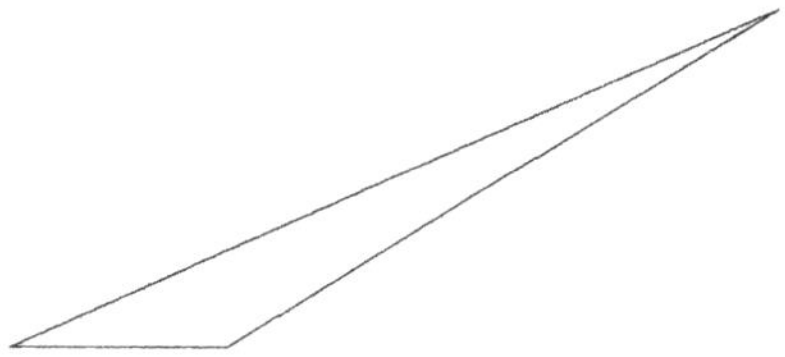

15. The class reconvenes and the groups report back.

16. The groups are given a sheet of shapes drawn on graph paper. The shapes should include a regular pentagon, a regular hexagon, an irregular convex polygon, a non-convex polygon, and a shape that has curved edges. The groups are asked to find the area of each shape. They are also charged with determining if they can come up with *formulae* for the areas of a regular pentagon and a regular hexagon. Can they come up with a formula for the area of a general quadrilateral?

17. The groups report back. In the discussion that follows the following points should be drawn out.

    a) We find the areas of other polygonal shapes by chopping them up into regions whose shape we know (triangles and rectangles).

    b) We *can* find formulae for the areas of regular pentagons and hexagons. The formula for a hexagon is quite nice; the formula for the pentagon includes the value of a trigonometric function. Since the shape of one of these objects is determined by knowing what kind of object it is and the length of one of its sides, it makes sense there should be formulae for their areas based solely on the length of one side.

    c) For a general quadrilateral it is harder to come up with a formula for its area. The reasoning behind this is that the shape and area of the quadrilateral are not determined solely on the lengths of the four sides, so any formula for its area would have to include some other lengths (not dissimilar to the way that the formula for the area of a triangle contains its height). However, it may be hard to describe what those other lengths might be. There *are* ways to do this but a) the formulae you get are not particularly memorable and b) interesting issues arise as to whether you can define what you mean by, say, the *height* of the quadrilateral. This discussion could touch on many interesting mathematical points including what it means to have a formula for something and properties of quadrilaterals (for example, every quadrilateral must have at least two acute angles; why?)

    d) To find the area of a shape with curved edges is much harder. Discussion should center around sandwiching the error between two numbers (an upper estimate and a lower estimate) and how we could get a better sandwich if we took finer graph paper, so that in theory we could estimate the area to any desired degree of accuracy. There could be a rich discussion here about what the difference is between this and actually knowing the area exactly. In particular, suppose the area is equal to π. What do we know about π? (The point is that we know the value of π only to a degree of accuracy.)

18. The moderator should show them the picture below that illustrates why the area of a circle is equal to πr^2. (Notice this derivation rests on the fact that the ratio of the circumference of a circle to its diameter is equal to π.)

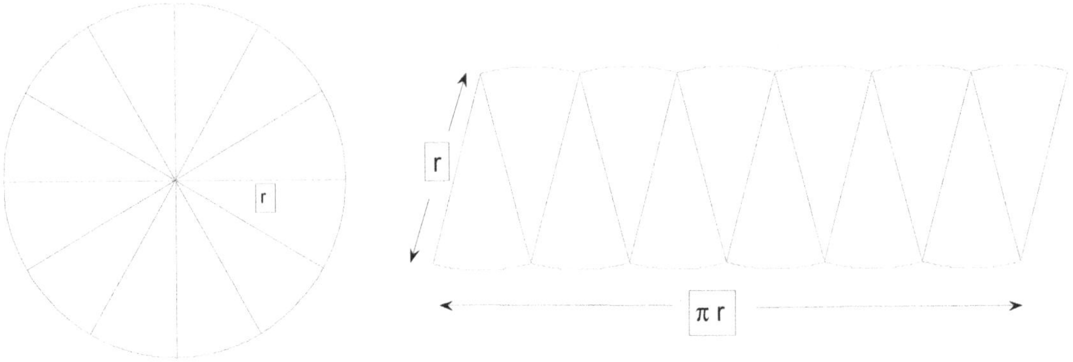

19. The students return to their groups to calculate the area of the region that is bounded by $y = x^2$, $x = 0$, $x = 1$ and the x-axis. Groups should use the ideas of area we've found so far and not use calculus. For students who have taken calculus this shouldn't be too hard; it will be a familiar Riemann sum calculation. For other students this is a very hard problem. They should be prodded to break the region up into n pieces and estimate the area in terms of n. Then look at what happens as n gets larger and larger.

20. The class reconvenes and the groups report back.

21. Now it is time to discuss what we mean by the area of a region that lies in three dimensional space and not in the plane. The groups are first charged with discussing what the surface area of the curved surface of a cylinder is and what the surface area of the curved surface of a cone is. Note: these two surfaces can be formed from planar regions, so this is simply a matter of bending a flat region in three dimensional space. Can the groups prove the familiar formulae for these two surface areas?

22. The groups reconvene and report back.

23. The groups are sent back to discuss what we mean by the surface area of a sphere and how we can find it. Can they find the surface area of a sphere *without* using calculus?

24. The groups report back. The discussion that follows should bring out the following points.

    a) Since a sphere cannot be formed by deforming a flat piece of paper, this is really a *new* sense of area. To relate it to areas in a plane and to the surface area of a cone, we approximate the sphere with many flat pieces (like a geodesic dome) and say that the sum of the areas of the flat pieces should be close to the area of the sphere. The smaller the pieces are (and therefore the more of them we use) the closer we should get to the actual area of the sphere.

b) How can we *find* the surface area of a sphere? The meaning of its area tells us how to find it in theory. However, is there some way we can see the area is equal to $4\pi r^2$? Here are some possible approaches.

- This method is similar to the picture of the circle. Take a hemisphere and cut it up in thin slivers that stretch from the top of the dome to its edge. Lay the pieces top to bottom. One might at first think that it would look a bit like a rectangle whose height is $\pi r/2$ and whose width is πr. From this you'd get the surface area of the sphere to be equal to $2(\pi r/2)(\pi r) = \pi\pi r^2$ instead of $4\pi r^2$. The problem with the argument is that the pieces are not *flat*, so when you lay them top to bottom, they do not fit snugly together, and even in the limit as they get thinner and thinner, the shape won't look like a rectangle.
- Archimedes calculated the surface area of a sphere by slicing it into thin strips each of which looks like the frustrum of a cone. His argument is close to the argument used in calculates today but doesn't involve evaluating an integral. This argument is easily found online.
- We can calculate the surface area of a sphere by considering the projection of the sphere onto the curved surface of a cylinder that has the same radius as the sphere and circumscribes it (see the picture below). The important observation is that this projection map is area preserving, so the surface area of the sphere is the same as the area of the curved surface of the cylinder. Below is a way to see this.

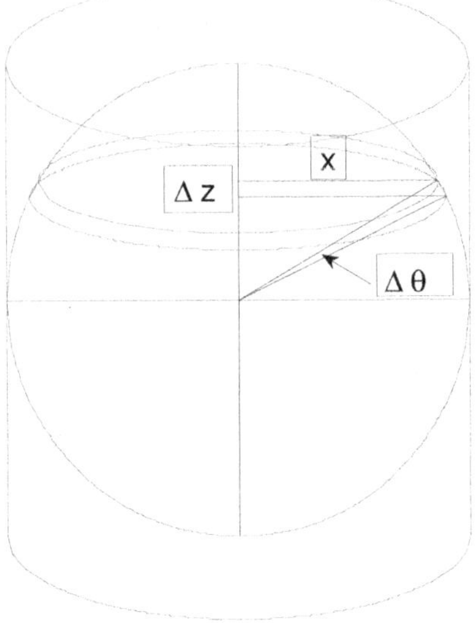

The surface area of the projection of the strip at height z onto the cylinder is exactly equal to $2\pi r \Delta z$. The surface area of the same strip of the sphere is approximately equal to $2\pi x r \Delta\theta$. But

$$\begin{aligned}\Delta z &= r\sin\left((n+1)\Delta\theta\right) - r\sin\left(n\Delta\theta\right)\\ &= r\sin\left((n+1)\Delta\theta\right) - z\\ &= r\left[\sin(n\Delta\theta)\cos(\Delta\theta) + \cos(n\Delta\theta)\sin(\Delta\theta)\right] - z\\ &= z\cos(\Delta\theta) + x\sin(\Delta\theta) - z\\ &= z\left(\cos(\Delta\theta) - 1\right) + x\sin(\Delta\theta).\end{aligned}$$

It follows that

$$\frac{\Delta\theta}{\Delta z} = \frac{1}{z\left(\frac{\cos(\Delta\theta)-1}{\Delta\theta}\right) + x\left(\frac{\sin(\Delta\theta)}{\Delta\theta}\right)},$$

so the surface area of the strip of the sphere is approximately equal to

$$2\pi x r \frac{1}{z\left(\frac{\cos(\Delta\theta)-1}{\Delta\theta}\right) + x\left(\frac{\sin(\Delta\theta)}{\Delta\theta}\right)}\Delta z.$$

As $n \to \infty$ (and the strips get narrower and narrower) the approximation gets better and better and $(\cos(\Delta\theta) - 1)/\Delta\theta \to 0$ and $\sin(\Delta\theta)/\Delta\theta \to 1$. Thus, the surface area of the strip of the sphere is closer and closer to $2\pi r\Delta z$ which is the same same as the surface area of the projection of the strip onto the cylinder. Thus, the projection of the sphere onto the cylinder is an area preserving map and the surface area of the sphere must be equal to the curved surface area of the cylinder. Since the cylinder has height $2r$ and radius r, its curved surface has area $(2\pi r)(2r) = 4\pi r^2$. Thus we get the formula for the surface area of the sphere without calculus (although the derivation did depend on the knowledge that $(\cos(\Delta\theta) - 1)/\Delta\theta \to 0$ and $\sin(\Delta\theta)/\Delta\theta \to 1$ as $\Delta\theta \to 0$).

Of course, finding the surface area of a sphere without using calculus is a very hard problem and we can't expect any of the students to come up with these arguments or even anything close to them. However, it is interesting for the students to *see* them and there are plenty of things to discuss. For example, why, exactly, is the area of the strip of the sphere at height z approximately equal to $2\pi x r \Delta\theta$? Why does this approximation improve as $n \to \infty$?

Variables

Mathematics covered: We use letters in mathematics in many different ways; here are *some* examples.

i) A letter can be used to represent a number or mathematical object that can take on any value in a given set of numbers or set of objects. This is how letters are used when we are describing properties that various mathematical constructs have. For example, we write

$$a(b+c) = ab + ac,$$

to state the distributive law for real numbers. In this statement, a, b, and c can be any real numbers. Similarly, when stating the distributive law in set theory, we write

$$A \cup (B \cap C) = (A \cup B) \cap (A \cup C),$$

where A, B, and C can be any sets that are subsets of the same universal set. In a more complicated example we write,

$$(kf + g)'(x) = kf'(x) + g'(x)$$

where f and g are any differentiable functions, x is any number that lies in both the domain of f and the domain of g, and k is any real number.

ii) A letter can be used to represent a specific mathematical object that is a known but complex object, in order to be able to refer to it by name. This is the case, for example, in statements such as 'consider the function f that is given by

$$f(x) = \frac{x^2}{3x-1}.'$$

Here, f refers to this very specific function. The letter x in this example, is being used in the sense of i); it represents any number that lies in the domain of the function f.

iii) A letter can be used to represent a quantity whose value is (presently) unknown. This is how letters are used in many elementary algebra problems. For example, suppose Abigail scored 72, 67, 82 and 79 on her first four math tests and her average score on these tests including the final was 77. To find her score on the final we may let x denote this score and write down the equation

$$\frac{72+67+82+79+x}{5} = 77$$

which we can solve for x.

iv) A letter can be used to represent the measurement of a feature associated with a set of objects. As such it is a function defined on a set of objects whose target space is a set of numbers (be they integers, real numbers, complex numbers etc.) This is how letters are used in many applied problems in calculus and in modeling problems.

In cases iii) and iv), the letters are often called *variables*, though this is something of a misnomer in the case of iii). (Sometimes letters that are used as in i) may also be called variables but in these modules we'll restrict the term 'variable' to refer to letters that are used as in iii) and iv).) In elementary algebra problems, variables generally refer to letters used in the sense of iii). The switch to using them as in iv) usually happens in algebra II or pre-calculus classes and is often not made explicit, so students can find themselves manipulating algebraic expressions in which variables in the sense of iv) are being used, but their concept of a variable is that of iii). This causes problems. To resolve the cognitive disconnect, they may start to think of a variable as something that varies. This is, indeed, true (usually), but it lacks specificity. In the following discussion we show how the variables we use in many problems in calculus should be interpreted as functions on an underlying set of objects as in iv). We shall also see how making this interpretation explicit can aid in solving problems and in understanding the method of solution.

In pre-calculus classes, students explore the relations between variables and, in particular, they explore *functional* relations. They learn y is a function of x if the graph of the relation between x and y passes the vertical line test. To test their understanding of this concept they are typically given various curves in the plane and asked to identify in each case if y is a function of x. However, this is completely abstract. If the variables x and y are 'unknowns,' what is meant by the *graph of the relation between x and y*? Indeed, in the most common examples of the relation between two variables that are not functionally related, the graphs are *regions* in the plane and not curves in the plane, so the examples that are given are not ones that encourage a concrete interpretation of the relation between variables. On the other hand, if you think of x and y as each measuring a feature associated with an underlying set of objects, then the graph of the relation between x and y has a very concrete meaning; (a, b) lies on the graph if and only if there is an object in the set for which $x = a$ and $y = b$. Functional relations are intimately connected with *causality* and *control.* In particular, suppose the underlying set is our universe and y is a function of x. If the value of x is known then this means that we are dealing with that subset of the universe consisting of all those elements for which x has the given value. Since y is a function of x the fact that we are dealing with this subset *causes* y to take on a particular value. Similarly, if we can *control* the value of x then we can control the value of y by choosing the value of x appropriately.

To clarify this, let's take a look at an example. Consider the set of all polyhedra that have a rectangular base and whose sides are triangles that meet at a point 5 units vertically above the center of the base (see the picture below).

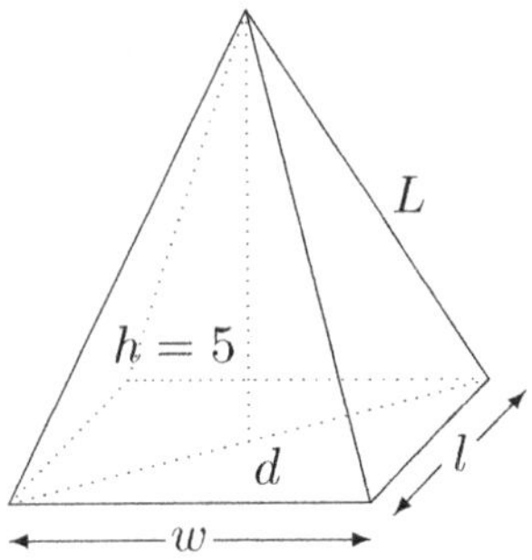

Some variables associated with this set include the width of the base w, the length of the base l, the length of the diagonal of the base d, the height h, and the slant height L. Notice each variable associates a number with each polyhedron; in other words, each variable is a function on this set of polyhedra. Since the height h is identically equal to 5 it may seem strange to call it a 'variable' but, as a function defined on the set of objects, it is a variable; just as we can have constant functions, so we can have variables that are constants. The graph of the relation between w and l includes everything in the first quadrant. The graph of the relation between l and h is a horizontal half-line at height 5 and restricted to $l > 0$. This passes the vertical line test, so h is a function of l (though it is rather trivially a function of l). The graph of the relation between d and L is that quarter of the hyperbola defined by $L^2 = d^2/4 + 25$ that lies in the first quadrant. This passes the vertical line test so the value of d determines the value of L and L is a function of d. The graph of the relation between w and L includes everything in the first quadrant that satisfies $L > \sqrt{(w^2/4) + 25}$. This graph does not satisfy the vertical line test, so L is not a function of w.

These kinds of variables in algebra are the same as variables in statistics and random variables in probability. In statistics there is a population of interest and variables are measurements of features associated with the set of interest; in other words, they are functions on the population. Random variables in probability are functions on the sample space of the experiment. Notice there is often freedom in exactly what you choose your population and/or sample space to be. Similarly we shall see that there is freedom in what you consider to be the underlying set of variables in algebra. Despite that, statisticians and probabilists still recognize the value in identifying *some* set as the population or sample set of the experiment and we argue that, similarly, when considering algebraic variables, we should have an underlying set in mind. The main difference between variables in statistics, random variables in probability and variables in algebra, is that in the former two cases we endow the population and sample space with the added structure of a probability measure. This means that the kinds of questions we ask in probability and statistics are a little different from the questions we ask in algebra; in the latter we are interested if one variable is a *function* of one or more of the others, in the former we are interested *to what extent* one variable is determined by one or more of the others. Another difference is many variables in statistics are not *numerical*; the target space of the function is not a set of numbers but instead a set of objects. (Examples include eye color or race.)

To explore the concept of variables and see how we can think of them as functions on sets of objects and why this is useful, let's look at a number of examples. Let's first consider the following problem that is a typical optimization problem covered in calculus classes. In this example, apart from seeing how various variables can be thought of as functions on an underlying set of objects, we shall also see that what we normally think of as *functions* are, in fact, variables. Indeed, any quantity that is a function of a variable, is itself a variable. Indeed, some of the confusion about variables no doubt arises from the common tendency to confuse a function with its output variable. Often this is benign, but there are many situations in which it can be problematic.

Problem A: A closed box with a square base is to have a volume of 12 m^3. The material for the top and sides costs \$2 per square meter and the material for the bottom costs \$4 per square meter. Find the dimensions of the least expensive box that can be constructed.

In this problem we are interested in the set of all boxes that might conceivably be constructed. These boxes all have square bases and volumes that are equal to 12 m^3. For example, one of the boxes in the set is 2 m by 2 m by 3 m and another box is 4 m by 4 m by 0.75 m. We are charged with finding the box in the set that is cheapest to construct.

There are many variables associated with this set, most of which are uninteresting in the solution of the problem. Here are a few of them.

a) The height h in meters.

b) The height l in inches.

c) The length of the base x in meters.

d) The cost of construction C in dollars.

e) The length of the diagonal d in meters. (In other words, the length from one corner of the box to the diagonally opposite corner.)

f) The girth g in meters. (In other words, twice the sum of the two shortest dimensions of the box.)

g) The rate at which the cost C changes as x is increased, measured in dollars per meter. This is the variable dC/dx.

h) The rate at which the length of the base changes as the height is increased, measured in inches per meter. This is the variable $r = 39.3701(dx/dh)$.

i) The rate at which dx/dh changes as x is increased, measured in meters per square meter. This is the variable $d^2x/dxdh$ (which is not, by the way, the same as $\frac{\partial^2 x}{\partial x \partial h}$).

j) The average cost of all the boxes as the height ranges from 1 meter to the height of the box, measured in dollars. This is the variable $\left(\frac{1}{h-1}\right)\int_1^h C\,dh$.

Every one of these variables measures a physical characteristic of the boxes in the set and is a function on the set of boxes; in other words, to each box it associates a number.

Notice that x, h and l are rather special; if we know the value of any of them then we can identify the particular box in the set to which that value corresponds. In one of the modules below we call these variables *characteristic* variables of the set. What it means is that every variable defined on this set of boxes can be thought of as a function of x or as a function of l or as a function of h. Indeed, this is how most people would think of most of these variables.

Let's take the variable x as an example. When we write another variable as a function of x, we are introducing the variable x as an intermediary, so that the variable is now a composition of its representations as a function of x and x as a function on the set of boxes. This is illustrated in the picture below.

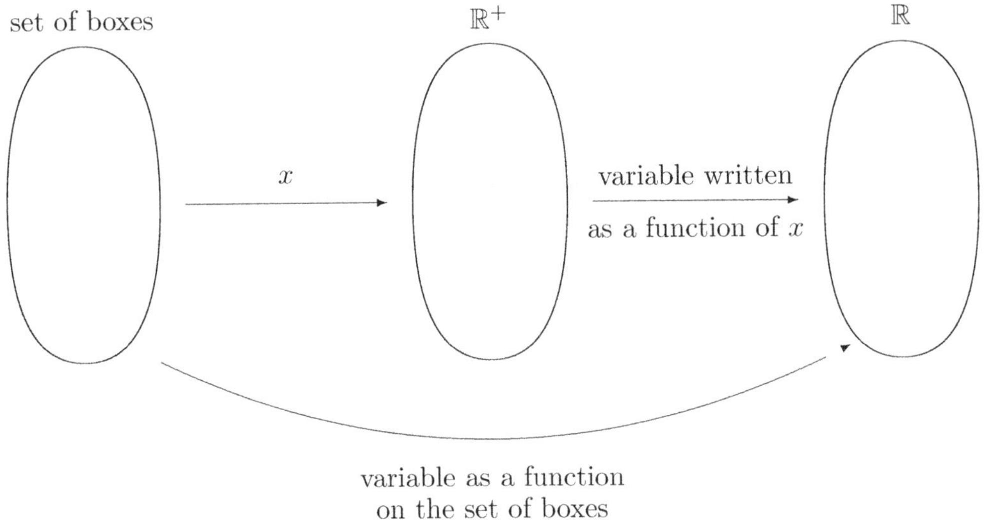

This process of introducing an intermediary variable is actually not an uncommon practice in mathematics. For example, when we evaluate an integral by substitution, we introduce a variable u that is a function of x and consider the quotient of the integrand divided by du/dx. When we write this quotient as a function of u in order to evaluate the integral, we are introducing the variable u as an intermediary variable, so the quotient becomes a function of u which in turn is a function of x.

To see how it can be useful to think of variables in this problem as functions on the set of boxes, imagine a student who has recognized that she wants to optimize the construction cost but is having difficulty writing the construction cost as a function of some other variable in the problem. If you can get the student to identify the underlying set that is of interest (namely, the set of boxes) then you can take an example of a box in the set, and find its construction cost. For example, consider the box that is $2 \times 2 \times 3$. Now you have a concrete box, the student will probably be able to find the construction cost for this box. Now take another example; say a box that is $3 \times 3 \times ?$. What is the third dimension? Well, you can calculate it because you know $3^2h = 12$ so $h = 3/4$. Now you have a concrete box again, so the student will probably be able to calculate the construction cost. Having done these two examples, you now take a box that is $x \times x \times ?$. The example shows you how to find the third dimension and, once you have that, the example again shows you how to find the cost. Notice this whole approach rests on the fact that the student has identified the underlying set of objects of interest and can consider examples of objects in that set.

There are other benefits to thinking about variables in optimization problems as functions on an underlying set of objects. In particular, it is easier to make sense of the process by which we perform the optimization (see the third module below) and it can help clear the fog created by the plethora of notations for derivatives and integrals.

The next example is a typical related rates problem that appears in calculus classes.

Problem B: An airplane, flying at an altitude of 6 miles, passes directly over a radar antenna. When the airplane is 10 miles from the antenna, the radar detects that its distance from the antenna is decreasing at the rate of 240 mph. What is the speed of the airplane?

The set of objects that should be considered in this problem is the collection of points in time before and after that point in time when the airplane is 10 miles from the antenna. There are many variables associated with this set. Here are some of them:

a) The (straight line) distance in miles of the airplane from the antenna, D.

b) The horizontal distance in miles of the airplane from the point that is 6 miles vertically above the antenna, x.

c) The time in hours from when the airplane is directly above the antenna, t.

d) The rate at which x is changing in miles per hour, dx/dt.

e) The rate at which D is changing in miles per hour, dD/dt.

f) The rate at which x is changing as D is increased in miles per mile, dx/dD.

g) The temperature at the antenna in degrees Celsius, T.

h) The vertical height of the airplane above the ground in miles, h.

Each of these variables is a measurement of a feature associated with the objects in the set. (Notice h is a constant variable.) The variable t is special, since the value of t is enough to determine the object in the set under consideration. In this problem, most people would think of all the variables above as functions of t rather than as functions of the underlying set of objects. Since t is so closely associated with the objects in the set, the two points of view are really equivalent.

To solve this problem, students need to do the following.

a) Identify the variable x whose derivative with respect to time is equal to the rate that is given.

b) Identify the variable D whose derivative with respect to time is the rate that we are trying to find.

c) Identify the relationship between x and D, namely $D^2 = x^2 + 36$. In particular, when determining this relationship a common mistake it to substitute $D = 10$ and write $100 = x^2 + 36$. This equation is not a relation; instead it is an equation that is satisfied by two objects in the set (those objects for which $D = 10$).

Students taking calculus are generally relatively comfortable with introducing a variable that is an unknown in the problem (as in standard algebra problems). However, they are less comfortable with introducing variables that are measurements of features of objects on an underlying set. This is the root of the difficulty doing a) and b). In particular, at this point in the course, students often think of speed as the derivative of distance without understanding there may be many different distances of interest. The root of the difficulty doing c) is that their notion of a varaible is that of an 'unknown,' so they do not realize equalities involving variables need to be interpreted contextually; some equalities refer to particular values of the variables whereas others refer to an identity of variables. For example, the statement, $D^2 = x^2 + 36$ is an equality between two variables; the variable D^2 and the variable $x^2 + 36$. For every element in the set these two variables have the same value. On the other hand, consider the statement 'When $D = 10$, $100 = x^2 + 36$.' This is shorthand for the more complete statement, 'For all those objects in the set for which the value of D is 10, the value of the variable $x^2 + 36$ is 100.' The equalities here are not equalities of variables, instead they are equalities of particular outputs of the variables (thought of as functions) for particular items in the underlying set. Notice when we talk about the value of a variable, we are using the same language that we use with functions (as in the value of a function). This is no surprise, since variables *are* functions.

The last example is a typical algebra problem. Unlike the two problems above, it is not clear thinking about variables arising in this problem as functions on an underlying set of objects is helpful for solving the problem. However, we are including the example in order to fully explore the concept of a variable and to see it is possible to think of variables that are unknowns as functions on an underlying set of objects. This is analogous to what happens to models we have for various arithmetic operations when we extend our number system. For example, when working with whole numbers, the primary model for the subtraction $a - b$ is that of taking b things away from a things. When we extend to integers this model breaks down and instead we think of $a - b$ as the distance and direction you need to go to get from the tip of vector b to the tip of vector a. This second model encompasses the first in that it still works with whole numbers though we do not *need* this model when we are just dealing with whole numbers. Similarly, thinking of a variable as a function on an underlying set of objects is a model that is important when discussing the relations between variables; it is still valid for problems that do not concern the relations between variables though it is not *necessary* in these problems.

Problem C: A painter has a paint mixture that contains 1% blue dye and another mixture that contains 6% blue dye. How many liters are needed of each in order to make 50 liters of a solution that contains 4% blue dye?

The standard way to solve this problem would be to let x denote the amount (in liters) that should be used of the 1% mixture and y denote the amount (in liters) that should be used of the 6% mixture. These 'variables' are not really variables at all. Instead, they are letters used in the sense of iii) above, each one represents a particular number whose value is (presently) unknown. We can think of x and y as functions on an underlying set, but the underlying set just consists of the single element that is the 50 liters of solution that contains 4% blue dye. Notice, since x and y take on single values only (indeed, $x = 20$ and $y = 30$), the graph of the relation between x and y consists of a single point (namely, $(20, 30)$).

To find the values of x and y we write down the simultaneous equations

$$x + y = 50 \quad (1)$$

$$\frac{0.01x + 0.06y}{x + y} = 0.04, \quad (2)$$

which we can solve by any number of methods. To help students gain an understanding of what they are doing when they solve these equations, we may have them graph each line and notice that the solution to the simultaneous system of equations is the point of intersection of the two lines. What is interesting here is the graphs of these lines do not correspond to anything physical in the problem, so by using this method of solution we are performing operations in our world of mathematics that are not physically meaningful.

Another approach to this problem is to consider the set of all possible mixtures (of any amount) that could be made from the 6% and 1% mixtures. Some variables associated with this set include:

a) The amount (in liters) used of the 1% mixture, x.

b) The amount (in liters) used of the 6% mixture, y.

c) The ratio of the amount of 1% mixture to the amount of 6% mixture, $r = x/y$.

d) The amount (in liters) of blue dye in the mixture, $D = 0.01x + 0.06y$.

e) The fraction of blue dye in the mixture, $f = (.01x + 0.01y)/(x + y)$.

f) The total amount of paint in the mixture (in liters), $P = x + y$.

These are all variables in the sense of iv). The variables x and y do not depend on each other in this set; knowing the value of x gives you no information about the value of y and the graph of the relation between x and y includes everything in the first quadrant. Together, the values of x and y constitute enough information to determine the object in the set. That is why all the other variables can be written in terms of x and y. We are interested in that object in the set for which $P = 50$ and $f = 0.04$. Each of these equations defines a subset of the set of objects in question; we are interested in that object which lies in the intersection of these two subsets. In each of these subsets, the values of x and y *are* related; equation (1) is the relation in the set $\{P = 50\}$ and equation (2) is the relation in the set $\{f = 0.04\}$. Now when we sketch the graphs of these lines, they are physically meaningful (provided we limit the graphs to those parts of the lines that lie in the first quadrant); each point (a, b) on the graph of, say, $x + y = 50$ corresponds to an object in the set $\{P = 50\}$ for which $x = a$ and $y = b$. The point of intersection corresponds to that object which lies in the intersection of the two sets and therefore its x and y coordinates constitute the solution to the problem.

In these modules students will explore the concept of a variable. They will see the term is used differently in different situations and the concept of a variable as an 'unknown' is insufficient to cover all of the ways in which variables are used. They will look at optimization problems and related rates problems in calculus where variables are used in the sense of iv) and explicitly determine the underlying sets of objects and see how interpreting the variables involved as functions on these sets can be helpful both in terms of solving the problems and in terms of understanding the method of solution. They will also see that variables as used in iv) are, in fact, the same as variables in statistics and random variables in probability.

Variables 1: Variables and Relations between Variables

Materials Needed: Ample supplies of chart paper, markers, and whatever is needed to attach the chart paper to the wall of the room for students to display their work. At least 30 cards with mathematical vignettes for item (2). Some story lines with pairs of variables prepared and ready to hand out if the students' own problems do not work well for item (7). One card for each group with a mathematical vignette with open-ended questions about variables for item (10).

Mathematics covered: In this module students will explore their own concept of a variable and of the relation between two variables. They will recall what we mean by a relation and explore different equalities involving variables and think about which of these equalities are relations between variables and which of them are not. From the exploration it will become apparent, and also made explicit, that algebraic variables measure features of objects and as such are functions on sets of objects.

Accessibility: This module is written assuming students have had at least one semester of calculus. However, the material is important for any student that has had high school mathematics, and the module could be adapted for students who haven't yet had calculus.

1. The moderator should tell the class the focus of the day will be *variables*: what we mean by them in mathematics and what we mean by the *relation between two variables*. There could be a short preliminary discussion about what students understand by the word variable.
2. The students are divided into groups. Each group is given, say, thirty cards. On each card is written a mathematical vignette that contains letters with one of those letters or an expression involving those letters written below the vignette. The students are asked whether or not they consider the expression to be a *variable*. Using these as examples they are asked to come up with a definition of what is meant by a variable in mathematics. Examples of vignettes with different expressions that could be used are given below.

Vignette: Consider the function f that is given by

$$f(x) = \frac{x}{x^2 - 4}.$$

By the quotient rule we see that

$$f'(x) = -\frac{x^2 + 4}{(x^2 - 4)^2}.$$

Examples of expressions include: f, f', x, $f(x)$, $f'(x)$, and $\dfrac{x}{x^2 - 4}$.

Vignette: Suppose $y = 3x^2 + 7x + 4$. Then

$$\frac{dy}{dx} = 6x + 7.$$

Examples of expressions include: x, y, $\frac{dy}{dx}$, $3x^2 + 7x + 4$ and $7x + 4$.

Vignette: The associative law of addition states that

$$(a + b) + c = a + (b + c)$$

for all real numbers a, b and c.

Examples of expressions include: a, b, c, and $a + b$.

Vignette: For all sets A, B, and C,

$$A \cap (B \cup C) = (A \cap B) \cup (A \cap C).$$

Examples of expressions include: A, B, C, $A \cap B$ and $B \cup C$.

Vignette: For all functions f and g, $(f \circ g)(x) = f(g(x))$.

Examples of expressions include: f, $f \circ g$, x, $g(x)$ and $f(g(x))$.

Vignette: Abigail scored 72, 67, 82 and 79 on her first four math tests and her average score on these tests including the final was 77. Let x denote her score on the final. Then

$$\frac{72 + 67 + 82 + 79 + x}{5} = 77.$$

Examples of expressions include: x, $79 + x$, and $\frac{72 + 67 + 82 + 79 + x}{5}$.

Vignette: A painter has a paint mixture containing 1% blue dye and another mixture containing 6% blue dye. In order to determine how many liters she needs of each in order to make 50 liters of a solution containing 4% blue dye, let x be the number of liters she needs of the 1% mixture and let y be the number of liters she needs of the 6% mixture. Then

$$x + y = 50$$

$$\text{and} \qquad \frac{0.01x + 0.06y}{x + y} = 0.04.$$

Examples of expressions include: x, y, $x + y$ and $0.01x + 0.06y$.

Vignette: A closed box with a square base is to have a volume of 12 m^3. The material for the top and sides costs \$2 per square meter and the material for the bottom costs \$4 per square meter. In order to find the dimensions of the least expensive box that can be constructed, let h denote the height in meters, let x denote the length of the base in meters, let C denote the cost in dollars, and let V denote the volume in cubic meters. Notice that

$$C = 2(4xh + x^2) + 4x^2 = 8xh + 6x^2.$$

Moreover, since $V = 12$, it follows that $x^2h = 12$, so

$$h = \frac{12}{x^2}.$$

Thus

$$C = \frac{96}{x} + 6x^2.$$

Examples of expressions include: x, h, C, V, x^2h, $8xh + 6x^2$, and $\frac{96}{x} + 6x^2$.

Vignette: An airplane, flying at an altitude of 6 miles, passes directly over a radar antenna. When the airplane is 10 miles from the antenna, the radar detects that its distance from the antenna is decreasing at the rate of 240 mph. In order to determine the speed of the airplane, let x denote the distance of the airplane from the point 6 miles vertically above the antenna and let D denote the straight line distance of the airplane from the antenna. Notice that

$$D^2 = x^2 + 36.$$

It follows that

$$D\frac{dD}{dt} = x\frac{dx}{dt}$$

and when $D = 10$, $x = 8$. Thus, when $D = 10$,

$$\frac{dx}{dt} = \frac{(10)(240)}{8} = 300,$$

and we see that the airplane is traveling at 300 mph.

Examples of expressions include: x, D, t, D^2, $\frac{dx}{dt}$, and $x^2 + 36$.

There should be at least 30 cards total including at least 5 different examples for each of the 5 kinds of expressions listed below. This will allow for a rich discussion about the ways in which we use letters in mathematics and a rich enough set of examples to be able to make comparisons and classify the different meanings associated with the word 'variable.' The same vignette can be used on more than one card with different expressions to be considered on the different cards.

i) An expression representing the name of a known and explicitly described mathematical object.

ii) An expression representing any number or object in a set of numbers or set of mathematical objects (as in the statement of a theorem).

iii) An expression representing a particular quantity that is an 'unknown.'

iv) An expression representing a measurable quantity that can assume different values.

v) An expression representing a measurable quantity that is constant on the set in question.

If the moderator notices some groups are having a hard time coming up with a definition of a variable, she could prompt the group by saying a) the definition could be multiple sentences long and b) the definition could encompass different things as long as all of those things are explicitly included.

3. The class reconvenes and the groups report back. There is material here for a very rich discussion and there is not a 'right' or 'wrong' answer. Discussion should center around whether or not each group has categorized the expressions correctly according to their definition. In particular, some students might insist a variable be a single letter, so that, for example, in the vignette about boxes, they may try to claim C, x, and h are variables but $8xh + 6x^2$ is not. The moderator should ask them if it makes sense for a variable to be equal to something that is not a variable? Different groups will probably come up with different definitions and this is okay. It would be good to bring this to the students' attention, however, since it means different people mean different things when talking about a 'variable.' After all the groups have reported back it would be good to try to come up with 3 groups of cards; those in which everyone agrees the expression is a variable, those in which everyone agrees the expression is not a variable, and those in which there is disagreement about whether or not that constitutes a variable. From this, perhaps the class as a whole can come up with a definition encompassing everything falling into the first group. The definition will hopefully encompass variables as in the sense of a fixed but unknown numerical measure and a quantity that can take on different values. The definition will probably also encompass some of the things that fall in the other two groups. These should be gone through one by one and re-evaluated as to whether or not they are variables. If students are resistant to calling, for example, $8xh + 6x^2$ a variable, the moderator should insist that it is a quantity that can take on different values and therefore should be considered a variable as well. If they are still resistant, ask them to give you the benefit of the doubt for a while and see how they feel at the end of the module. If the definition includes expressions whose value is non-numerical (e.g. the value may be a function or it may be a set) then the moderator should explain that in what follows we'll only be interested in variables whose values are numerical. Having explored everyone's notion of

a variable, the next focus will be on determining what we mean by the relation between two variables.

4. Students are divided into groups and asked to come up with examples of 'word' problems in which they would use at least *two* variables to solve the problem. Tell the students to be creative; the problems could come from algebra or calculus or differential equations etc. For each example, the students should write down the example, define the variables that they would use in its solution, and *very briefly* say how they would use those variables to solve the problem. (It is not important that they solve the problem; indeed, most of the problems probably won't have 'nice' numbers). The moderator should walk around to the different groups and make sure there is a rich set of examples. It would be helpful if there was at least one regular problem in algebra, at least one regular calculus I optimization problem and at least one calculus I related rates problem. It would be great if there were other types of problems as well; for example, there could be modeling problems, problems from differential equations, problems from geometry etc. Depending on how many groups there are, the moderator may want the groups to come up with more than one example each. It would be good to have at least 10 examples total.

5. The class reconvenes and the groups share their problems. Hopefully some of the examples will include variables having fixed numerical values and variables that are more clearly functions on an underlying set and can assume many different values. If this is not the case, the moderator should include some problems of his or her own. When the groups share their problems, the moderator should press them, if necessary, on the exact definition of the variables they use. (The variable x is the length *of what*? Starting where and ending where? Notice the question *of what* is priming students to realize that x is measuring a feature *of an object*. This set of objects is the underlying set.) Discussion should center around whether or not everyone agrees with the presenters that the 'variables' that they used are, indeed, variables as the class had decided.

6. The moderator should ask students if they recall what we mean by the *relation* between two (numerical) variables and the graph of the relation. The goal of the discussion that follows is to have students understand that the relation between two variables x and y is the set of all pairs of values of x and y. In particular, the relation between x and y is a subset of the plane; (a, b) lies in the relation if and only if we can simultaneously have $x = a$ and $y = b$. If students have no recollection of what is meant by the relation between two variables, the moderator should prod them to remember back to their pre-calculus class when this was discussed. If that doesn't work, ask them what it means to say that y is a *function* of x. They will probably remember this but there are different things that they could say. They may say that the graph of x and y should pass the vertical line test. If they say this, the next

question should be what is meant by the graph of x and y; hopefully they will say something to the effect that it consists of all the possible pairs of values of x and y. The moderator can then explain that this is what is meant by the relation between x and y. Another thing they might say is that the value of x determines the value of y. In this case the moderator should affirm this and ask them to come up with examples of variables x and y where y is *not* a function of x. (If they are stymied here the moderator may have to send them back to their groups to discuss this between themselves.) Having the example(s), the next question is why is y not a function of x? Hopefully they will say there is more than one possible value of x for some values of y. How do they know this? They may give an example of two objects in the underlying set for which the value of x is the same but the value of y is different. From this it will be easy for them to understand that the relation is all pairs of values of x and y. Or possibly, they will write down an equation that describes the relation between x and y. In this case the moderator can explain that this equation is the relation between x and y and the graph of the equation is the graph of the relation. In particular, the relation is the set of all pairs of values of x and y.

However the discussion goes, the outcome should be that students understand the relation between x and y is a subset of the plane and (a, b) lies in the relation if and only if it is possible to have both $x = a$ and $y = b$. The goal in what follows will be for students to be able to say explicitly what it means to be 'possible to have both $x = a$ and $y = b$.' In particular this means x and y are measurements of features of objects in an underlying set and there is an object in the set for which both $x = a$ and $y = b$.

7. The students divide into groups. Each group is given a story with a collection of pairs of variables and is asked to sketch the relation between the two variables in each case and to determine in which cases the relation is a functional relation and in which it is not. The moderator could use the students' own examples or give them other examples. Below is an example of what could be given to one of the groups.

   Boxes with square bases cost \$2 per square meter for the material for the sides and top and \$4 per square meter for the material for the bottom.

   a) The length of the base and the height of the boxes.
   b) The length of the base and the surface area of the boxes.
   c) The surface area and the construction cost of the boxes.
   d) The length of the base and the area of the bottom of the boxes.

   Notice the pairs of variables include two cases where the relation is a region in space; one in which knowing the value of one variable gives *no* information about the value of the other (as in a) above) and one in which it gives *some* information but not enough to determine its value (as in b) above). It also includes a case where there is a functional relation between the two variables (as in d) above). Ideally the fourth pair would be a case where the graph of the relation is a curve but the relation is not a functional relation. However, it is hard to come up with such examples. Failing that, the fourth example could be a pair of variables where the graph is more of a challenge to determine (as in c) above).

8. Class reconvenes and groups report back. Discussion that follows should center around how they worked out whether or not a point (a, b) was or was not in the relation. The moderator should press them to say what they mean by it is possible for x to be equal to a and y to be equal to b. In each case they should end up saying there is an object in the underlying set for which $x = a$ and $y = b$. Moderator summarizes the discussion by noticing in each case there is an underlying set of objects under consideration and the two variables under consideration are both measurements of some feature of those objects; as such they are *functions* defined on the set of objects. A point (a, b) is in the graph of the relation between x and y if there is an object in the set for which $x = a$ and $y = b$.

9. The next item is to discuss different kinds of equalities involving variables. The moderator presents the following equalities to the class.

    - $(x-1)(x+1) = x^2 - 1$
    - $x^2 + 1 = 2x$
    - $\sqrt{x^2} = x$
    - Let $y = x^2 - 1$.

    Discussion centers around what the equalities mean. However the discussion goes, it should come out that the first equality is an identity of variables; for every value of x the left-hand side is the same as the right-hand side, the second equality is only true for a particular value of x, the third equality is true on a large subset of the domain of each expression, and the fourth equality is being used to *define* the variable y. The moderator should ask students for which of the equalities does it make sense to add 3 to both sides, to multiply both sides by x, to compose each side with another variable t (supposing x is a function of t), to differentiate both sides, and to integrate both sides? Since the equalities are not appearing in context, the discussion is necessarily open-ended; there are not right and wrong answers but just an exploration of what each of these might mean in some context. The idea is simply to prompt students to think about the nature of equalities; for complex objects, such as variables that are functions on an underlying set of objects, there are different ways that an equality can be interpreted, since there are different aspects of the objects that can be equated, and operations we perform may depend on the nature of the equalities involved.

10. The class is divided into groups and each group is given a mathematical vignette that contains equalities involving variables and open-ended questions about the nature of those equalities. Different groups are given different cards. Each group is asked to prepare a presentation for the rest of the class discussing the issues raised on their group's card. Here are some examples for the cards.

Vignette: Consider the curve in the plane that is described by the equation

$$x^3 - 3y^3 - x + e^y = 1.$$

Notice the point $(1, 0)$ lies on this curve. To find the slope of the curve at the point $(1, 0)$, we differentiate both sides of the equality with respect to x to obtain

$$3x^2 - 9y^2\frac{dy}{dx} - 1 + e^y\frac{dy}{dx} = 0.$$

Substituting $x = 1$ and $y = 0$ we then get

$$3 - 1 + \frac{dy}{dx} = 0.$$

It follows that the slope is -2.

Questions: In what sense is $x^3 - 3y^3 - x + e^y$ equal to 1? Is this only true when $x = 1$ and $y = 0$ or is it true more generally? Is this the relation between the variables x and y? What does it mean to differentiate both sides with respect to x; if $x^3 - 3y^3 - x + e^y = 1$ is the relation between x and y, why do we say that y is a function of x? Does this graph pass the vertical line test? What is the meaning of the variable dy/dx? In what sense is $3x^2 - 9y^2\frac{dy}{dx} - 1 + e^y\frac{dy}{dx}$ equal to 0? Is this true only when $x = 1$ and $y = 0$ or is it true more generally? How generally? Think about the two equations

$$x^3 - 3y^3 - x + e^y = 1$$

$$\text{and} \quad 3x^2 - 9y^2\frac{dy}{dx} - 1 + e^y\frac{dy}{dx} = 0$$

in the three variables x, y, and dy/dx. Do these two equations describe the relation between the three variables? What do we mean by the relation between three variables anyway? Consider the equation

$$3 - 1 + \frac{dy}{dx} = 0.$$

What is the nature of this equality? If we wanted to find the concavity of the curve at $(1, 0)$ could we differentiate both sides to get

$$\frac{d^2y}{dx^2} = 0?$$

Vignette: A painter has a paint mixture that contains 1% blue dye and another mixture that contains 6% blue dye that she mixes to make other shades of blue. Let x be the number of liters that she uses of the 1% mixture and let y be the number of liters that she uses of the 6% mixture. If she wants to make 50 liters of a solution that contains 4% blue dye, then she'll need

$$x + y = 50$$
$$\text{and} \qquad 0.01x + 0.06y = 0.04(x + y).$$

Questions: In what sense is $x + y = 50$? In what sense is $0.01x + 0.06y = 0.04(x + y)$? Do these equations show the relation between x and y? What is the relation between x and y? What does the graph of the line $x + y = 50$ represent physically? What about the graph of $0.01x + 0.06y = 0.04(x + y)$? Would it make sense to differentiate both sides of $x + y = 50$ to obtain

$$1 + \frac{dy}{dx} = 0?$$

Why or why not?

Vignette: An airplane, flying at an altitude of 6 miles, passes directly over a radar antenna. When the airplane is 10 miles from the antenna, the radar detects that its distance from the antenna is decreasing at the rate of 240 mph. In order to determine the speed of the airplane, let x denote the distance of the airplane from the point 6 miles vertically above the antenna. Since the airplane is 10 miles from the antenna,

$$100 = x^2 + 36.$$

Differentiating, we get

$$0 = 2x\frac{dx}{dt} + 0$$

so the airplane is not stationary.

Questions: What mistake did this student make? How could you help this student understand her error? Under what circumstances are we allowed to differentiate both sides of an equation and under what circumstances is it nonsense to do this? Can you come up with examples to explain this to the student?

Vignette: A closed box with a square base is to have a volume of 12 m^3. The material for the top and sides costs \$2 per square meter and the material for the bottom costs \$4 per square meter. In order to find the dimensions of the least expensive box that can be constructed, let h denote the height in meters, let x denote the length of the base in meters, let C denote the cost in dollars, and let V denote the volume in cubic meters. Notice that

$$C = 2(4xh + x^2) + 4x^2 = 8xh + 6x^2.$$

Moreover, since $V = 12$, it follows that $x^2h = 12$, so

$$h = \frac{12}{x^2}.$$

Thus

$$C = \frac{96}{x} + 6x^2.$$

Questions: What is the nature of the equality $C = 8xh + 6x^2$? Is this a relation between variables? Compare this equality with the equalities $V = 12$ and $x^2h = 12$. Can we differentiate these equalities to obtain

$$\begin{aligned} \frac{dC}{dx} &= 8\left(h + x\frac{dh}{dx}\right) + 12x \\ \frac{dV}{dx} &= 0? \\ 2xh + x^2\frac{dh}{dx} &= 0? \end{aligned}$$

What about the following. Are they meaningful?

$$\begin{aligned} \frac{dC}{dh} &= 8\left(\frac{dx}{dh}h + x\right) + 12x\frac{dx}{dh} \\ \textit{or} \qquad 1 &= 8\left(\frac{dx}{dC}h + x\frac{dh}{dC}\right) + 12x\frac{dx}{dC} \end{aligned}$$

Does it make sense to talk about $\frac{dC}{dx}$? What about $\frac{dC}{dh}$? What about $\frac{d^2C}{dxdh}$?

11. The class reconvenes and groups report back. Discussion follows.

Variables 2: Comparing Variables in Algebra with Variables in Statistics and Random Variables in Probability

Materials Needed: Ample supplies of chart paper, markers, and whatever is needed to attach the chart paper to the wall of the room for students to display their work. Collections of word problems for item (6). Each group needs a collection of problems.

Mathematics Covered: Variables in statistics and random variables in probability are functions defined on the population and sample space of the experiment respectively. Variables in algebra can also be thought of in this way. In this module, through a comparison of variables in statistics and random variables in probability, students will gain an understanding of how variables in algebra can be thought of as functions defined on a set of objects. They will also contrast variables in algebra with random variables in probability and variables in statistics. They will see that it is the additional structure of a probability measure on the underlying set that leads us to ask different kinds of questions in probability and statistics.

Accessibility: This module is designed for students that have taken a class in probability and therefore know what a random variable is. Since random variables in probability are defined as functions on the sample space of the experiment, students will be primed for thinking about a variable in algebra as a function on a set of objects. This will allow a rich discussion and comparison of these two kinds of variables.

1. The moderator should tell students in this module the class will be comparing variables in probability, variables in statistics and variables in algebra. In particular, the goal is to understand what variables *are* in these different contexts and to determine in what ways they are the same and in what ways they are different in these different contexts.

2. In groups, the class is charged with discussing what *random* variables in probability are. They should be asked to remember the definition (if they can), come up with a rich set of examples illustrating the different *types* of variables that there are, and to compare random variables in probability with non-random variables in statistics.

3. The class reconvenes and groups report back. Hopefully at least some of them will have remembered the definition of a random variable. If not, the moderator should make sure that the definition emerges from the discussion. Most of them will probably be able to come up with examples, though the majority of their examples will probably be of discrete numerical variables, so the moderator should ensure examples of categorical variables and numerical continuous variables are also presented to the class. They may have difficulty articulating what a non-random variable in statistics is, because this is often not discussed explicitly. (A variable in statistics is a function defined on the population of interest. Different kinds of variables are characterized by the target space of the function; if it is a set of objects then the variable is said to be categorical and if it is a set of numbers with which it makes sense to do arithmetic then the variable is said to be numerical.) If nobody is able to come up with a definition and examples, the moderator should present some examples. To contrast these examples with random variables in probability, the moderator should have students consider the experiment of taking a sample from the population and consider some *random* variables associated with that experiment. It is important the following things emerge from the discussion and the students are relatively comfortable with them.

   a) The fact that a random variable X is a *function* whose domain is the sample space of an experiment and that $\{X = 5\}$, for example, is a subset of the sample space.

b) Examples of variables including both numerical continuous, numerical discrete, and categorical variables and an explicit discussion of how variables are categorized according to the kinds of values they assume.

c) The fact that non-random variables in statistics are functions on the population of interest.

d) A comparison of random variables and non-random variables where the experiment is 'take a sample' from the population of interest. For example, if we consider the set of all voters, a non-random variable is the party affiliation of each voter. If we take a random sample from the set of all voters, then a random variable is the number of republicans in the sample. (Notice, by the way, the party of affiliation of voters in the sample is not a random variable since this is a function on the specific sample obtained and not on the sample space of the experiment.)

4. Students return to their groups charged with comparing variables in algebra with random variables in probability and variables in statistics. They should come up with examples of problems in which (algebraic) variables are used in the solution. What do the variables mean? Is the meaning of the variable the same as the meaning in probability and statistics? As the moderator visits the groups she should listen to the kinds of examples they are coming up with. If all the examples are of variables that represent an 'unknown,' the moderator should encourage students to think about optimization problems and related rates problems in calculus in which the variables used are more obviously similar to variables in probability and statistics.

5. The class reconvenes and groups report back. Hopefully they will have come up with a rich set of examples for discussion. They may or may not have come to the conclusion that variables in algebra are the same as variables in probability and statistics. In the discussion that follows the moderator should use their examples to encourage them to think of variables as a function on a set of objects. Take the easier examples first, and ask them if there is some way that the variable that is used *could be thought of* as a function on a set of objects. Go through each example in this way, so that in every case they can see that, indeed, a variable is a function on a set of objects.

6. Students return to their groups. Each group is given a collection of 'word' problems in algebra and calculus and they are asked, in each case, to identify the underlying set of objects of interest and to define 5 variables associated with this set of objects. Ask them to be as creative as possible when defining the variables. Can they come up with variables that are undoubtedly variables but are irrelevant to solving the problem at hand? Can they come up with variables that are a little different from the usual things that they think of as variables? Here are a collection of problems that could be used along with some examples of variables.

a) A closed box with a square base is to have a volume of 12 m^3. The material for the top and sides costs \$2 per square meter and the material for the bottom costs \$4 per square meter. Find the dimensions of the least expensive box that can be constructed.

b) An airplane, flying at an altitude of 6 miles, passes directly over a radar antenna. When the airplane is 10 miles from the antenna, the radar detects that its distance from the antenna is decreasing at the rate of 240 mph. What is the speed of the airplane?

c) A painter has a paint mixture that contains 1% blue dye and another mixture that contains 6% blue dye. How many liters does she need of each in order to make 50 liters of a solution that contains 4% blue dye?

7. The class reconvenes and the groups report their discussions.

8. The next step is to contrast variables in algebra with variables in statistics and random variables in probability. The students return to their groups and are asked to formulate six questions total that we might be interested in; two involving algebraic variables, two involving random variables, and two involving variables in statistics. Examples of questions in the former might be simple algebra I or algebra II problems, optimization problems and related rates problems. Examples of questions in probability might have to do with finding the probability a random variable takes on a particular value or finding the distribution of a random variable. Examples of questions in statistics might be the degree to which two variables are correlated or the degree to which one variable can be used to predict the value of another.

9. The class reconvenes and groups share their stories. Discussion should center around the kinds of questions we are interested in about algebraic variables, the kinds of questions we are interested in about random variables, and the kinds of questions we are interested in about variables in statistics. As the class brainstorms, someone should be assigned the task of writing down on chart paper or the blackboard (where everyone can see) the kinds of questions that are asked in each case. Here are examples of things that might be said.

   - *Algebraic variables:* What is the relation between two variables? Is one variable a function of another and, if so, can we find a formula for that function? What is the largest possible value of a variable? How fast does a variable change as time moves on?
   - *Random variables:* What is the probability that the variable assumes a particular set of values? What is the distribution of the variable? What is the joint distribution of two variables? What is the correlation between two variables? What is the expected value and variance of a variable?
   - *Statistical variables:* What is the distribution of the variable? How much information does one variable give us about another? What is the expected value and variance of a

variable?

10. The next question is whether it makes sense to take a question that we ask about algebraic variables and ask it of a random variable or a variable in statistics? For example, does it make sense to ask what the largest value of a random variable or a variable in statistics is? Does it make sense to ask what the relation between two random variables in probability or two variables in statistics is? Does it make sense to ask how fast a random variable in probability or a variable in statistics is changing? These questions can be asked concretely by referring to the student's stories. The moderator should summarize that most of these questions certainly make sense but they may not be very interesting questions.

    Similarly, does it make sense to take a question that we ask about random variables or variables in statistics and ask it of an algebraic variable? If we have a variable in algebra, does it make sense to ask what its distribution is? Does it make sense to ask the probability of it assuming a particular value? If you have two variables in algebra, does it make sense to ask to what extent one of the variables is determined by the other? Again, these questions can be asked more concretely by referring to the students' stories. The moderator should summarize that most of these questions do not really make sense, because they require the additional structure of a *probability measure* on the underlying set on which the variables are defined. It is the assumption of this additional structure that drives the kinds of questions that are asked.

Variables 3: The Power of Thinking of Variables as Functions on Sets of Objects in Optimization Problems

Materials Needed: Ample supplies of chart paper, markers, and whatever is needed to attach the chart paper to the wall of the room for students to display their work. Optimization problems written out with pictures and a wire frame (depending on the problems chosen) for item (1). Optimization problems for each group for item (7).

Mathematics covered: This module is about optimization. Optimization problems encountered in first semester calculus problems are typically very challenging for students. The main challenge is setting up an expression for the variable to be optimized in terms of another variable in the problem. Some of the reasons for this difficulty are the following.

- Students have a poor understanding of what variables are and that a variable is the measurement of a *feature* associated with a *set of objects*. In particular, this means a variable is a *function* whose domain is a set of objects.

- The choice of independent variable is left to the student. Since students have a poor understanding of what variables are, they have difficulty pulling variables out of thin air (so to speak).

In this module students will reflect on optimization as a process. They will look closely at the variables involved and how the nature of these variables and their relationship to the underlying set determine the method of optimization to be used.

When optimization problems are covered in first semester calculus classes there is a natural underlying question that is rarely addressed directly; why is it that writing the variable to be optimized y as a function of another variable x, helps us find the maximum and minimum values of y? When you first think about this it may, indeed, seem a little mysterious. However, when you consider a variable is actually a function on a set of objects, it becomes less mysterious; by writing y as a function of x (where the values of x range over some interval), rather than thinking of it as a function on a set of objects that has no structure, we are endowing the domain space with the structure of the real numbers. This makes it easier to perform the optimization, because it allows us to systematically run through all the values that y can attain by letting x increase from its lowest value to its highest value.

In these optimization problems we are interested in the optimum value of y and/or the object in the set for which that optimum is attained. The independent variable x is chosen by the person solving the problem as a *tool* for solving the problem and is generally not of interest in its own right. As a tool, it has some important properties. First of all, knowing the value of x is enough to identify the object in the set being considered. In particular, this means there is a one-to-one relationship between the values of x and the objects in the set. Moreover, in calculus I problems, the values of x range over an interval. In this module we will say variables having these properties are *characteristic* of the set. Not all sets have characteristic variables; for example, if the cardinality of the set is greater than the cardinality of the real numbers no such variable can exist. Calculus I optimization problems are problems where the set of objects under consideration has at least one characteristic variable. Optimization problems encountered in multivariable calculus classes are problems where the underlying set is best characterized by more than one variable and other kinds of optimization problems in which one might use, say, simulated annealing, or problems in the calculus of variations are concerned with more complex sets that are not easily characterized by even a handful of variables.

In contrast to the independent variable x, in optimization problems studied in first semester calculus classes, there generally is not a one-to-one relationship between the values of the variable to be optimized y and the objects in the set. (In those rare cases where there is a one-to-one relationship, it is far from obvious.) This variable is generally not characteristic of the set, so the range of values it takes on are more difficult to determine.

In this module students will think deeply about what we mean by a variable. They will practice identifying the underlying set of objects being considered in various kinds of problems involving variables, will practice identifying variables associated with these sets, and will learn to classify variables as characteristic or uncharacteristic. They will explore the optimization process used in Calculus I optimization problems and gain some appreciation of what it is about those optimization problems that means they can be solved using the technique that is generally employed.

Accessibility: This module is written for students who have taken a calculus class. Moreover, it is assumed in this module that students already have an understanding that variables can be thought of as functions on an underlying set of objects.

1. The students are divided into groups. Each group is given one or more optimization problems to do. (The moderator shouldn't tell them they are optimization problems; they should simply be asked to do the problems.) Some of the optimization problems should be typical Calculus I optimization problems but others should involve optimizing a characteristic variable and others should involve optimization over a set that has less structure. Here are some sample problems that could be used. The moderator should ensure there are at least 2 problems from each type of problem. It might be appropriate to give each group three problems; one of each type. Notice it is quite unlikely they will be able to solve the Type C problems, but that's okay; this exercise is not about finding the solutions, it is about the processes we use to find the solutions.

   Type A: Optimizing a characteristic variable.

   a) If a taco costs \$1.25 and you have \$62, what is the maximum number of tacos you can buy?
   b) A company wants to package its product in a cylindrical can whose volume is 16π cubic inches and whose radius is between 1 and 2 inches. What is the smallest possible height of the can? What is the smallest possible radius?
   c) A church is organizing free thanksgiving meals for the homeless. One meal consists of 4 ounces of turkey, 1/4 cup of cranberry sauce, 1 cup of mashed potatoes, 1/2 cup of yams, 5 brussel sprouts, a cup of green beans and a cornbread muffin. If the church has 280 lb of turkey, 600 pints of cranberry sauce, 1250 pints of mashed potatoes, 500 pints of yams, 6000 brussel sprouts, 400 pints of green beans and 1200 cornbread muffins, what is the maximum number of people that the church can feed?

   Type B: Typical Calculus I optimization problems.

   a) What are the dimensions of the box of largest volume that has a square base and can be wrapped with 106 cm of ribbon if 10 cm are left for the bow? (See the picture below.)

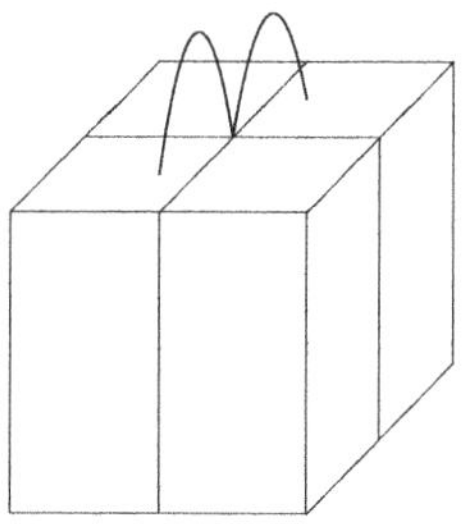

b) A cable is to be run from a power plant on one side of a river 900 meters wide to a factory on the other side 3000 meters downstream. If it costs \$5 per meter to run the cable under water and \$4 per meter to run the cable over land, what is the most economical route by which to run the cable?

c) You operate a tour service that can be booked by private parties. For each excursion you require a minimum of 50 people and you can accommodate up to 80 people total. For 50 people you charge \$200 per person and for each additional person beyond 50 the rate is reduced (for everybody) by \$2 per person. It costs you \$6000 plus \$32 per person to conduct the tour. What are the maximum and minimum profits you can make on each tour? How many people do you need to maximize your profit?

Type C: Optimizing over a set with less structure.

a) Consider the set of all quadrilaterals that are completely contained in the figure below and all of whose vertices lie on the edges of the figure. What is the largest perimeter that such a quadrilateral could have? What is the smallest perimeter that the quadrilateral could have?

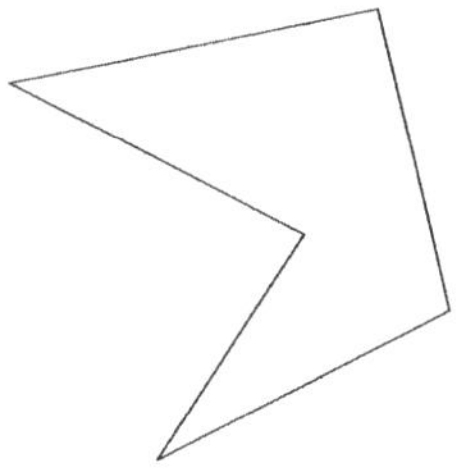

b) The group should be given a wire frame living in three dimensional space. They should consider all those surfaces that include all edges of the frame. Which one of these surfaces has the smallest possible area? What does that surface look like?

c) Shown below is a graph with 'costs' associated with different edges. Find the least expensive route through the graph that starts and ends at A and visits every node at least once.

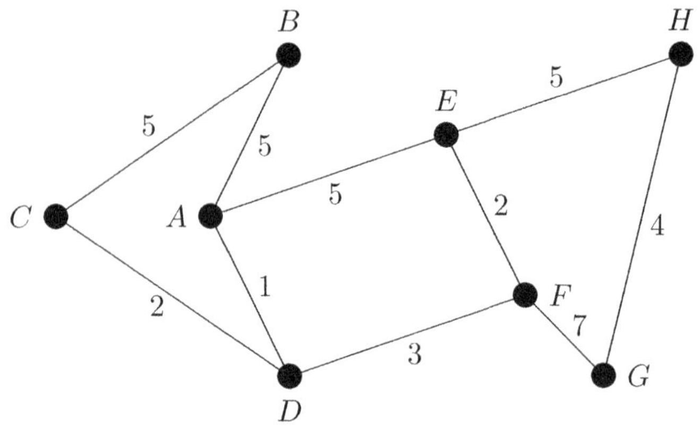

2. The class regroups and students share their solutions with each other. The moderator asks the class to think about what's similar about all the problems. Hopefully it will emerge in all of the problems we are trying to make some quantity as large or as small as we possibly can; in other words, these are optimization problems.

3. The students go back to their groups and are asked to discuss what's different about the different problems. In particular, what is it about the problems that requires the *method* by which the optimization is done to be different in the different cases? How do the methods differ? Can the students categorize all the problems according to the method used to solve them? Students should try to articulate in what ways the methods of solution for the problems in each category are similar and what it is about the problems in each category that means such an approach would work.

4. The class reconvenes and the groups report back. In the discussion that follows the moderator should try to steer the class to noticing some problems were solved by looking at simple inequalities, some problems were solved by writing the variable to be optimized as a function of another variable (this is the standard Calculus I technique), and other problems were not solved at all or were solved by ad-hoc methods. The class should discuss what it is about the problems and variables in the first group that means they can be solved by looking at simple inequalities, what it is about the problems and variables in the second group that means they can't be solved by looking at simple inequalities, and what it is about the problems and variables in the third group that means they can't be solved using Calculus I techniques. The moderator should summarize the discussion as follows. All the problems are concerned with making some *quantity* either as small or as large as we can. Since this quantity has different possible values, it is a *variable*. Depending on the nature of the variable to be optimized and its relation to the underlying set , different approaches may be needed to solve the different problems.

5. Students return to their groups. For each problem they should first identify the underlying set of objects under consideration, the variable that is to be optimized, and any other variables that are used in solving the problem. Then they should explore how all these variables are related to the underlying set on which they are defined; in which problems and for which variables is there a one-to-one relationship between the values of the variables and the objects in the set? If you know the value of the variable, can you identify the object in the set to which it corresponds?

6. The class reconvenes and the groups report their discussions. The moderator should summarize that in the Type A problems (the ones that were easy to solve), the variable to be optimized had a one-to-one relationship between its values and the objects in the set. In the Type B problems (the calculus I problems), the variable x that was used as a tool for the optimization had a one-to-one relationship between its values and the objects in the set but the variable to be optimized did not. In the Type C problems there were no easily identifiable variables having a one-to-one relationship between its values and the objects in the set. Let's label those variables having a one-to-one relationship between their values and the objects in the set as *characteristic*.

   The next question on the table for discussion is *why* in the Type B Calculus I optimization problems, it was helpful to write the variable to be optimized as a function of another variable x. In other words, what is it about doing this that makes it possible to determine the maximum and minimum values of y? Why can't we determine the maximum and minimum values of y the same way that we determined it in the other problems? In the discussion that follows, the moderator should ensure that it emerges that writing y as a function of x allows us to systematically explore all the values of y by running through the values of x from its minimum value to its maximum value. In the harder optimization problems there are no natural characteristic variables so this is not a viable method of solution.

7. Students return to their groups. Each group is given a bunch of optimization problems and asked, not to solve the problems, but in each case, to identify the underlying set being considered, identify at least 5 different variables, and for each variable determine if it is a characteristic variable or not. Ask them to be creative when identifying variables.

8. The class reconvenes and groups report back.

9. Now the focus will be on using this new way of looking at optimization problems as a way to help others solve optimization problems. To this end, the groups are asked to *solve* the optimization problems they just considered by doing the following.

   a) First they should choose a characteristic variable for each problem and find the range of values of that variable.

b) Then they should choose a value of the variable in the range, consider the corresponding object in the set, and determine the value of the variable to be optimized, y, for that object.

c) Repeat with a different value of the characteristic variable.

d) Use their work in the two concrete examples to now write the variable y as a function of x.

e) Complete the optimization.

Part V

Center Assessment

Internal Evaluation And Reporting

© Tiana Bosley,
Wayne State University

1 Introduction

This chapter discusses some elements of internal evaluation of a Quantitative and Mathematics Support Center (QMaSC) from the prospective of a center director. To understand how to conduct an internal evaluation, it is good to have a clear understanding of what internal evaluation involves. After explaining this information, the chapter discusses a specific method of evaluation that can be applied to a mathematics or quantitative center that is in the early stages of acquiring data. This means that the internal evaluation refers only to the operation of the center; comparisons with other centers are addressed in the chapter involving external evaluation.

J. L. Fitzpatrick defines evaluation as "the identification, clarification, and application of defensible criteria to determine an evaluation object's value, its merit or worth, in regard to those criteria" [1]. Evaluation consists of several levels requiring preparation that may include focusing the evaluation on decision-making criteria, specifying the type of data for collection, organizing the format and design of the information, choosing appropriate methods of analyzing the information, specifying the type of information, identifying recipients of the reports, and administering the evaluation. It can be performed by internal, and/or external evaluators.

"Internal evaluation traditionally has been viewed — and largely practiced — as a staff function informing operations, management, and/or strategic planning" [2]. Thus, a QMaSC's evaluation can be performed by the director or coordinator of the center or by staff appointed by a manager

Suggested Citation: Tiana Bosley (2016), "Internal Evaluation And Reporting", http://dx.doi.org/10.5038/9780977674435.ch16 in G. Coulombe, M. O'Neill, M. Schuckers (Eds.) *A Handbook for Directors of Quantitative and Mathematical Support Centers*, Neck Quill Press, http://scholarcommons.usf.edu/qmasc_handbook.

This material is based upon work supported, in part, by the National Science Foundation under Grant DUE-1255945. Any opinions, findings, and conclusions or recommendations expressed in this material are those of the author(s) and do not necessarily reflect the views of the National Science Foundation

to operate as an evaluation committee or group. Center directors or coordinators can determine their method of internal evaluation by considering the size of their center and the budget allowed for the evaluation.

There are both benefits and disadvantages to internal evaluation. "Internal evaluators are likely to know more about the program, its history, its staff, its clients, and its struggles than any outsider" [1]. This type of evaluation can be conducted by the center director and staff. It can be a budget-friendly option. However, an internal evaluator may be too close to the program and subject to bias, so an external evaluation may be necessary.

2 Basic Evaluation Roles

Evaluation can take on two roles, formative or summative. An evaluation is categorized as formative if the results of the evaluation are to be used to improve the program. "Formative data collected during the early stages of a program can help identify problems in the program model or theory or in the early delivery of the program that can be modified or corrected" [1]. This data can contribute to the creation of new or additional programs. The summative evaluation approach is often used to assist in making decisions about the continuation or adoption of a program. Initially, in a mathematics or quantitative center, a formative evaluation may be the most suitable method. Summative evaluation can be used in the future if the center's coordinators or directors plan to expand the program beyond their institution.

Before a formative evaluation begins, the role of the evaluator must be negotiated and the goals of the program should be clearly defined. For example, the main goal of a research institute's QMaSC may be to help WSU students develop mathematical skills necessary to be successful learners through the creation of a safe and friendly learning environment. It is the job of the internal evaluator to determine if these goals are being met. It can be useful to implement a decision-oriented and descriptive formative evaluation plan.

Prior to his position as an emeritus professor and director at the Western Michigan University Evaluation Center, Daniel Stufflebeam created the CIPP (Context, Input, Process, and Product) Evaluation Model that can be applied to help in decision-making and quality assurance. The Context of the program guides planning, Input relates to structuring decisions, Process is for implementing decisions, and Product is for recycling decisions. J. L. Fitspatrick calls Stufflebeam "a leader in developing an approach oriented to decisions" [1]. The CIPP Evaluation Model Checklist [3] can be a useful guide for conducting a math center's evaluation. An example of the steps in this method as applied to a QMaSC are presented in Table 1.

The descriptive design of the evaluation should include elements of a cross-sectional design to survey and observe the program as it relates to students and tutors utilizing the center. Eventually, this form of evaluation will provide outcomes that will help management to modify this program and continuously re-evaluate any adjustments in future stages. Thus, a descriptive design will assist with the four stages of evaluation implementation in the CIPP Evaluation Model.

Table 1: Example of the CIPP Evaluation Model for a QMaSC

Context Evaluation	Input Evaluation	Process Evaluation	Program Evaluation
The QMaSC will use these findings for review and revision of their goals to properly target assessed needs. The QMaSC will use this step throughout the entirety of its tutoring program to assist in assessment of its effectiveness and significance.	The QMaSC will maintain records of program events, problems, costs, and allocations. The QMaSC will maintain an up-to-date profile of the tutoring program offered by the QMaSC.	The QMaSC will use these findings to coordinate and strengthen staff activities. The QMaSC will use these findings to report on the center's progress to the financial sponsors, the college or university.	The QMaSC will use these findings to judge the extent to which its tutoring program is serving or did serve its tutees. It will also use these findings to judge the extent to which its programs have addressed or are addressing important needs of the university community.

There are some limitations to this formative evaluation approach. This evaluation can only provide information for evaluating a specific center based on the goals of that center. It will be necessary to repeat the evaluation over several terms to collect enough data for comparisons with other tutorial programs. Moreover, the purpose is to assess quality assurance, and the program coordinator or director will have more explicit decision-making responsibilities than other stakeholders (tutees and staff) in the program. This could lend itself to bias in the collection and interpretation of data unless an external evaluator is also utilized or different methods of data collection are used to assess the same item to discover if the results of the analysis are comparable.

3 Components of the Evaluation

Stufflebeam further explains, "Sound evaluations are grounded in clear and appropriate **values** (principles, attributes, or qualities held to be intrinsically good, desirable, important, and of general worth) and **criteria** (standards on which to base judgments)" [4]. An internal evaluation should be guided by questions that address the overall quality of the center. These questions can focus on environmental factors, tutor subject and skill proficiency, tutoring benefits and resource availability. The center should be the main source for data collection because it can be used to observe the interactions of the tutors and students and to supply students with the survey instruments.

Several methods can be used to collect data. They can include documents, records, observations, surveys, interviews, focus groups and/or tests. Some of this data can be collected directly from individuals. For example, a center director may require tutors to submit their transcripts to validate their mathematical background. Data can also be collected via an assessment test given to tutors each academic year to assess their current mathematical skills. Equally important, the program director/coordinator can use face-to-face interviews to observe the tutor's overall method of tutoring through questions and role-play situations. An additional data collection source is the information collected by the center on a regular basis. This can be a useful and inexpensive source of information. Attendance records are a good example of this type of data. They are collected on a daily basis and can provide the frequency of the visits of each tutee and listings of courses that require the most assistance for students. This is important when a center coordinator or director decides to make improvements in the services of the center such as having tutors review subject areas that cause the most confusion or operating the center during hours that are most needed by its client population.

The internal evaluator must decide on the types of instruments (surveys, tests, etc.) to be used to carry out the evaluation. Most often, the program coordinator or director creates the necessary instrument for evaluations. This can be a tricky task because evaluation questions need to meet certain standards and be clearly written to assess whether the center's goals are being met. Another option would be to locate some instruments through an online search engine like Google and to contact the authors for permission to use them. Lastly, some instruments for evaluation can be purchased from an established evaluation company, but this option can be expensive.

The following are examples of instruments that can be used in a QMaSC center:

1. *Student Survey (Tutors)* This survey can consist of several items using a five-point Likert scale: 1-Strongly Disagree, 2-Disagree, 3-Neutral, 4-Agree, 5-Strongly Agree. It can also contain a few demographic items. The purpose of this instrument would be to measure student attitudes toward tutors' ability. The topics covered can include Quality of Services, Academic Benefits, Environment, Knowledge of Content (Tutor) and Teaching Ability of Tutors. An example is provided in Appendix A.

2. *Assessment Test* This instrument can be developed to measure tutor proficiency in mathematics. It can test major concepts and computational skills from the main courses covered by tutors in the center. For example, tutors may be given a timed test of 15 - 20 problems in which they must score at least 80 out of the maximum 100 points to be considered proficient.

These problems can come from old final exams. An example of questions used for assessment of the tutors is provided in Appendix B. These questions were taken from an old Precalculus group final.

3. *Skill Assessment Interview* An instrument can be developed to measure tutor proficiency in instruction during various situations. It can be conducted as a face-to-face interview between each tutor and the internal evaluator. The interview can consist of questions regarding tutoring skills and the purpose of tutoring as well as role-play situations in which the tutor has to respond to the behavior of the internal evaluator. For example, the internal evaluator can act the role of an irate student and then ask the tutor how he or she would respond in this situation. An example is provided in Appendix C.

4. *Checklist* This instrument can be developed using the "Guidelines for Developing Evaluation Checklists: The Checklists Development Checklist" by Daniel Stufflebeam [5] to determine if sufficient resources, such as textbooks, worksheets, etc. are available to students who use the center.

5. *Observation Instrument* Random 20-minute observations can be conducted of each tutor throughout the term. An instrument can be developed to summarize these observations. The data collected can be organized in the following categories: Instructional Strategies, Organization, Communication Skills and Student/Tutor Rapport.

6. *Tutor Transcripts* Updated academic transcripts of tutors can also be used to determine if tutors have demonstrated proficiency in the courses in which they will tutor.

These instruments use mixed methods of data collection (qualitative and quantitative). They can be administered randomly by paper or online in the forms of open ended questions or surveys. Randomization is common for many evaluations because it can provide better material for analysis and control for bias. Also, multiple methods of analysis are helpful in supporting the validity of the evaluation being performed.

An example of an evaluation plan is presented in Table 2 on the following page. This table can be adapted for use by any mathematics or quantitative center. SPSS refers to the statistical software package offered by IBM. However, other packages and software with statistical data analysis methods can also be used.

4 Reporting

To create a good evaluation report, an evaluator should consider several factors. It is important to consider the audience that will receive the report because reports and presentations will need to

Table 2: Example of an Evaluation Plan Summary for a QMaSC

Evaluation Question	Source of Data	Method of Data Collection	Method of Data Analysis
Is the QMaSC a comfortable learning environment ?	Students	Survey	Descriptive statistics and Chi-Squared Test (SPSS)
	Tutors/Students	Observation	Observation instrument data summary
How proficient are tutors in calculus courses and below?	Tutors	Request Tutor Transcripts	GPA in relevant math courses will be computed
	Tutors	Assessment Test	Scored out of 100 points (80 points + considered proficient)
How skilled are the tutors at providing supplemental instruction?	Students	Survey	Descriptive statistics and Chi-Squared Test (SPSS)
	Tutors	Interview	Summarize major themes of interviews
	Tutors/Students	Observation	Observation instrument data summary
Is supplemental material, such as textbooks, worksheets, etc., available in the QMaSC?	QMaSC	Checklist	Categorical (Yes/No)
Do students feel they benefit from the services of the QMaSC?	Students	Survey	Descriptive statistics and Chi-Squared Test (SPSS)
	QMaSC	Sign-in sheets/Attendance Records	Frequency Analysis (SPSS)
Does tutoring improve student understanding of the material?	Students	Student survey	Descriptive statistics and Chi-Squared Test (SPSS)
	Students	Students self-report test/quiz/homework scores with Access ID, run of grade reports for mathematics courses at the level of first-semester calculus and below	Multiple regression with final course grade regressed on test, quiz, homework scores (SPSS)

be custom-designed for that particular audience. For example, the tutors do not need to receive a report on the budget of the center. This type of report would be more suitable for the business manager or dean of the center's college. Also, the report must be delivered in a timely fashion that will allow the information to be used.

Evaluation findings can be written in a report or provided during an oral presentation. Written information can be given at meetings, sent through email, listed in a blog or posted on the center's web site. With all of these methods, the internal evaluator must consider the audience and only list pertinent information. For example, budget allocation information would not be provided on a center's web site. Also, a final report should be presented to the stakeholders invested in the success of the center. "Final reports include an executive summary, an introduction, a description of the evaluation object and the questions to be answered" [1]. They should also include a brief description of the evaluation plan, methodology, results and recommendations. Technical information should also be included in the appendix of the report. Oral presentations of the findings, if required, can provide a nice platform for conveying the results of the evaluation with the use of PowerPoint, handouts, and even posters.

5 Summary

Internal evaluation can be a useful and cost-friendly method of evaluation for a QMaSC. It can be performed by a program coordinator, director, or a group of staff members specifically assigned to the task of evaluation. The internal evaluation should be created to assess the overall quality of the program as it relates to each center's goal. This can be done through use of the CIPP Evaluation Model during the formative stage of the evaluation and during center re-evaluations. The weaknesses of this method can be minimized by the use of an additional external evaluator or by using several data collection methods for analysis. Clarity, timing and style of presentation must be considered for reporting to a specific audience. Ultimately, internal evaluation can be useful in the improvement of the center's program. It can provide data that can be used to make management decisions for improvement of the structure of the center, budget allocations, and confirmation that the center is succeeding in meeting its goals.

6 Bibliography

[1] J. L. Fitzpatrick, J. R. Sanders, and B. R. Worthen, *Program evaluation: alternative approaches and practical guidelines*. 4th edition ed., 2011.

[2] R. T. Torres, "Improving the quality of internal evaluation: the evaluator as consultant- mediator," *Evaluation and Program Planning*, vol. 14, pp. 189–198, 1991.

[3] D. Stufflebeam, "CIPP evaluation model checklist." https://www.wmich.edu/sites/default/files/attachments/u350/2014/cippchecklist_mar07.pdf, 2015.

[4] D. Stufflebeam, "Evaluation values and criteria checklist." https://www.wmich.edu/sites/default/files/attachments/u350/2014/values_criteria.pdf, 2015.

[5] D. Stufflebeam, "Guidelines for developing evaluation checklists: the checklists development checklist." https://www.wmich.edu/sites/default/files/attachments/u350/2014/guidelines_cdc.pdf, 2013.

7 Appendix A

QMaSC Student Satisfaction Survey – A few sample statements

Directions: Please circle the number that best fits your response where
1 = Strongly Disagree, 2 = Disagree, 3 = Neutral, 4 = Agree, 5 = Strongly Agree

1. The Center provided the services that I needed.

   1 2 3 4 5

2. The quality of tutoring meets my needs.

   1 2 3 4 5

3. The staff makes me feel comfortable.

   1 2 3 4 5

4. The Center has suitable hours of availability.

   1 2 3 4 5

5. The tutors are knowledgeable and help in the understanding of my coursework.

   1 2 3 4 5

6. I would recommend the Center to others.

   1 2 3 4 5

7. Overall, the Center is a good source for mathematics help.

   1 2 3 4 5

Please provide any additional questions or comments regarding the Mathematics Resource Center on the lines below.

____

____

____

8 Appendix B

QMaSC Tutor Assessment for Precalculus

Read the directions to each problem carefully. ALL WORK MUST BE SHOWN. Each problem is worth 5 points. DO NOT USE A CALCULATOR.

1. Find the domain of the function $f(x) = \frac{e^x + \ln(1-\frac{3}{2})}{\sqrt{4x+1}}$.

2. Find the average rate of change of the function $f(x) = -x^2 + 2x - 3$ from $x = -2$ to $x = 5$.

3. Let $f(x) = 2 - \frac{3x}{3x-1}$. Find $f^{-1}(x)$, where f^{-1} is the inverse function of f.

4. Let $f(x) = \frac{1-x}{1+x}$ and $g(x) = 1 - \frac{2x}{x+1}$. Find the function $(f \circ g)(x)$ and **simplify** your answer.

5. Graph the polynomial $p(x) = -x^2(x+2)^3$, finding and labeling all intercepts.

6. Graph the function $f(x) = \begin{cases} (x+1)^2, & x < 0 \\ 2 - x, & 0 \le x \le 2 \\ 5, & x > 2 \end{cases}$

7. Find all the zeros of the polynomial $p(x) = 2x^3 + 10x^2 + 9x + 4$. Please express any non-real zeros in the form $a + bi$.

8. Graph the function $g(x) = \frac{x^2+x-6}{x^2-25}$, labelling all intercepts and asymptotes.

9. Solve the equation $2e^x - 1 = 3e^{-x}$.

10. The general function $n(t) = n_0 e^{kt}$ is used to model the number of bacteria in a certain culture, where the time t is measured in hours. Suppose the culture initially contains 630 bacteria. After 3 hours the bacteria count is 1050. find the exact time it took for the bacteria count to grow to four times its initial size. **Simplify** your answer.

11. Find all the values of x, if any, such that $\log_4(x+6) - 2 = \log_4(5-x)$.

12. Simplify the expression $\ln(e^{3\ln(e^2)}) + 4^{-2\log_4(1/2)}$.

13. Find the exact value of each trigonometric function at the given real number, if it exists.

    a. $\csc(-5\pi/3)$ b. $\cos(15\pi/4)$.

14. Find and state the amplitude and period length of the function $f(x) = 4\sin(\frac{1}{3}x - \frac{\pi}{12}) - 1$ and graph one complete period. Be sure to label the highest and lowest points on the graph.

15. Given that $-\pi/2 \le \theta \le \pi/2$ and $\csc\theta = -5/2$, find the value of $\cos(\pi/6 - \theta)$.

16. Find the exact value of each expression, if it exists.

    a. $\sin(\sin^{-1}(\pi - 3))$ b. $\cos^{-1}(\cos(11\pi/6))$.

17. Find the exact value of $\cot(\sin^{-1}(12/17))$.

18. Verify that the trigonometric equation is an identity.

$$\frac{\cot x - \tan x}{\sin x \cos x} = \csc^2 x - \sec^2 x$$

19. Find all primary solutions (i.e. $0^\circ \leq x < 360^\circ$) to the equation $\frac{1}{2}(\sin 2x)(\cos 2x) = \sin x \cos x$.

20. Two strings stretch from a balloon to two posts on the ground which are 100 feet apart. String 1 is attached to a post that lies to the left of the balloon and string 2 is attached to a post that lies to the right of the balloon. String 1 forms a 57° angle with the ground and string 2 forms a 79° angle with the ground. How high up is the balloon? **Simplify** your answer.

9 Appendix C

QMaSC Tutor Interview – Sample guided questions and role play situations

Open Ended Questions:

1. What do you feel your purpose is as a tutor?
2. How confident are you in your ability as a tutor?
3. What do you do when you are presented with a problem that you cannot solve/explain?
4. How does tutoring affect your current situation? Classes? Time?
5. Are there any issues that you are concerned with as a student/tutor?
6. Overall, how would you rate yourself as a tutor and why?
7. Do you have any recommendations for the center?

Role Play Situations:

1. Evaluator pretends to be an irate student that yells and complains about his/her math course in hopes of observing the tutor's honest reaction in dealing with the attitudes and behaviors of a student frustrated in mathematics.
2. Evaluator pretends to be a thoroughly confused student on a specific mathematics problem in hopes of observing the tutor's ability to dissect the situation and offer alternative examples and explanations of the material.

Reporting Assessment Findings to External Audiences

© Rachel M. Dunwell, Grace Coulombe, Michael E. Schuckers,
Rhodes College, Bates College, St. Lawrence University

1 Introduction

Appropriate assessment of the performance of a QMaSC is an important task. The presentation of assessment results to external audiences is one job that QMaSC directors may be required to perform. Here we will refer to any group outside the QMaSC staff as an external audience. The key to good communication of these results is to understand the purpose for which the audience will use an assessment report. One of the most demanding audiences for an assessment of a QMaSC is an accreditation body. Because an evaluation might be likely to occur during the parent institution's accreditation review (AR), this chapter will address the expectations of a typical accreditation body. While our emphasis in this chapter is on the rigorous endeavor of reporting for accreditation bodies, the general ideas and principles will apply to other external audiences.

Not every review of a QMaSC will be as extensive as that done by an accreditation body. However, the various functions of the QMaSC should be assessed regularly. Included in such a review should be assessments of the QMaSC's performance in terms of the quantity and quality of interactions; a catalog of center accomplishments and additional services such as workshops or special projects that arise between reviews; and appropriate summaries regarding the performance of the staff as a whole as well as individual staff members. It is typical for a QMaSC to create a

Suggested Citation: Rachel M. Dunwell, Grace Coulombe, Michael E. Schuckers (2016), "Reporting Assessment Findings to External Audiences", http://dx.doi.org/10.5038/9780977674435.ch17 in G. Coulombe, M. O'Neill, M. Schuckers (Eds.) *A Handbook for Directors of Quantitative and Mathematical Support Centers*, Neck Quill Press, http://scholarcommons.usf.edu/qmasc_handbook.

This material is based upon work supported, in part, by the National Science Foundation under Grant DUE-1255945. Any opinions, findings, and conclusions or recommendations expressed in this material are those of the author(s) and do not necessarily reflect the views of the National Science Foundation

set of yearly goals and objectives and then assess which of these are met during the annual review process. Depending upon the reporting structure at a given institution, assessment results may be presented to an advisory board or academic dean in brief at the end of each semester, followed by a more extensive annual review at the close of each academic year.

Accreditation for an institution of higher education is conducted by one of six regional bodies (see [1] for a more complete description of accreditation in the U.S.). Each body has its own set of standards. While some variation exists in the way that these bodies view academic support centers, such differences exist at the micro level. Appendix A includes links to the web pages for these bodies. The requirements of all bodies are satisfied by a set of documents that includes the following:

- a mission statement for the center and a description of how this mission supports that of the parent institution,
- a breakdown of the QMaSC mission into a set of specific objectives,
- a detailed description of how each objective will be or has been assessed,
- a description of how assessment results are or will be converted into action (often called "closing the loop"),
- a schedule for the assessment of each objective,
- a set of reports for every assessment exercise that has been conducted (these are often called tracking reports),
- a full review of the QMaSC after every planned assessment exercise has been conducted,
- a set of annual reports showing how the assessment is integrated into the decision-making process of the QMaSC, and
- a narrative explaining how the results of the QMaSC's assessment have informed wider decisions made by the parent institution.

Such a set of documents would, for example, satisfy the Southern Association of Colleges and Schools Comprehensive Standard for Academic and Student Support Services (see page 51 of [2]), it also matches the "Developed" or "Highly Developed" descriptions in the relevant sections of rubrics used by visiting teams from the Western Association of Schools and College ([3]).

The standards of the accrediting bodies are written with the intention that they can be satisfied only if best assessment practices have been followed consistently over an extended period of time. There are many resources to assist someone who wants to learn about these practices, two of which include: *Assessment Clear and Simple: A Practical Guide for Institutions, Departments, and General Education* [4] and *Assessing Student Learning: a common sense guide* [5]. Workshops suitable for directors of QMaSCs are held regularly by the accrediting bodies and are advertised on their websites. The people working at the parent institution who are tasked with coordinating the institution's accreditation process will be the most valuable source of help; in fact, it is essential for the director to work with this group. This group may be identified as the Accreditation Liaison, Institutional Effectiveness, Institutional Research, or by a similar name. In this document, such a group will be identified as the Accreditation Liaison.

The rest of this chapter gives more detailed advice on each of the parts of the accreditation documentation described above. While not every QMaSC will undergo a formal review as part of an accreditation process, the information that follows is still wholly relevant to performing any external review process as well as the production of the subsequent report.

2 Mission Statement

An important part of documentation for the accreditation body is the mission statement of the center. These statements help guide the leadership of the center in steering long-term decision making. Typically an institution will have a hierarchical structure of missions and strategies for attaining the mission. The accrediting body will check every level of this hierarchy for compatibility to determine if all constituent parts of the institution work to support this mission. Crafting a mission statement for a QMaSC is complicated by the fact that the center services are a vital component of more than one of the institution's strategies. In particular, it is common for QMaSC services to support:

- the institution's general education curriculum,
- the learning outcomes for the academic program of one or more departments,
- the institution's student support services,
- the institution's experiential learning outcomes (the work of QMaSC student staff may serve as the basis of experiential learning), and
- the pedagogical initiatives of individual faculty and/or of academic departments (e.g., by assigning tutors to work with groups of students, a faculty member may be able to assign more challenging projects).

Therefore, the mission statement for the QMaSC should reflect the full range of its activities. By working with other members of the institutional administration, directors can develop a mission statement appropriate for their institutional culture. In the case of an AR, the QMaSC director should work with the institution's accreditation committee to discuss exactly how the center supports the mission of the parent institution and determine the most appropriate way of documenting this for the relevant accrediting body. One way of recording a mission statement and linking this to a wider mission is to write a document called an assessment plan. An adaptation of the template used at Rhodes College for its assessment plans is provided in Appendix B. Chapter 2 of [4] also details how assessment works at the institutional level and provides QMaSC directors with insight into the world of the Accreditation Liaison.

3 Setting Goals for Fulfilling a Mission

One method for meeting the QMaSC mission is to establish a regular plan, which is most often accomplished by developing goals and objectives alignment with each element of the mission. Each goal must be specific enough to allow for the identification and implementation of the means of measuring its attainment. Appendix B includes a sample assessment plan template, designed to document the statement and measurement of achievement for each goal as stated in the QMaSC mission. Thus, if every goal is attained, the QMaSC has fulfilled its mission. Chapters 1 of [4] and 8 of [5] offer advice on developing goals. Appendix A of [4] includes a template for a "curriculum map" which may be used to verify that the goals relate directly to the mission and are designed to allow for a measure for attainment.

One difficulty that is particularly acute for QMaSCs is the temptation to phrase every goal in terms of improvement. Certain benchmarks should measure improvement, but it is not necessary for them all to specify this criterion. It is natural for an academic support service to have goals that start "Students attending the support sessions regularly will improve their ability to ...." What makes such goals problematic is the need for a baseline from which to measure the improvement. Without such a baseline, the goal cannot be assessed. Every goal that concerns improvement should be weighed against a goal that specifies a standard for students who interact with the QMaSC.

4 Assessment

For each objective itemized by the QMaSC leadership, a method must be developed for determining whether or not the center has achieved the goal. The accrediting body requires detailed descriptions of these methods and an explanation of how each assessment tool was used. Assessment is a process through which missions are fulfilled, so accrediting bodies take a dim view of data being collected and analyzed but not converted into action. They seek evidence to verify "how the loop is closed" for each individual assessment exercise. Each assessment exercise should gather a sufficient amount of data for conclusions to be drawn while allowing for the possibility of translating these conclusions into action items. Appendix B provides a sample assessment plan template for documenting such methods and anticipated actions.

Gathering such data is no simple task. Doing so in a way that facilitates converting conclusions into action is even more challenging. This process is described in depth in chapters 9 and 10 of [5] and is the subject of sessions at most assessment workshops. Forethought and planning are required for any sort of assessment. For example, suppose a QMaSC identifies as one of its goals, "Students who regularly attend sessions at the center will be able to undertake successfully the basic calculations of their courses." Possible measures might include the following:

- asking faculty members to assess their students' competency in basic calculations by examining a selected question from one of their exams and providing student scores on that question to the QMaSC director,
- surveying students by asking them to rate their competency on a specific question or set of questions,
- asking faculty members to assign an extra credit review homework containing questions on basic calculation (it may be necessary for QMaSC staff to grade this assignment), and
- asking QMaSC staff to rate the competency of student visitors in answering questions that require basic calculations.

It is hard to conclude reliably from data derived from any one of the above metrics either that "students can successfully undertake basic calculations" or that "students cannot successfully undertake basic calculations." This is because there are multiple circumstances in which students could score high on a single measure (only one of which is that they have basic calculation skills) and many more circumstances in which they would gain a low score (only one of which is that they do not possess basic calculation skills). Suppose, for example, all the students who attended multiple sessions in the QMaSC had good scores on the exam question selected and assessed by their teachers. The circumstances that would give rise to this data might include the following:

- the question was one that could be answered using skills other than calculation,
- the question was inadvertently answered in a review session before the exam, and
- the question could be answered using only one calculation skill that is more elementary than most other basic calculations in the course.

The circumstances in which students score high on the specified exam question, rate their own abilities highly, and in which those working in the center rate the students' abilities highly, are limited, however, to students' having the ability to perform basic calculations. This suggests that perhaps more than one measure might best be applied to assess the attainment of certain goals.

A set of measures that allows a reliable conclusion to be drawn is not necessarily an adequate basis on which to act. In the case of the objective used in the example above, it would be unwise to act in any way without establishing the extent to which students possessed the relevant calculation skills prior to the start of the course, possibly using a pre-test. If it is found that students possess the desired skills prior to the start of the course, then the action must be to change the goal. If, in anticipation of this, the plan was to raise the minimum standard in the goal so that it becomes "Students who regularly attend sessions in the center will be able to successfully undertake all the basic calculations for their course and several of the more advanced calculations," then it would be sensible to collect data on students' ability to perform calculations that faculty judge to be "more advanced" (this would include collecting relevant pre-test data). If, however, the intended action is to change the instructions for QMaSC workers to focus on the concepts involved in calculations rather than the mechanisms (or vice versa), then additional data is needed regarding the areas on which tutors focus when working with students calculations. In short, the loop should not only be closed after the event, but it should be designed from the outset to close.

The question of whether students improve as a result of visiting the center is fraught with challenges. Unfortunately, it is not enough simply to compare the performance of QMaSC visitors with those enrolled in the same courses who do not visit the center. The issue of center visitation may not be the sole contributing factor in determining the difference in performance for these groups. For example, students who visit the QMaSC may be more motivated to do well in their courses, whereas some students might perform well in the course regardless of whether or not they use center resources. Though difficult to implement, the approach known as student matching may be used to assess the effectiveness of student visitation to the center on course performance. In particular, one can compare every student who visits the center with one or more students in the same course who have similar characteristics (SAT scores, high school GPA, class year, etc) but who do not choose to do so. Alternatively, a more complicated methodology, requiring the consultation of a statistician, involves using data from every student in a particular course to build a statistical regression model, which accounts for the possible confounding factors with terms in the regression model. Such a model would include an indicator variable to address whether or not the student used QMaSC services.

Lastly, the assessment of learning outcomes may not be appropriate, especially in the case of those students who use services only once or twice. Consequently, a director might choose to assess the effectiveness of certain services by considering qualitative reports by students regarding their experiences, counting the proportion of repeat visitors, or other approaches that use feedback and data that can be easily acquired.

Documenting Assessment

Whenever a goal is assessed, the process should be formally documented. There are several templates publicly available for creating this kind of documentation, also known as a tracking report. Some institutions have a template specific to that institution, if so, the Accreditation Liaison will have a copy on file. Appendix C contains a tracking report template adapted from the template found on page 32 of [6]. In addition to assessing the degree to which the QMaSC attains its goals, it is also important to assess the suitability of facilities (including physical space) and resources used by the center. These assessment exercises should also be documented with a tracking report.

Scheduling Assessment

For assessment to be effective — i.e., resulting in positive change — it needs to be conducted with care and sufficient time for thorough reflection. A well-done assessment is a rigorous and time-consuming process, especially in the case of an accreditation. As such, it is not unreasonable to complete only one assessment exercise per semester or per year. Accrediting bodies will expect to see a detailed schedule of assessment that gives a timeline for the assessment of every goal over a multi-year span. Many publicly available templates exist for such schedules. The assessment of facilities and resources should also be included in this schedule. A sample of how the timeline might be incorporated into the assessment plan is provided in the template found in Appendix B. As with tracking reports, some institutions have their own assessment schedule template. If so, the Accreditation Liaison will have a copy on file.

5 Review of the QMASC

After a full cycle of assessment exercises has been completed, i.e., when every goal has been measured once, the QMaSC will possess a rich set of data upon which to draw. These data can be useful for and can provide a great deal of insight into the operation of the QMaSC. As with any data analysis, it is important to recognize that any conclusions that are drawn by center leadership as well as individuals outside of the center are only as good as the data collection and analysis methodology. Consequently, it is worth reiterating that a good deal of time should be spent at the beginning of the assessment process thinking about the type of data that should be collected.

Accrediting bodies will look to see that the assessment data is fully utilized. Only after a full assessment cycle is it possible to determine whether or not the center is productive, fulfilling its mission and supporting the mission of the parent institution. Appropriate actions should be devised based upon the external review, and the associated results should be communicated to interested parties. Tracking reports are a suitable way to document these reviews. Upon completion of an assessment cycle, the mission statement should be revisited and, if necessary, revised when previous stated objectives are deemed no longer suitable. When such changes are made, the feasibility of the assessment schedule should also be examined and appropriately altered to reflect any changes in the mission. This review and the implementation of any changes should also be documented.

The ways in which the center functions and the facilities it uses are also subject to review. A review of center operations should involve QMaSC staff and the faculty members whose students use the center. Making changes to QMaSC operations requires cooperation and tact. A review of the facilities should include not only the space it occupies but also the resources (including financial) that are necessary for the center to function successfully. Changes, whether in the way the QMaSC operates or improvements in facilities, are often made at the end of an assessment cycle when the assessment report and mission review provide the director with documented rationale for prospective change, the addition of resources, or the improvement in facilities.

6 Annual Report

An annual report is an ideal document to bring together all of the center's activities. A sequence of annual reports demonstrates to external reviewers evidence of the longitudinal accomplishments of the QMaSC. In addition, writing a summative report every year provides administrators and other interested parties with accurate information regarding QMaSC activities and services, and at the same time it helps the director to stay focused on fulfilling the QMaSC mission. Moreover, including the QMaSC staff in the preparation of the report or providing to them a version of the report educates them on the QMaSC's mission and how their work contributes to a wider set of goals. It is common for institutions to require that annual reports be written in a specific format, and, if so, the person to whom the QMaSC director reports can provide information regarding such a requirement. For those whose institutions have no required format, a suggested template is provided in Appendix D.

Directors who hesitate to perform the ambitious sort of assessment required for accreditation but still desire to craft an annual report may want to include some of the following information, when appropriate: the numbers of drop-in visitors and individual appointments with tutors; a list of center accomplishments and additional services such as workshops or special activities; populations served; and appropriate summaries regarding the learner satisfaction and the performance of the staff. When preparing an annual report, directors might want to refer to the associated chapters in this handbook on internal evaluation, reporting, and staff feedback.

7 External Review Panels

As mentioned in the introduction, not every review of a QMaSC is as extensive as that done by an accreditation body. Some institutions and QMaSC directors opt to have an external review of the QMaSC and its services conducted by an external team of experts. Often, these experts are directors or administrators of academic support centers from peer institutions. The QMaSC director may work with the institutional administration in identifying a list of potential candidates to serve on an external review panel. The director, or perhaps an academic dean or provost, will invite at least two, ideally three, candidates to serve on the review panel and establish a schedule of relevant meetings for the panel's campus visit. Once the panel is finalized, arrangements must be made for the transportation, lodging, and suitable compensation for the external reviewers.

The preparation of a self-study is a crucial element for such an external review. The self-study, assembled by the director, should contain materials similar in nature to the assessment documents described in detail for the accreditation body. In particular, the study should introduce the external review panel to the college and its mission; the QMaSC mission, objectives , services, and workshops; the most recent assessment of QMaSC services performed by the director; a frank appraisal of features and services when objectives are not met; a list of potential changes to meet those objectives; and ideas for expanding and enhancing services.

The self-study must be provided to the review panel prior to its arrival on campus. Once on campus, the review panel should meet with relevant faculty, QMaSC staff, students, and any other constituents who use QMaSC services. It is important that the panel's visit include considerable unscheduled time to allow for reflection. The panel needs time prior to meetings with relevant parties to discuss the study and to generate questions to ask of each constituency. Likewise, the review team needs a block of time after scheduled meetings to discuss their findings and to compose a report. In some cases, the report may be completed before the team members return to their home institutions. In other cases, the team may need time to devise a plan for compiling the report.

8 Conclusion

As stated at the outset, this chapter focuses primarily on compiling a full set of documents that may be required by an accreditation body. Most other audiences will not need to see such an extensive set of documentation, but it should be possible to construct a useful report to any external audience from this body of documents regardless of whether that external audience consists of institutional administrators or an outside review panel. Directors seeking additional resources regarding the main principles of effective communication about the services of their center, especially in the context of assessment results, may choose to refer to chapter 17 of [5].

9 Bibliography

[1] J. S. Eaton, "An overview of u. s. accreditation." `http://www.chea.org/Research/index.asp#overview`, 2012. Internal document for Council of Higher Education.

[2] Southern Association of Colleges and Schools - Commission on Colleges, "Resource manual for the principles of accreditation: Foundations for quality enhancement." `http://www.sacscoc.org/handbooks.asp`, 2012.

[3] Western Association of Schools and Colleges, Accrediting Commission for Senior Colleges and Universities, "Assessment rubrics." `http://www.wascsenior.org/document-list`.

[4] B. E. Walvoord, *Assessment Clear and Simple: A Practical Guide for Institutions, Departments, and General Education.* Jossey-Bass, 2nd edition ed., 2010.

[5] L. Suskie, *Assessing Student Learning: a common sense guide.* Jossey-Bass, 2nd edition ed., 2009.

[6] "Academic program assessment workshop." Internal document of Southern Association of Colleges and Schools, 2012.

10 Acknowledgements

Rachel Dunwell is very grateful for considerable assistance she received from Associate Dean Anita Davis, the Accreditation Liaison for Rhodes College, in the preparation of this chapter.

11 Appendix A

Table 1: List of Accreditation Bodies and Associations

Name	Full Title	Home Address	States
MSA	Middle States Association of Colleges and Schools	`http://www.msche.org/`	New York, New Jersey, Pennsylvania, Delaware, Maryland, the District of Columbia, Puerto Rico, and the US Virgin Islands
NEASC	New England Association of Schools and Colleges	`http://cihe.neasc.org/`	Connecticut, Maine, Massachusetts, New Hampshire, Rhode Island, and Vermont
NCA	North Central Association of Colleges and Schools	`http://www.ncahlc.org/`	Arkansas, Arizona, Colorado, Iowa, Illinois, Indiana, Kansas, Michigan, Minnesota, Missouri, North Dakota, Nebraska, New Mexico,Ohio, Oklahoma, South Dakota, Wisconsin, West Virginia, and Wyoming
NWCCU	Northwest Commission on Colleges and Universities	`http://www.nwccu.org/`	Alaska, Idaho, Montana, Nevada, Oregon, Utah, and Washington
SACS	Southern Association of Colleges and Schools	`http://www.sacscoc.org/`	Virginia, Florida, Georgia, Kentucky, Louisiana, Mississippi, North Carolina, South Carolina, Alabama, Tennessee and Texas
WASC	Western Association of Schools and Colleges	`http://www.wascsenior.org/`	California, Hawaii, Guam, American Samoa, Micronesia, Palau, and Northern Marianas Islands
CHEA	Council for Higher Education Accreditation	`http://www.chea.org/`	NA

12 Appendix B

Assessment Plan

Last Updated:

Organization: (name)

Mission Statement:
(description)

Table 2: Example Template for Assessment Plan

Mission	Link to Institution's Mission	Outcome	Measurements and Actions	Responsible Party	Date/Frequency of analysis
(portion of own Mission supported)	(Portion of institution's mission that is supported)	(goal)	1. (means of measuring) (area of intended action) 2. (means of measuring)(area of intended action)	(who will collect/analyze measurements)	(date last measured/ analyzed) (next date to measure/analyze)

(Full descriptions of the measurement used should be given after the table, and cross-referenced to the appropriate row of the table.)

13 Appendix C: Example Template for Tracking Report

An adaptation of the template found on page 32 of [6].

Goal. What goal did you assess? How does this goal support the mission?

Evidence. What assessment evidence did you collect? Where and how did you obtain the evidence? How many artifacts were collected and assessed?

Assessment Process. How did you assess the evidence? Include a copy of the rubric, if one is used. Who did the assessment? Describe any statistical analyses that you preformed. Describe how assessment findings and analysis were shared with interested parties.

Conclusions. What did you learn about the goal?

Closing the Loop. What actions did you take? Discuss consequences (including financial implications) of these changes.

Assess the Process. Did the data collection, analysis, and implementation of changes go well? What suggestions do you have for next time this goal is assessed?

14 Appendix D: Example template for Annual report

Annual Report: Suggested Template for QMaSCs. (Adaptation of Template used by Rhodes College for Academic Departments)

This template organizes the report into three parts: an executive summary, a narrative section and four appendices.

Executive Summary: This should be a brief, one or two paragraph, summary of the document that follows.

Narrative :

- Services. Begin your narrative with a description of all the services that your Center provided for students using the Center, and an account of how widely the Center's services were used. Link these to the institution's curriculum, and other relevant areas of the institution's mission. Include descriptions of area of strength and weakness, with reference to supporting data, describing how the strengths can be maintained and weaknesses addressed. In the narrative consider the following questions:
  - How many students used the various services that you offered? What proportion of students used the services? What proportion of the students made repeated use of services?
  - How frequently all people working for the Center met and discuss the services provided? What was the focus of the discussion? Where any follow up actions taken?
  - How frequently have you met with faculty to discuss how the Center's services support their teaching, and the learning outcomes of their students? What follow up actions were taken?
  - Where any new services offered this year? Did you change any of your procedures? Did these changes result in positive changes?
  - What are the most significant challenges that the Center faces? How do you plan to address these challenges in the coming year?
  - Are then any new services, or changes of procedure that you plan to introduce in the coming year?
- Staff. Begin this narrative with a consideration of the people who work in the Center and their various roles. Follow this with a consideration of the following questions:
  - How effectively have you been able to staff the services? Is your current faculty staffing level sufficient to accommodate next year's projected needs?
  - Is the work load spread evening across the staff? Should differences that exist be addressed?
  - How does working in the Center help students attain learning outcomes? How do these outcomes link with and support the institution's experiential learning curriculum?
  - What do you consider the most significant challenges in regard to your Center's staff? How might these be addressed?
- Facilities (including space). Begin this section with a consideration of your current facilities and an account of how well those are (or are not) meeting your needs. What are the issues

related to facilities that we should try to address as soon as possible, and what should we plan for in the near future (say five years)?

- Departmental Activities. Describe any activities that your department carried out during the year, including participation in workshops, collaborating with other learning support centers, holding seminars, etc.
- Additional Information. Provide any other information about your Center that you would like to bring to our attention.

<u>Appendices: Organize your supporting documents into appendices as follows:</u>

I. Appendix 1. Attach the following assessment documents as Appendix 1, with sub-sections 1A- 1D:

   - Appendix 1A: Your department's current assessment plan, including your mission statement and goals.
   - Appendix 1B: An account of how attaining all the goals will allow the Center to fulfill its mission.
   - Appendix 1C: A schedule of services offered by the Center. This should indicate the maximum number of students who could utilize the service, and the number or proportion of students who did so.
   - Appendix 1D: Activity Tracking forms for each goal measured during the past academic year.

II. Appendix 2. List all noteworthy faculty and staff accomplishments for the year. This should include any academic credit that students earned through their work for the Center, any successful application for funds that support the work of the Center, and any publications relating to activities of the Center.

III. Appendix 3. Attach a list of services provided by each staff member in the Center during the year.

Center Staff Feedback

© Dennis Lewandowski,
Michigan Technological University

1 Communication Between Director and Staff

Staff to Director Feedback

The tutors of the QMaSC have the most direct contact between the student-clients and the learning center organization. Thus they are in the position to understand what is working and what is not. It is important for the director of the center to have a system in place that can glean this information from the staff of the center. Three systems are commonly in use: regular written evaluations of the center's operations, open staff meetings where all aspects of the center are fair game for discussion, and a combination of the two.

With regular written evaluations, simple one-page, four-question forms of this type should be sufficient to encourage reflection upon the operations of the center:

a What do you like best about the center?

b What things about the center would you like to change?

c What are the center's strengths?

d What are the center's weaknesses?

The evaluations may be printed on paper or available online (e. g. as a Google form), and the evaluations may be collected on a regular basis (weekly or monthly) or once each semester (mid-semester assessment or end of semester). Since the QMaSC staff is commonly composed of

Suggested Citation: Dennis Lewandowski (2016), "Center Staff Feedback", `http://dx.doi.org/10.5038/9780977674435.ch18` in G. Coulombe, M. O'Neill, M. Schuckers (Eds.) *A Handbook for Directors of Quantitative and Mathematical Support Centers*, Neck Quill Press, `http://scholarcommons.usf.edu/qmasc_handbook`.

This material is based upon work supported, in part, by the National Science Foundation under Grant DUE-1255945. Any opinions, findings, and conclusions or recommendations expressed in this material are those of the author(s) and do not necessarily reflect the views of the National Science Foundation

students who have recently taken the course(s) being tutored, the composition of the staff changes quite often. It ordinarily takes eight weeks or so for new tutors to start feeling comfortable enough with processes to be able to offer feedback.

Open staff meetings where the entire staff of the center is present and can discuss everything openly can be useful both for identifying strengths and weaknesses of the center and for brainstorming solutions to issues. The director of the center must be able to accept all input and be open to discussion, and the director should be prepared to ask open questions that can lead to successful group problem solving. Obtaining staff input about solutions also helps increase the likelihood that the staff will take ownership of new procedures and policies.

If both written evaluations and open staff meeting are used as feedback mechanisms, the results of regular written evaluations can be used as the starting point of the staff meeting discussions, possibly improving the efficiency of the staff meeting. Additionally, staff members should be encouraged to speak privately to the director about any matter as it arises. These individual conversations can help identify topics to discuss at staff meetings, too.

Director to Staff Feedback

In an ideal world, the feedback from the center director to the staff would consist of an individual evaluation of each tutor's job performance. This is sometimes difficult because of several factors:

- The high turnover rate of tutors.
- The high turnover rate of student-clients.
- The lack of a consistent tutor-to-student pairing for more than one academic term
- The constraints on the learning center director's time that may make individual evaluation of each tutor's performance problematic at best.

However, at the very least, a director should be alert for indications of the quality of the tutors' work, such as comments, compliments, and complaints from student clients.

One feedback mechanism is the student-client evaluation of each individual tutor. This is a standard evaluation given near the end of the academic term that attempts to gauge the tutor's job performance. This type of evaluation is most helpful when the client has had a standing appointment with the same tutor throughout the academic term. It is less useful, but still appropriate, for a drop-in tutoring center. The director can use the results of these evaluations when judging the effectiveness of the tutors.

Ultimately, the QMaSC director must use the overall operation of the center as the best feedback tool to and from the staff. The number of visits to the center is the best gauge of center operations. If the number of visits rises from one term to the next, it is a strong indication that students are satisfied with the services they have received and that the center is a welcoming place.

2 Incorporating staff input into management decisions

The use of staff input in management decisions works best if the following principles are observed:

1. Hire very good staff. The hiring process must be rigorous and thorough.
2. Clearly communicate operational rules to the tutors.
3. Give the staff the learning tools they need.
4. Treat the staff as colleagues more than employees.
5. Let them do their jobs.

3 Conclusion

All of the feedback between the learning center director and the staff is most meaningful if it results in some visible change to the center's operation. The change must be visible to the tutors, the student-clients, the director of the center, and the academic administration. By spelling out the steps tutors can take in order to communicate with each other, and with the director, open and honest dialogue can take place, thus enhancing the ability for all constituents to focus on a positive tutoring process and on preparation of the tutors.

Part VI

Starting a Center

Defining Staff Roles

© Darry Andrews, Carolyn Johns,
The Ohio State University

1 Introduction

Staffing in a QMaSC assumes many structures depending on the size of the institution, budget constraints, number of students served, and the availability of qualified tutors. In the case of smaller institutions, the QMaSC may be part of a more encompassing academic learning center. Typically, at a minimum, the center or lab is staffed with a director or lab coordinator, and undergraduate or graduate student tutors. Tutors are also referred to as coaches, learning assistants, quantitative tutors, and more. The center may also employ a dministrative personnel or a pedagogy assistant. No matter what configuration of staff roles is determined to fit the needs of the institution best, a set of essential duties must be performed by the QMaSC staff.

Philosophy of Center

One of the first responsibilities of the staff is to determine the philosophy or mission statement of the QMaSC. The center philosophy should complement the institution's philosophy. In addition, the philosophy must be student-centered. Although some centers focus only on helping students answer homework questions, a better mission is to help students become self-motivated, independent learners, who eventually no longer need tutoring services. Tutoring sessions should be focused on the lasting effects of the session for the students.

While the staff may wish to create the center's philosophy, the tutors themselves, once hired, can be responsible for creating the philosophy through a training activity. This approach not only

Suggested Citation: Darry Andrews, Carolyn Johns (2016), "Defining Staff Roles", `http://dx.doi.org/10.5038/9780977674435.ch19` in G. Coulombe, M. O'Neill, M. Schuckers (Eds.) *A Handbook for Directors of Quantitative and Mathematical Support Centers*, Neck Quill Press, `http://scholarcommons.usf.edu/qmasc_handbook`.

This material is based upon work supported, in part, by the National Science Foundation under Grant DUE-1255945. Any opinions, findings, and conclusions or recommendations expressed in this material are those of the author(s) and do not necessarily reflect the views of the National Science Foundation

allows tutors to reflect on their tutoring role but also to create ownership. [1]

2 Tutoring Models

The staff must also decide how to organize the center. What tutoring model will the center adopt? There are many options and combinations available. There are drop-in or appointment-based models. There is group tutoring vs. one-on-one tutoring. Tutoring may occur with all subjects being tutored in one common room, or tutoring may occur with separate rooms or spaces for each course. Each combination has its advantages and disadvantages. The needs of the students, budget, and space available will determine the best model.

One-on-One Appointment-Based Model:

When to use: This model is best for students who need significant help in a course due to gaps in their knowledge, poor study skills, or low self-confidence. Example of Model: Students contact the center to make one-on-one appointments: either for a single appointment or a weekly appointment. Ideally, students can have a weekly appointment to see the same tutor.

Advantages: Students get individual, uninterrupted attention to focus on their needs. Students know they can get help at a specific time and don't feel as though they are wasting time waiting for a tutor. Weekly meetings are easy for students to remember; they are responsible for keeping their appointments; and it is relatively easy to monitor student progress and evaluations. Tutors can work on study skills, assign small tasks, and build rapport.

Disadvantages: A 1:1 ratio is less cost-effective than group tutoring; missed appointments waste tutors' time; it is harder to schedule last-minute appointments; and it makes no provision for students to have short questions answered.

Comments: Given a large budget, this is an ideal model to help students who struggle significantly.

One-on-One Drop-in Model:

When to use: This model is best for students who need significant help with a particular topic and for courses with lengthy homework problems. Example of model: When students arrive at the center, they are placed in a queue to meet with a tutor. As each tutor becomes available, he or she takes the next person in the queue to a private room for tutoring.

Advantages: Students get individual, uninterrupted attention to focus on their needs.

Disadvantages: Students may have to wait; tutors may not be able to meet with drop-ins. This model is less cost-effective than appointment-based one-on-one tutoring. Because students may see a different tutor each session, this model may prevent rapport building or hinder working toward long term goals.

Comments: This model is somewhat impractical from a budget standpoint. Using appointments, either one-on-one or with groups, would be a better option.

Group Appointment-based Model:

When to use: This model is available when students need significant help in a course due to prior knowledge gaps, poor study skills, or low self-esteem, and the center's budget constraints do not allow for one-on-one tutoring. Example of Model: This model is similar to supplemental instruction. Students sign up for a weekly group meeting with a tutor. The tutor works with the group as a whole to handle the most common questions, encourages the group to answer their own questions, and provides help with study skills.

Advantages: This model is cost effective. It imposes no wait time for students; students can learn from each other; and having the same tutor allows rapport to build. Student progress can readily be tracked, and tutors can be evaluated.

Disadvantages: This model is not able to handle quick homework questions; students must have an appointment, and individual attention is limited.

Comments: Given a medium budget, this model may be the most beneficial way to balance both cost-effectiveness and regular appointments with the same tutor.

Group Drop-In Model:

When to use: This model is useful if the tutoring budget is low and when students mainly want a place to study with an expert around to answer quick questions. Example of model: Students work in a common room, tutors circulate and answer questions as needed.

Advantages: Tutors can help many students in a short amount of time, students can work in groups and on other problems while they wait for the tutor.

Disadvantages: Individual attention and student-tutor rapport are limited. It is difficult to work on long-term goals and difficult for students to work on study skills.

Comments: This model may be cost-effective in that tutors may be able to help many students quickly; however, it becomes less cost-effective when fewer students attend tutoring than anticipated in the schedule of tutors. Overall, this model is the least effective in helping students become independent learners. Tutors must be excellently trained and supervised for continuation of trained skills in order to ensure students are not simply getting answers to homework questions. In this model, all courses may be tutored in one common room, with distinct tables for each course, or courses may be grouped and placed into separate rooms.

A combination of these tutoring models may be used to meet the needs of all students and stay within budget. In addition, the center director should check with other campus tutoring services to see where gaps in tutoring need to be filled.

3 Center Roles and Responsibilities

Center Director

In the case of the stand-alone center, the director or coordinator may be a faculty member or non-tenured staff member with a Master's or Ph.D. in a Mathematics/quantitative field. In the case of a larger academic learning center, the director or coordinator may have alternative degrees. In any case, the main duties of the director are to establish and execute the mission of the center, provide leadership, and manage all aspects of the center operations, including budgeting and staffing. Some centers may employ an assistant director to assume a subset of duties that are traditionally performed by the director. This may occur at larger research institutions where the sheer volume of students mandates this arrangement. Additionally, at some institutions, a director may have teaching obligations that take him/her away from the center. At smaller institutions where the tutoring occurs under the umbrella of an academic learning center, the director should have a liaison with departments or an assistant director to ensure the quality of the tutoring.

Center Assistant Directors

Some institutions may choose to hire an assistant director to fulfill some of the duties of the center. The assistant director may take on an administrative role of tracking appointments and tutors, as well as scheduling. The assistant director may be in charge of hiring, training, and evaluating tutors. The assistant director may hold a part-time position in charge of only one of the duties such as marketing, budgeting, record-keeping or processing of tutor timecards. How the director and assistant director divide tasks will depend on the size of the institution and each staff member's strengths and weaknesses. The assistant director should preferably have a master's degree in mathematics or another quantitative discipline, or in math education; however, the role could be filled by an individual with a bachelor's degree or a graduate student. The assistant director reports to the director of the center.

Pedagogical Assistant

The role of the pedagogical assistant may be filled by the director or assistant director or may be a stand-alone position. The pedagogical assistant may offer workshops and feedback to instructors on best math pedagogy practices. The pedagogical assistant may also be in charge of tutor training or creating/overseeing online materials for courses, and ideally should have a master's degree in math, in a quantitative subject, or in math education. The pedagogical assistant reports to the director of the center or to the appropriate administrator, if the position is a stand-alone one.

Center Tutors

Center tutors are responsible for the tutoring of students. Depending on the institution, tutors may be undergraduate or graduate students. In some instances, outside professionals serve as tutors.

Tutors may have some administrative responsibilities within the center. In the case of appointment-based tutoring, they may be responsible for keeping appointment records such as student progress, material covered, completing learning style inventories, and making appointments with their students. Tutors may be involved in marketing the center services to students through word of mouth and classroom visits. Tutors are responsible for reporting hours worked for payroll.

Center tutors may be senior tutors and junior tutors. Senior tutors possess extensive experience, while junior tutors are just beginning their career as a tutor. A third category is master tutor, which ranks above senior tutor and may be added in the exceptional tutor category. Tutors with excellent evaluations may be given higher pay, more hours, or preferred working hours. Master tutors may also assist in training and mentoring of junior tutors. All tutors may report to the director, assistant director, or center assistants.

As mentioned above, tutoring may take on many different forms. Regardless of the model used, center tutors must understand the mathematical content, have pedagogical knowledge of best tutoring practices, possess excellent customer service skills, and demonstrate use of study skills in their own studies. Tutoring should never be a homework service. Tutors must learn to put down their chalk or pencil and allow the students to take control. Tutors must learn to listen to students and use questions to find the misunderstandings as well as lead the students to the correct answer. A student who learns to find the correct answer on his/her own, rather than being given the answer, will learn how to problem-solve and eventually become an independent learner.

Center Assistants

If the math center assistant is a full-time employee, the assistant may perform clerical tasks and ensure correct record-keeping, as well as processing hired tutors' time sheets. However, the assistant's role may also be filled by student workers who perform such tasks as checking students in for appointments, ensuring tutoring rooms have supplies, and printing fliers. The math center assistants may also be responsible for tracking loan materials such as books and calculators if such services are offered by the center. The assistant reports to the director or assistant director. In some cases these assistants may be tutors with additional responsibilities.

4 Other Staff Duties

Budgeting

Budgeting for a QMaSC includes non-personnel costs for the physical tutoring space, cost of furniture, room decorations (such as posters and bulletin boards), office equipment, computers, software, and paper supplies. Personnel costs to be accounted for include the pay of the director, assistant director, tutors, administrative staff, and other personnel. Centers working with limited budgets can hire students with Federal Work Study Financial Aid, employ faculty on a volunteer basis, employ graduate students as part of their graduate student duties, or seek student volunteers. Another option would be to create tutoring as a course component of an education course. However, certain cost-cutting measures may leave a center with fewer high-quality applicants for tutor positions. A center may employ some combination of these types of employees to balance quality tutoring with budget constraints. When budgeting for hourly paid tutors, it is important to factor in time paid for training as well as extra staffing needed during exam periods.

Scheduling

The administrator must determine not only the hours the center will be open, but also how the center should be staffed. Budgeting, as well as the needs of the students, will greatly effect both of these facets of scheduling. Additionally, the model of tutoring being used will effect scheduling.

For appointment-based tutoring, the only constraints on hours of operations are the hours the tutors are available. Additionally, a supervisor should be available to assist during sessions. However, weekend and evening tutoring sessions might take place in a library, allowing a library staff member to serve as a supervisor during those times. TutorTrac software [2] is a great tool for managing appointment-based scheduling.

For drop-in tutoring, the hours will depend on the budget of the center and tutor availability. On a small budget, it is advisable to offer tutoring in the evening when it will be accessible to the majority of students. For residential campuses, evenings are a particularly advantageous time to offer tutoring. For non-residential campuses, evenings may not be advantageous, as many students will have left campus. In addition, residential campuses may offer tutoring on weekends, particularly Sundays, whereas non-residential campuses may not offer weekend tutoring. In any case, a center should make sure the hours meet the needs of the students. On some campuses, it may be feasible to poll students for the best times. If the center can only offer limited hours, the center should refrain from offer tutoring the same hours each night. On non-residential campuses and when budgeting allows, it is beneficial to offer daytime tutoring. This gives students a space to study while waiting for their next course and allows students to attend tutoring right before or after a class. Furthermore, student athletes and those involved in extracurricular activities may find it difficult to attend tutoring during the evening. Daytime tutoring hours greatly help these students.

Scheduling of tutors will vary depending on the tutoring model being used. With appointment-based tutoring, tutors will be assigned to students, and the scheduling will be determined by the tutor and students. For drop-in tutoring, care should be given to staff more tutors during the busiest times of day. Ideally, the center will have a usage tracking system such as TutorTrac. For smaller colleges offering only a few sections of a course, one would expect the tutor room to have very few students while the course is taking place. Similarly, at large universities, many sections may be offered at the same time, rendering those times less popular. Additionally, extra staff will be needed on homework due dates and prior to exams. It is important to coordinate with instructors to determine when due dates and exams are.

Once the numbers of tutors needed during each hour of the day are determined, tutor availability needs to be collected. This can be collected through a survey tool such as Surveymonkey or Doodle. The staff will then place tutors according to center needs and tutor availability. For scheduling, one may choose software such as Microsoft Excel [3] or purchase an online scheduling software such as Shiftboard [4]. The center may find it beneficial to have a tutor schedule which repeats weekly while allowing for additional shifts added during exam weeks. However, for centers with tight budgets, it is more economical to use past semester student usage data, trends in the current semester's usage data, and course schedules to have a schedule which varies weekly. Additionally, for centers working with multiple rooms, it is important for the director to visit the rooms periodically as tutors may be asked to change rooms to help out in a particularly busy room.

Marketing the Center

Marketing, or publicizing the center, is one of the most important duties for a director and for the staff members of a QMaSC.

The necessity of marketing the center cannot be overstated. Often students and instructors are unaware of the existence of a center unless its services are properly advertised. Like any business, a center needs customers in order to prosper.

There are many ways a center can market itself. Starting at the local level, the center should place an abundance of signs around the center's location, instructors' offices, and classrooms announcing the center's presence and hours. Additionally, emails can be sent to students enrolled in supported courses as well as the instructors for those courses. A representative from the center needs to inform the instructors of the services before syllabi are made, and request that center information be printed on the syllabus or posted on the course website. Additionally, center staff or tutors, depending on the number of courses, should visit each course individually to introduce themselves and announce the center's services. They may take along fliers to hand out. Furthermore, the center can create a website and/or Facebook page with information regarding services. Before the start of the autumn term, some centers' staff meet with targeted student support groups, staff (particularly advisors), and faculty to inform them of the various services offered.

Marketing of the center does not stop after the beginning of the term. Emails, announcements, and website/Facebook updates must continue throughout the semester. These outreach efforts should increase student awareness of services offered by the center.

Hiring Tutors

There are many different ways to hire tutors. Depending on the program, the hiring of tutors may be subject to a variety of considerations and constraints. In some cases, hiring may be part of a lecturer's or graduate assistant's duties, so that tutors are not hired directly by the center. In others, budget constraints may limit the center director to hiring Federal Work Study students. Budgets may allow for students to be selected at will. In some programs, tutors may be handpicked after completing selected coursework, or tutors may be selected after they have completed a tutoring course.

The staff must seek to hire exceptional students who have demonstrated excellence in mathematics and other quantitative courses, or students who are referred by their math instructors, and who have interpersonal skills,. Such tutors typically major in mathematics, statistics, actuarial science, and engineering. However, students from all majors should be considered.

The first step in hiring is to have applicants fill out an application that includes past course grades, references, and their responses to questions such as why they wish to become QMaSC tutors. Some centers may have specific criteria for undergraduate tutors such as achieving A's and B's in upper division mathematics courses and a minimum GPA of at least a 3.0. In contrast, other centers prefer students with only A's or B+'s in the course they are tutoring or the next course in the sequence. (See the chapter on "Staffing and Hiring" by Christine Tinsley.)

Once the pool of applicants has been screened by performance in previous course work, the QMaSC staff conducts interviews to determine if undergraduate students possess the interpersonal and communication skills necessary to work with many different types of students. At the time of the interview, a diagnostic quiz may be administered. This quiz consists of problems students typically struggle with and is used to evaluate the applicant's ability to assist other students in these topic areas. At some colleges, administrators use the information from this test to place tutors into categories of expertise (e.g., introductory calculus, advanced statistics, etc.) and tutors work with students only in specific categories.

The final portion of the interview process consists of a tutoring scenario. Prior to the interview, applicants are given a few short readings on tutoring pedagogy to read to help them prepare for the interview. During the interview, the applicant is asked to help a confused student with a homework problem. During this portion of the interview, the staff observes to see how tutors interact with students, their capabilities in explaining concepts, and their pedagogical knowledge of the tutoring process. These scenarios also help the director determine the training needs of the tutors.

In some cases, the staff may choose not to use the application process model for hiring. The director may offer an open tutoring training course for all interested applicants. The applicant goes through several sessions of training and at the end, is asked to perform tutoring scenarios. This model may be preferred, as only applicants who are serious about tutoring will attend the training. In addition, tutors are now trained when they begin, and it allows the staff to see tutors' potential after training, rather than judging tutors prior to being trained.

Training Tutors

Tutor training can take many different forms. Directors may choose to train tutors as part of the hiring process, create a tutor training course, administer initial training, or administer on-going training. In addition, these trainings may take place in-person, online, or consist of readings and discussion posts. In any form, training should be interactive and follow educational best practices. Tutors will need to be trained on the essence of tutoring: how is it different from teaching? In addition, tutors will need to learn center policies and procedures, as well as the philosophy of the center. Tutors learn how to manage sessions and provide excellent customer service. Tutors need training in greeting students, defining goals with students, and closings. Tutors learn how to help students develop problem-solving skills, rather than providing answers directly to students, by asking probing questions and listening to the students as they talk through their problem-solving process. Tutors learn how to accurately identify the part of a concept that confuses students. For returning tutors, training on more advanced topics such as difficult students, learning styles, and additional goal development may be offered. In addition, center directors may choose to have excellent returning tutors mentor new tutors. This can be done through shadowing opportunities.

Evaluating Tutors

Tutors should be evaluated at least once, if not twice, a year. For an evaluation, the staff should sit in on a tutoring session and observe the tutor at work. Other sources, such as student evaluations of their tutoring experiences should also be taken into consideration. Ideally, the evaluation period should occur after a majority of the tutor training has occurred.

There are many reasons for evaluating tutors. Evaluations determine the effectiveness of each tutor. This may determine their future employment or hours at the center. Areas which need addressed by further training for all tutors can also be brought to light during evaluations. Most important, the staff can give tutors feedback on how to improve their sessions and become better tutors. Through these evaluations, excellent tutors may be identified who can serve as mentors for new tutors.

Tracking Tutor Attendance

Daily tutor attendance must be tracked to ensure tutors are present and on time for their shifts. A system of tutor notification and ongoing tracking for missed or late arriving shifts should be in place. In addition, a tardy and absent policy should be in place for tutors. Late or absent tutors undermine the mission of the center.

Tutor Payroll

Staff must also ensure tutor payroll is tracked and human resource needs are met. It is important to track payroll throughout the term to ensure the center stays within its budget. In addition, all proper work documentation and tax forms need to be filed. Usually this paperwork is performed by staff in the institution's human resources department.

5 On-going Needs Assessment

The staff must be aware of the on-going needs of the center. This includes daily monitoring of rooms to ensure the quality of tutoring and adherence to center policies. In addition, the staff needs to be aware of the attendance in the tutoring rooms to determine when additional tutors or fewer tutors are needed. Low attendance may indicate that further marketing is needed to make the students aware of the center or may indicate that the center's hours are inconvenient to students. Software such as TutorTrac is helpful in monitoring center usage. Usage data can also be used to justify the center's existence. In cases where students are receiving appointment-based tutoring, grades may be tracked using GradesFirst [5] software. Evidence of improvement in grades or success in courses can be used to build support for the center.

Staff should also monitor students' appraisal of the center. Dissatisfied students spread the word, and the center will soon lose its clientele . Staff must work to ensure the center has happy customers. Student feedback on center performance should be collected via periodic surveys or by using a suggestion box.

6 Supporting Courses with Other Materials

In addition to providing tutoring services, a QMaSC may choose to support courses with other aids. For example, a center may offer courses in study skills, provide exam review workshops, or provide additional resources for particularly hard topics in courses. A center may hold workshops that provide extra explanations and problems for students to do while an instructor or center staff member is present for questions. A center may furnish a set of lecture videos online for student use. In addition, the center may house a library of textbooks and calculators to be checked out. At some institutions this role may also be filled by the library. Center staff may work with course instructors to ensure best teaching practices are used. Center staff may also assist instructors through grading or printing services.

7 Conclusion

The center staff configuration is up to the discretion of the director and institution. The number and type of staff a center employs depends greatly on the size of institution, available budget, and level of demand for tutoring and other services. In any case, the essential duties of the center must be carried out by center staff. The division of labor will depend on the skill set of each staff member. Once hired, the director may choose to define the scope of the staff roles by defining their duties and may perform a search for additional staff with the duties of each role in mind. On the other hand, a director may be given an existing staff and must assess the duties each staff member and determine their level of competency based on their existing duties. The various configurations for reporting depend on the center's needs.

8 Bibliography

[1] K. S. Agee and R. Hodges, *Handbook for Training Peer Tutors and Mentors.* Cengage Learning, 2012.

[2] Redrock Software Corporation, "Tutortrac." `http://www.tutortrac.com/`, 2013.

[3] Microsoft Corporation, "Microsoft Excel version 2013," 2013.

[4] Shiftboard Inc., "Shiftboard." `http://www.shiftboard.com/`, 2013.

[5] GradesFirst, "Gradesfirst advising management system." `http://www.gradesfirst.com/`, 2013.

Hiring a Director

© George Recck,
Babson College

1 Introduction

Directors of quantitative and math support centers take on many roles, each of which requires specialized skills. A director needs a strong quantitative background, the ability to teach/tutor in various quantitative subject areas, and strong management skills. Because QMaSCs often serve many constituencies, patience is a necessary quality in a successful director. Those served by the QMaSC include students, some with math anxiety; faculty, with unique needs specific to their courses; and administrators. This chapter seeks to arm administrators and search committees with sufficient information about the role of a QMaSC director to enable the identification of the best candidate for the job.

This chapter addresses the concerns of hiring a QMaSC director through discussing aspects of the director's role that inform the hiring process, budget considerations, and search committee composition.

2 Aspects of the Director's Role that Inform the Hiring Process

Position within the Academic Organization

Suggested Citation: George Recck (2016), "Hiring a Director", `http://dx.doi.org/10.5038/9780977674435.ch20` in G. Coulombe, M. O'Neill, M. Schuckers (Eds.) *A Handbook for Directors of Quantitative and Mathematical Support Centers*, Neck Quill Press, `http://scholarcommons.usf.edu/qmasc_handbook`.

This material is based upon work supported, in part, by the National Science Foundation under Grant DUE-1255945. Any opinions, findings, and conclusions or recommendations expressed in this material are those of the author(s) and do not necessarily reflect the views of the National Science Foundation

The director's position within the overall organization varies greatly depending upon the culture within the academic institution. Who is hired is often determined by where the director sits in the organizational structure.

Some schools centralize resources by creating an academic "supercenter" in which all resource centers are housed. In this way, the institution gains some economy of scale with one senior member at the head and lower-level individuals in specialized subject areas. In this case, the supercenter director may hold a position equivalent to that of an associate dean or vice-president. Such a reporting structure is demonstrated in the diagram below.

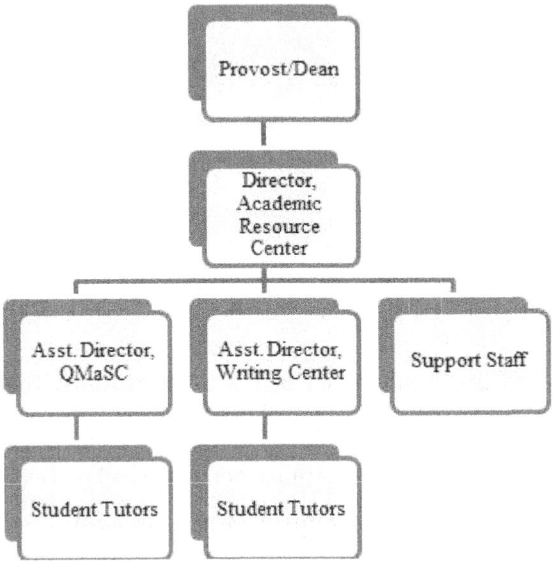

Figure 1: Sample Reporting Structure for a "Supercenter for Academic Resources"

Some schools may look to a stand-alone QMaSC to save money due to mounting fiscal pressures. In this situation, schools may choose to hire someone with fewer academic credentials or less experience. For example, schools may consider candidates with a master's degree even though a previous search required a candidate with a Ph. D. Interested readers may find specific examples of different reporting structures for QMaSC directors as well as their specific role within the institution in the Case Study section at the end of this handbook.

The changing external environment calls for individuals with strong quantitative résumés. The demand in the corporate job market for candidates with strong quantitative skills is rising and it appears that this trend will continue in the foreseeable future. This change in the competitive landscape will likely make it harder for schools to hire well-qualified individuals. Some schools may need to either be creative in terms of organizing their resource centers or lower their expectation level for a director.

Once the specific positional level within the academic organization is established, the next

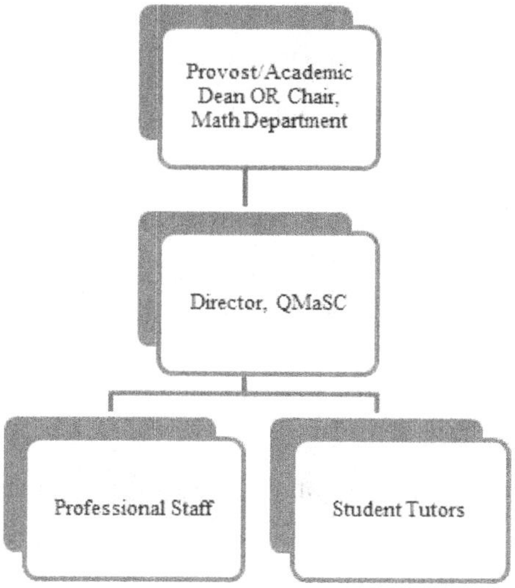

Figure 2: Sample Reporting Structure for Stand-Alone QMaSC

question is whether or not the director will teach classes. Is the director a member of an academic department or only tangentially related to a department?

Consider the three main resource centers at Babson College. In recent years, the four people who have held the leadership positions of these resource centers consisted of a senior tenured faculty member, a full-time lecturer, a part-time lecturer, and a part-time staff member. Each is highly regarded by most faculty members. Recently, the part-time staff member replaced the senior faculty member in the leadership position. While this sequence may not be representative of all institutions, it reveals two key issues. First, many full-time faculty members shy away from leading a QMaSC because it takes them away from their primary responsibilities of teaching and research. Second, the institution may feel that the return on the budget item is better spent on someone who is not a member of the faculty.

Hiring Full- and Part-time Staff Members

Hiring staff, whether full-time support staff or student help, is an important function of every QMaSC director. Larger institutions will likely have a larger academic support staff. In such cases, the larger environment will require a management-oriented director who may rarely interact with QMaSC patrons. Such institutions may choose to expand the responsibilities of the director. Many of the sections below will address some of these opportunities.

At smaller institutions the director plays a much different role, often wearing many different hats. The director may go from teaching a class to attending a curriculum-planning meeting to covering an understaffed shift in the resource center.

As schools strive to have increasingly more diverse student bodies, consideration should be given to hiring staff who can work with a wide variety of cultures and ethnicities. For example, if the QMaSC serves a large Spanish-speaking population, consideration should be given to hiring some individuals who can speak Spanish while assisting such students. Because quantitative material can be viewed as a language unto itself, students may find that having a difficult concept explained in a more familiar language can often lead to a breakthrough in their overall understating of the material. For more information, refer to the chapter of this handbook on Addressing Diversity.

In any case, the director will likely need to establish relationships with relevant departments for regular consultation regarding the recruitment of students with the requisite knowledge to serve as tutors.

For more information refer to Section IV of this Handbook on staffing, hiring, and training QMaSC personnel.

Training

Training is another vital responsibility of the director. Traditionally, the director ensured each staff member was up-to-date in the subject matter supported, prepared to work with students, and, perhaps, knowledgeable in several software packages. As previously noted, QMaSC staff members may also need to be trained to handle a host of different issues. As schools look to diversify their student population, new challenges may emerge that have little to do with being good at teaching math. Directors and tutors might need to learn new technologies in terms of delivery of their services. For example, remote access, on-line support and even things like Skype and social media might be used to deliver support services. For more information on how technology may be incorporated into the management and services of a QMaSC refer to Section II of this Handbook.

As quantitative data analysis permeates the academic institution, QMaSC directors might provide assistance or training to faculty and interested staff members in quantitative techniques as well as the latest versions of available software.

Adaptability to Emerging Trends in Curriculum

Many of the changes in the academic environment are forcing schools to review and update their curriculum, particularly in the quantitative areas. One megatrend is the transition of state-mandated curricula (K-12) to focus more on statistics and data analysis and less on traditional pathways into calculus [1]. Babson, a small business college, introduced a new curriculum in the fall of 2013 to address this trend. The College is prepared to revise the curriculum again in the near future, given the expectation of seeing as many students arriving with AP statistics as currently arrive with AP calculus. As such, a new QMaSC director should be up-to-date with these curricular trends as well as the latest advances in quantitative software and delivery technologies.

Massive On-line Open Courses (MOOCs) are perhaps the biggest megatrend in academics in recent years. Many colleges and universities now offer MOOCs. Additional information regarding a current list of schools that offer MOOC programs may be found by searching online for "Colleges that offer MOOCs."

It is unclear how MOOCs will affect QMaSCs. If the integration of MOOCs into large-scale curricula is successful, it follows that better online support systems will not be far behind. Could these replace QMaSCs and other resource centers altogether? Some might argue that MOOCs will have a bigger impact on larger institutions, whereas smaller, more specialized schools will be less likely to see an impact. In particular, students at smaller institutions are likely to demand the personal attention that can only be provided at the QMaSC. A cursory survey of a few online quantitative courses shows the potential for remedial and lower level teaching and support. However, as the material becomes more complex, the online courses become less effective at the present time. While there is still a lot of uncertainty in this area, the QMaSC director will need to pay close attention to the impact of MOOCs in quantitative learning.

Many publishers bundle textbooks with electronic homework. The software often comes with dynamic parameterized problem sets, which allow students to complete the same kind of problem many times without encountering the same exact example. This prevents students from memorizing answers for the second and third attempts. Faculty at many institutions use these tools, viewing them as an added resource for the students. Students might see those same questions again at the QMaSC, but that possibility results in increased familiarity with the material as opposed to a redundancy. QMaSC directors need to understand how the use of these tools can be integrated into their overall support strategy as well as ensuring that the QMaSC staff is familiar with how the tools work.

Assessment

The goal of assessing the QMaSC should be to ensure that the center reaches its intended "market" of students and appropriately serves those students by providing the necessary assistance for the students to be successful in all aspects of developing their quantitative skills. Several years ago, accreditation bodies began requiring an assessment component to ensure that college and universities "measure" their success in terms of achieving their learning objectives.

Achieving the assessment goal involves determining how to measure and record results of the QMaSC services in a meaningful way. Assessment should not be taken as a measure of failure but as a constructive means to identifying areas in which the QMaSC is not fully achieving its objectives. An assessment may also provide a mechanism for making improvements.

While assessment of QMaSCs is formally addressed in Section V of this Handbook, the following example demonstrates why considering future assessments is relevant to hiring a director. One measure of success is to demonstrate that students who routinely visit the QMaSC, after performing poorly on an exam early in the semester, show improved performance on assignments and exams later in the semester. Such a measure requires the collaboration of faculty, the Registrar, an academic dean, and the QMaSC director, coupled with various electronic tools, to provide a record of the student's performance over time. Consequently, a successful QMaSC director will need to establish strong relationships with relevant faculty and administrators while possessing knowledge and skills for using technology and data analysis in performing assessments.

3 Budget Considerations

The QMaSC budget is another important aspect to consider when hiring a director. Budgets vary widely and are dependent upon the overall organizational structure around the QMaSC. QMaSCs staffed mostly by students often have small budgets. Budgets will be higher for QMaSCs staffed by professionals. In the budget-conscious world of academia, one question of importance is how much control the director will have over the QMaSC budget.

The strategic role of the QMaSC, its place in the institution's organizational structure, and the budget allocated to the QMaSC should all be considered collectively. An underfunded center that plays a key strategic role in the institution is likely doomed to failure. Savvy candidates for the position of director are likely to understand that and will shy away from the position. In such a case, the institution may end up losing the better candidates.

Appropriate compensation for student workers may be another challenging issue for a QMaSC director. The director may need to fight for higher hourly wages for the student tutors on the basis that only a select few students are qualified to hold such a position. Competition for the best students is also challenging, as they tend to have other equally lucrative opportunities on and off campus. By contrast, some institutions may find just the opposite for several reasons. Some schools, like large schools, may simply not have sufficient funding to pay tutors at a premium rate. In addition, the demand for good positions may outstrip the supply of available positions. Hopefully, the schools that find themselves with limited budgets will have the advantage of the supply being greater than the demand.

Other budgetary considerations include money for the purchase of new technology, small hardware items, software, travel funds, paper, copies, print and digital resources, food for training, training materials, etc. In short, a QMaSC director needs to be assertive enough to lobby for budget increases to cover not only student wages but a myriad of other line items.

4 The Search

Before starting, the search committee should check with the institution's office of human resources for guidelines to follow in a job search. Developing a job description, or updating an existing job description, is the first step in the process of hiring a QMaSC director. This task may be primarily the responsibility of the individual who will supervise the director, but ideally it should be done in consultation with others, such as a committee. If, for example, the director reports to a program dean, the department head (or division chair) responsible for teaching quantitative methods courses should be consulted. Conversely, if the director reports to a department head, then a program dean should be consulted. The job description should be written by those familiar with the student body as well as other resource centers and support services provided by the institution.

The job description will reflect the culture of the institution and its services while addressing whether or not the position is that of a founding director. Schools starting up a new QMaSC will seek a founding director. In such a case, the new director influences the success or failure of the center. Therefore, careful consideration should be given to the experience level of the candidates. The founding director should be someone with considerable experience working at or running a resource center and teaching at a similar institution. Conversely, because most established QMaSCs already have well developed policies, procedures, and routines, candidates for the director position may not require as much experience as a founding director. Schools seeking a non-founding director might prefer to hire a less experienced candidate in favor of one possessing a wealth of technology skills and a fresh perspective to take the center to a new or different level.

Once the job description is finalized, the recruitment process for the new director can begin. The search committee should consist of an academic Dean, three to five faculty members, and may include some experienced student tutors.

Determining where to "place the ad" is often a consequence of the QMaSC's position relative to the institution's organizational structure and the budget allocated to the search. The search process may take up to one year. Institutions with considerable resources may opt to employ a search firm. Search firms, usually very expensive, can reduce both the duration of the search as well as the amount of time spent by the search committee. Institutions not employing search firms typically advertise their search nationally. For example, the *Chronicle of Higher Education*, a popular destination for academics seeking positions, has two offerings. In particular, as of 2014, a 2" x 3" advertisement costs about $1,500 and comes with a 60-day free posting on their web site, whereas, to post the ad only online for 60 days costs about $300. Given that today's candidates are likely to start their job search for a new position online, search committees may also want to consider other online offerings such as higheredjobs.com and mathjobs.com as alternatives to the *Chronicle*. In addition, some online sites, like monster.com (jobs.com), offer free or low-cost job postings.

5 Conclusion

A QMaSC director needs to possess a diverse set of skills and attributes. The individual may also need to be nimble enough to adapt to emerging trends in pedagogy and technology. The following is a list of some of the skills and attributes that are desirable in a QMaSC director: strong quantitative skills, good organization skills, good interpersonal skills, strong oral communication skills, good written communication skills; be a community and team builder, a competent manager, and possess the ability to work in a diverse environment.

Finally, careful consideration should be given to identifying the particular goals, needs and constraints of the institution when hiring the director. The director will encounter unexpected demands and face many challenges. Finding the individual who is the best fit for a given institution will not only enhance many students' experiences in their quantitative courses but also allow faculty to achieve more in their classes. This will enrich the overall quantitative portion of the institution's academic mission.

6 Bibliography

[1] Common Core Standards Initiative, "Mathematics Standards." http://www.corestandards.org/Math/. Retrieved on October 22, 2013.

7 Appendix

Position Descriptions:

For examples of other job postings, see `http://www.higheredjobs.com/admin`.

Position Description I

The following is a sample of the position description for the director of the Math Resource Center at Babson College.

Director of the Babson College Math Resource Center
One of three academic resource centers at Babson College, the Math Resource Center (MRC) serves students in the undergraduate programs and to lesser extent graduate programs.
The MRC is a dedicated facility that offers:

- Walk-in consultation and tutoring in the core quantitative competencies
- Use of physical space and equipment for practicing quantitative skills using the most up-to-date technology and software tools
- Outreach to the community (custom-designed events integrated with existing curricula in the UG and MBA programs).

Reporting Structure: to Chairman of Math & Sciences Division while keeping the UG Program Office fully informed.

The Director of the Math Resource Center has responsibilities in three related areas (in addition to maintaining professional competence in field):

A Administration & Staffing

B Curriculum Design/Outreach

C Day-to-Day Activities & Public Relations

A. Administration & Staffing

Staffing:

- Hire, train, and assess undergraduate student tutors who help to staff the Center.
- Design job descriptions; manage work schedules and compensation for the undergraduate student tutors.
- Interface with instructors of three quantitative foundation courses to identify qualified student tutors.

Manage Physical Facility and Equipment:

- In order to enhance this permanently dedicated space, the Director interfaces with CITG, ITSD, UG Program Office, and the Math & Sciences Division to help re-design and renovate the technology and software available to the students.
- Manage relationships with those same constituencies on an ongoing basis in order to maintain the physical attributes of the facility.

- Write summary reports to UG Dean's Office to keep them informed about goals of the MRC, condition and use of facilities, service to community, etc.

B. Curriculum Design/Outreach

- Take a lead role in the evaluation, development and implementation of new technology in the three quantitative foundation courses.
- Lead training sessions in new software.
- Serve as a resource to other faculty on certain software programs and packages.
- Work with new faculty on the use of technology as it pertains to the three quantitative foundation courses.
- Work closely with ITSD on incoming student needs as they pertain to quantitative methods and in particular to the standard laptop software load.

C. Day-to-Day Activities & Public Relations

- Manage schedule of student tutors.
- Track MRC usage by course and instructor.
- Arrange for and oversee security of facility, which contains expensive equipment
- Ensure (with help of ITSD/Media Services) that equipment is maintained in good working condition.
- Create awareness of services available to the college community by:
  - designing brochures for distribution
  - arranging for announcements placed in The Free Press
  - composing tailored e-mail letters to UG & MBA students and to faculty
  - placing notices in UG and MBA handbooks

Position Description II

The following is a sample of a position description for a QMaSC Director at Hamilton College. Director of the Quantitative/Symbolic Reasoning Center. Hamilton is one of the nation's top liberal arts colleges. It is independent, highly selective, coeducational and residential. Originally founded in 1793 as the Hamilton-Oneida Academy, Hamilton is today composed of 1,812 diverse and talented students from nearly all of the 50 states and approximately 40 countries. More information can be found at www.hamilton.edu/about.

Hamilton College invites applications for a full-time academic-year director of its Center for Quantitative and Symbolic Reasoning. The Center provides resources and tutoring for students in courses designated as fulfilling Hamilton's Quantitative and Symbolic Reasoning requirement and for faculty teaching such courses. We expect the Director will also teach one credit-bearing quantitative skills course each year. We seek an engaging professional leader adept at working both independently and with multiple constituents on campus to provide superior academic support in the areas of quantitative and symbolic reasoning. The Director oversees the daily operation of the QSR Center facilities and is responsible for recruiting, training, and supervising student tutors. In addition, the Director coordinates the Peer-Tutoring program for students enrolled in quantitative/symbolic courses as well as in courses in other areas of the curriculum. Other responsibilities include the development of tutorial materials, conducting faculty workshops on quantitative/symbolic reasoning in the curriculum, participation in professional communities, and publicizing the activities and resources of the QSR Center.

Experience in an administrative or leadership capacity, teaching experience, and an advanced degree in a quantitative/symbolic field required. Characteristics of a successful candidate may include a PhD. Equivalent combination of education and experience will be considered.

Please submit a résumé, cover letter, and the names and contact information of three references to apply@hamilton.edu. Address questions about the position to Human Resources at (315) 859-4302. Review of applications will begin on May 13 and continue until the position is filled.

Hamilton (www.hamilton.edu) is a residential liberal arts college located in upstate New York. Applicants with dual-career considerations can find other Hamilton and nearby academic job listings at www.upstatenyherc.org. Hamilton College is an affirmative action, equal opportunity employer and is committed to diversity in all areas of the campus community. Hamilton provides domestic partner benefits. Candidates from underrepresented groups in higher education are especially encouraged to apply.

Designing a QMaSC Space

© Eileen M. Phillips,
Keene State College

1 Introduction

The design, location, and layout of a quantitative or mathematics support center (QMaSC) are key factors to a successful center. A positive experience at the center will encourage students to return and continue to seek help to understand the math and quantitative reasoning in their courses. "It is not just a space that is being created, but an environment that encourages and supports learning of a subject that leaves many wanting for understanding" [1].

In the design of a space for a quantitative or mathematics support center, many questions arise. Should the center stand alone or be with other support services in a large learning center? How large a space is needed for the center? Where on the campus should the center be located? What type of furniture is needed? How can the space attract students to use the center? All centers are different and reflect the culture and needs of each campus.

Much research has been conducted on the effectiveness of mathematics tutoring, but very little research has focused on the best practices in designing a tutoring center. Therefore, studies of space planning for effective learning spaces at institutions were consulted and adapted for use in this chapter. Most spaces are designed by those who will supervise the space rather than those who will use the space. If at all possible, a director should involve the potential users of the QMaSC early in the process to help inform decisions [2].

Suggested Citation: Eileen M. Phillips (2016), "Designing a QMaSC Space", http://dx.doi.org/10.5038/9780977674435.ch21 in G. Coulombe, M. O'Neill, M. Schuckers (Eds.) *A Handbook for Directors of Quantitative and Mathematical Support Centers*, Neck Quill Press, http://scholarcommons.usf.edu/qmasc_handbook.

This material is based upon work supported, in part, by the National Science Foundation under Grant DUE-1255945. Any opinions, findings, and conclusions or recommendations expressed in this material are those of the author(s) and do not necessarily reflect the views of the National Science Foundation

Two principal components will be addressed within this chapter: space planning and designing a center. The first component, space planning, involves those aspects of the center that relate directly to the physical space on campus that will be used for the center. The second component, designing a center, involves those arrangements needed to ensure a comfortable and fully functioning center environment.

2 Space Planning for a QMaSC

When designing a QMaSC space, a committee, including the director, the academic dean, representatives from the physical plant, and other stakeholders should start by looking at the overall culture of the campus. Every institution will create its space in a unique way based on its strengths and weaknesses as well as the resources and infrastructure available [3]. To help create that space, visit centers at other institutions and then create a list of pros and cons to determine what features would work best given the space and resources of the new center.

An important step in planning for a center is to survey potential users to determine what spaces they find most beneficial to learning. This may not always be possible, particularly if there isn't enough time or money to complete and analyze a survey. A 2008 survey completed by the Society for College and University Planning (SCUP) and Herman Miller, Inc. can be used to identify the most important characteristics for learning spaces on a campus. These characteristics emerged: spaces need to be flexible, to promote social interaction and healthful living, to use available assets and be sustainable, and to be attractive and stimulate creative thinking [2]. Special consideration must be given to individuals with special needs. A helpful resource that contains information about the legal issues involved in creating a space that is accessible to everyone as well as a checklist that can be used when designing a space can be found at: `http://www.washington.edu/doit/programs/accesscollege/student-services-conference-room/accommodations/tutoring-and-learning-centers` [4].

The location of the center is another important consideration in the planning phase. Each director will need to think about the culture of the campus to determine which location is best. Is the QMaSC in a central location, such as a library or a learning center with a coordinator and all support services involved, or is it located in the mathematics department [1, 5]? Housing the center within the mathematics department puts tutors close to faculty members who can assist in answering questions and shows the commitment of the department to the quantitative needs of the campus. However, centers within departments may not be open for extensive hours, and the mathematics department may not be located in a central space on campus. Some institutions may find it more attractive to have the center in a centralized location such as the library, that is open many more hours of the week and is in the center of the campus where it is easily accessible to students. Students may also be more willing to visit a center where they won't run into their mathematics professor [3]. The downside to housing the center away from the mathematics department is that mathematics faculty members may not have time to visit frequently or to get to know how the center is functioning.

Visibility of a QMaSC on campus is critical. Support centers tucked away are hard to find, and they provide an excuse for students not to visit. Students who are having difficulty understanding the quantitative components of a class often feel anxious about their lack of understanding and feel awkward about getting help. According to Halcrow and Iiams, students felt a sense of intimidation asking for help from peers who excelled in mathematics, and they were also reluctant to visit an unfamiliar place with unknown people. They found students were "more willing to risk their success in the course than to risk being perceived at a lower status in mathematics than their peers" [6]. They also found that lower-ability students were the most hesitant to seek help, even though they were the population who needed it most. Centers must be very visible on campus so that when students find the courage to visit the center, they can do so quickly and easily.

The size and functions of the center should also be considered. Will more than just tutoring take place in the center? Will there be testing, study sessions, group work, meetings, and classes also be using the space? How will all of these functions work in the allotted space? Many centers will be located in existing spaces and will have to manage with what is given, but if new construction is a possibility, try to allow for the most versatile spaces. Versatility will allow for the space to get the most use possible as well as allow growth of the center through programmatic changes. The director of the center will need to produce a schedule that allows all of the functions of the center to coexist harmoniously.

The director or designer should decide at the onset whether or not there will be separate spaces for mathematics majors to work and for non-majors to be tutored. If there is only one room for tutoring, there is the risk that math majors will visit their friends who are on tutoring duty to work together on assignments. Although such visits might help build a community of learners, it is a problem if students who seek help at the center are overwhelmed by the level of mathematics that is being discussed or distracted by the noise created during the process of working on higher level mathematics problems. Apprehensive students do not want to enter such an environment. If two rooms are not established at the beginning, students may become comfortable in one space and not want to leave. It is best to start out with two separate rooms and make the purposes of the spaces known from the outset. Croft [3] cites a center that dedicated one room to be used for tutoring only, and an adjoining room to be used for individual or small group work, rather than trying to repurpose a space later for majors to complete their work. Solomon [1] found that when second and third year math majors were asked to use a new space to make more room for the first-year math majors and non-majors in the tutoring center, it disrupted the "strong collaborative working system" that had developed. This change caused a polarizing effect in that some students worked in the new room provided, while others stopped visiting the center altogether. Hunley and

Schaller [7] also found that students are less likely to use a space for a new purpose if it has not been physically changed in some manner. They also found that students wanted permission to use the space in a different way before they felt comfortable changing the established rules of a prior space.

3 Designing Center Space

Hunley and Schaller [7] used multiple methods to collect data to discover what the best spaces were for encouraging student engagement. Two of their findings were related to students and their workspaces: (1) students reacted best to a tutoring experience that was serious, to tutors who treated them with respect and encouraged interaction with other students, and (2) students were more engaged in the learning process if the physical characteristics of the space were welcoming. Carroll and Gill [8] also found in their research that a center needed to create an environment "where students feel no question is a stupid question" and that students valued the center for its ability to help them increase their confidence and comfort with mathematics.

It is critical to consider the physical environment of a support center because it contributes to the improvement or hindrance of students' learning. There are many factors that may go unnoticed such as the temperature, lighting, and noise level until there is a problem [9]. Through photographic studies, surveys, and focus groups, Hunley and Schaller [7] found that students were more engaged in learning spaces that were "comfortable, open, flexible, and appealing to the emotions." Engagement also occurred when the settings encouraged interactions with others, had appropriate technology, comfortable furniture, adequate air circulation and temperature, good lighting, and limited distractions.

To ensure all aspects of the center will fit in the allotted space, it is suggested that the design of the center be created on paper using a scale model. In 2003, a renovation was done on the building that housed the Mathematics Department and the Math Center at Keene State College in Keene, New Hampshire. Figure 1 shows the scale model that was created to show what elements were needed in the new space and how they would work together.

A more recent renovation of a QMaSC space can be found at the Quantitative and Symbolic Reasoning Center at Hamilton College in Clinton, New York [10]. Refer to the case studies section at the end of this handbook, where O'Neill writes about the development of the new center space.

Another idea to consider while planning a QMaSC is the use of satellite centers, where tutors provide help for students outside the four walls of the center. Tutors can be given a mobile cart that contains all the necessary items they will need while tutoring. They can hold drop-in times at the library on the weekends or in the residence halls during the late evening when the center is closed. These satellite centers are also helpful for centers that are newly formed and do not have a designated space yet or while a center is under renovation.

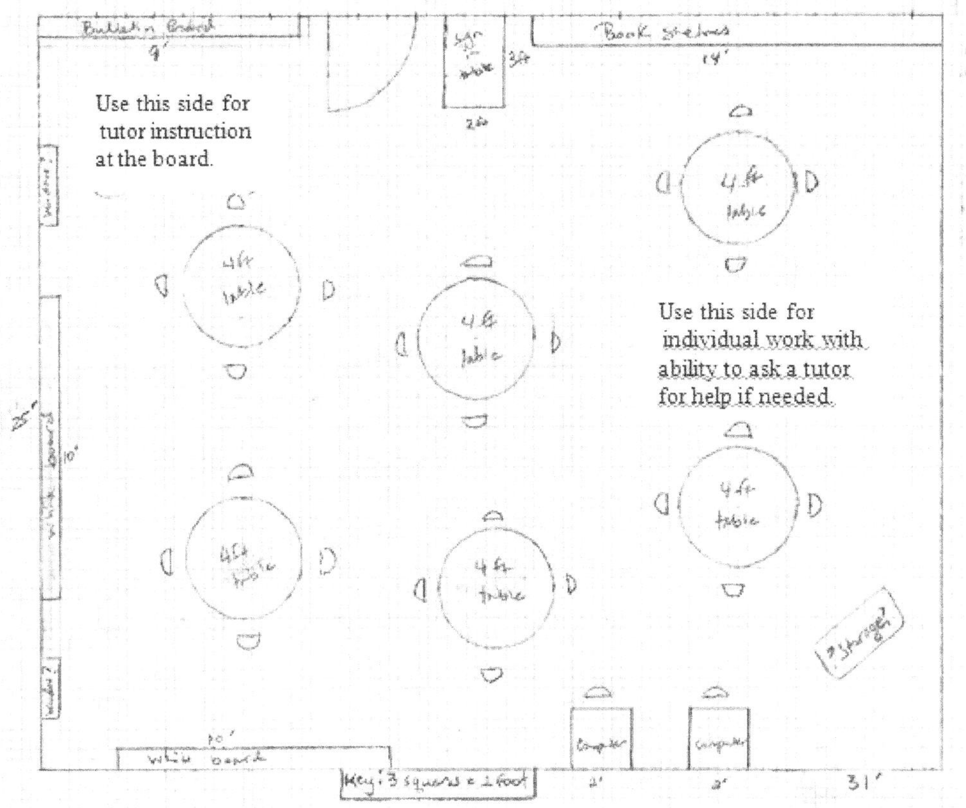

Figure 1: Scale Model for the Math Center at Keene State College

Creating the physical space that will be used on a day-to-day basis requires detailed thought. The remainder of this chapter identifies several components for directors to think about while designing a functional space: tables and chairs, board space, displays, storage, supplies, computers, a reading area, additional rooms, and finishing touches. Each component is explained below.

<u>Tables and Chairs:</u> Many types of tables and chairs can be selected for use in tutoring centers. The director needs to decide what type of furniture will work best for the purposes of the center. Some directors prefer circle tables because they give the students more space to put their textbooks, notes, and other personal items while they work on homework or a project. Circle tables also allow several students to sit comfortably around the table for discussions or group work. Other directors prefer square or rectangular tables so that students can move them around and make different sized tables to fit their needs. Types of chairs also vary from center to center. Some directors prefer rolling chairs for easy movement, whereas others prefer non-fabric chairs for ease of cleaning. As furniture can be very expensive, directors also need to consider the cost of the tables and chairs when making selections. Furniture also needs to be accessible and comfortable to all students, especially larger students or handicapped students [9]. Hunley and Schaller suggest furniture that

is flexible, "lightweight and easily movable, with a ratio between furniture and free space that allows for alteration of the room setup" [7].

Board Space: Some sort of writing surfaces should be installed to create areas where students can work together on solving problems. Once again, many options are available, including whiteboards, chalkboards, whiteboard or chalkboard paint that goes directly on the walls, glass whiteboards, or even easels. Preference really depends on the individuals using the space. If classes or study sessions will be offered in the center, the needs of those teaching in the space should be taken into consideration as well. For example, will instructors need a projector? If so, then perhaps the writing surface chosen can serve as a screen for the projector.

Displays: Bulletin boards should be placed in the center to display such things as times of operation, announcements to the students and tutors, tutor photos and names, rules of the center, funny math cartoons, and anything else that needs to be visible to the users of the center. Try to keep this area current and changing to hold students' interest.

Storage: Storage space should be incorporated into any center. As they say, one can never have too much storage! Many types of storage should be considered: filing cabinets, bookshelves, display cases, mailboxes, vertical hanging storage, and literature organizers. These organizing units can house such items as past and current textbooks, manipulatives, schedules, handouts on mathematical topics, answer keys, journals, signs for each tutor to put out while on duty, surveys for completion after tutoring, and anything else the tutors may need to perform the duties of their job. Figure 2 shows a photograph of a corner of the Math Center at Keene State College where most of the storage is housed for convenient access. The textbooks shown in the photo are out in the open for students and tutors to use for extra problems or research. Current textbooks are housed in a drawer in a filing cabinet so that they are more secure. Another model to ensure that current textbooks are not taken is described by Croft [3] that textbooks can be "physically secured to the walls or desks." Directors need to provide locked storage for items that are only for the tutors to use and that they do not want to leave the center. Storage should also be provided for the tutors' belongings while they are on duty.

Supplies: Create a work area that contains all the supplies students may need to use at the center (see Figure 2). These items would include a pencil sharpener, stapler, 3-hole punch, tape, pencils/pens, scrap paper, rulers, spare calculators, protractors, compasses, and anything else they may need to execute an assignment.

Computers: Although many centers are equipped with wireless internet, and students will bring

Figure 2: Example Work Area

their own laptops or tablets, it is still important to have at least a couple of desk top computers for students to use at the center. Tutors may need to use the computers to help students with technology-based assignments using mathematical software often not economical for students to have on their personal laptops, or tutors may want to use the computers for their own assignments if there is no one at the center being tutored. Students may also use the computers to complete projects assigned in math or quantitative courses, knowing that someone is there to help them should they run into a problem. Some centers have a computer located near the entrance to use for tracking attendance and survey purposes. For example, according to Cooper [11], the Tutoring Center at Western Washington University used TutorTrac software to electronically track the number of students who visit the center, how many hours they spend at the center, and for what courses they received tutoring help. Other centers use laptops or tablets to record this information. Whatever type of computers the center uses, it is important to keep them secure.

Reading area: If the center has an unused corner, try to create a reading area that is more relaxed with comfortable chairs or a sofa that students can lounge in and read an assignment, catch up on email, or chat with a friend.

Additional rooms: Many centers are housed within the walls of one room. If a center has the luxury to have more than one room, the rooms can be assigned for different purposes. Some ideas might be: a quiet area for studying, a testing room, a room for small groups to meet, or a room for majors to hang out.

Finishing touches: Finally, after all the research has been done about the space needs of the center and what items to include in the center, the finishing touches need to be done to make the center an inviting and welcoming place. Here are some ideas: paint the center a calming color, put up colorful and informative posters, paint a border with magnetic and/or chalkboard paint so that colorful magnets can be placed around the room (see Figure 2) or inspirational sayings may be written, have colorful signs throughout the room stating what tutors can and cannot do, post mathematical cartoons, paint the value of π on the wall, or have manipulatives displayed in glass display cases to add some color.

4 Conclusion

Creating a QMaSC space can be an exciting and challenging endeavor with many factors to consider, from the planning phase to the day-to-day functioning phase. A director should take the time to visit other QMaSC spaces or inquire about the ways in which other QMaSC spaces are designed to find the best practices and use that information to create a center that fits the culture and needs of the individual institution. Center directors should create a welcoming and inviting space that is visible on campus and encourages students to seek the help they need. QMaSC spaces should be flexible so the physical space can be ever-changing with the programmatic needs of the center.

5 Bibliography

[1] D. Lawson, Y. Solomon, and A. C. Croft, "Safety in numbers: Mathematics support centres and their derivatives as social learning spaces," *Studies in Higher Education*, vol. 35, no. 4, pp. 421–431, 2010.

[2] P. Grummon, "Best practices in learning space design: Engaging users," *EDUCAUSE Quarterly Magazine*, vol. 32, no. 1, 2009. Accessed from `http://www.educause.edu/ero/article/best-practices-learning-space-design-engaging-users` June 13, 2013.

[3] A. C. Croft, "A guide to the establishment of a successful mathematics learning support centre," *International Journal of Mathematical Education in Science and Technology*, vol. 31, no. 3, pp. 431–446, 2000.

[4] S. Burgstahler, "Tutoring and learning centers." DO-IT, University of Washington, Access College: Student Services Conference Room. Accessed from http://www.washington.edu/doit/programs/accesscollege/student-services-conference-room/accommodations/tutoring-and-learning-centers on March 1, 2016.

[5] J. Landsberger, "E-Learning by Design: Academic Support Centers: quo vadis?," *TechTrends*, vol. 49, no. 4, pp. 8–11, 2005.

[6] C. Halcrow and M. Iiams, "You Can Build It, but Will They Come?," *Problems, Resources, and Issues in Mathematics Undergraduate Studies*, vol. 21, no. 4, pp. 323–337, 2011.

[7] S. Hunley and M. Schaller, "Assessment the key to creating spaces that promote learning," *EDUCAUSE review*, vol. 44, no. 2, pp. 26–35, 2009.

[8] C. Carroll and O. Gill, "An innovative approach to evaluating the university of limericks mathematics learning centre," *Teaching Mathematics and Its Applications*, vol. 31, no. 4, pp. 199–214, 2012.

[9] WGBH Educational Foundation, "Diversity and classroom management: Building a physical space for learning." Retrieved from http://special.league.org/gettingresults/web/module1/diversity/build_space_learning.html on March 1, 2016, 2006.

[10] M. O'Neill, "A new tutoring center: Challenges, opportunities, and surprises." Retrieved from http://www.hamilton.edu/qsr on August 14, 2013, 2013.

[11] E. Cooper, "Tutoring center effectiveness: The effect of drop-in tutoring," *Journal of College Reading and Learning*, vol. 40, no. 2, pp. 21–34, 2010.

Part VII

Case Studies

Case Study: Symbolic and Quantitative Resource Center at Lewis & Clark College

© Margot Black ,
Lewis & Clark College

1 Introduction

The Symbolic and Quantitative Resource Center (SQRC) is an informal and free drop-in tutor center, staffed by undergraduate peer tutors, open to the entire Lewis & Clark student body. Until the spring of 2013, the SQRC was known as the Math Skills Center (MSC). This case study will serve both as a study of the new center, but will also address the logistical and philosophical elements and challenges involved in its transition from MSC to SQRC.

Lewis & Clark College is a private, selective, residential liberal arts college located in Portland, Oregon. In addition to offering Bachelor of Arts degrees in 28 majors in the College of Arts and Sciences, Lewis & Clark is also home to a Law School with a top ranked Environmental Law program, and a Graduate School of Education & Counseling. The College of Arts and Sciences is comprised of approximately 2000 students, most in the top quarter of their graduating class, representing all 50 states and 66 different countries. Popular majors are in the social sciences, but close to a quarter of all students major in the natural sciences, including mathematics and computer science. The college does *not* offer an education major (nor any pre-professional major), however the graduate school offers an M.A.T. degree in Education, with endorsements in Math, as well as Science, Art, Social Studies, and English. Undergraduate students choose to come to

Suggested Citation: Margot Black (2016), "Case Study: Symbolic and Quantitative Resource Center at Lewis & Clark College", `http://dx.doi.org/10.5038/9780977674435.ch22` in G. Coulombe, M. O'Neill, M. Schuckers (Eds.) *A Handbook for Directors of Quantitative and Mathematical Support Centers*, Neck Quill Press, `http://scholarcommons.usf.edu/qmasc_handbook`.

This material is based upon work supported, in part, by the National Science Foundation under Grant DUE-1255945. Any opinions, findings, and conclusions or recommendations expressed in this material are those of the author(s) and do not necessarily reflect the views of the National Science Foundation

Lewis & Clark for its small class sizes, extensive opportunities for overseas study, undergraduate research opportunities, beautiful campus, and dedicated faculty, well known for their commitment to excellent teaching.

The Math Skills Center was created in 1980 in response to a growing need for dedicated space and personnel to address and manage math remediation. Its creation coincided with the establishment of a remedial math course, Math 055 — Review of Intermediate Algebra, of which the Director of the MSC was the sole instructor of record. The center and the course were closely intertwined in a symbiotic and mostly static relationship for the better part of 30 years. Math 055 was taught *in* the MSC, either by the MSC director (for the lecture formatted sections) or facilitated by MSC tutors (for the self-paced sections). The MSC director was even in charge of placement testing to determine which students would take Math 055 and which were proficient enough to move onto college level math and science classes. At the time, "Mathematical Proficiency" at the intermediate algebra level was an official *graduation requirement* and the MSC was the place on campus designated to help students establish that proficiency and hence meet that requirement. Secondary to this, the MSC provided a space for students in lower division math classes to get help with their math homework.

Though the Math Skills Center was a valuable and well utilized resource, it was located in a hidden basement corner of campus and regarded as primarily a resource for weak students. Thus, many students were reluctant to visit the MSC even if they could find help there and faculty, wary of its stigma as a center for remedial math help, were reluctant to refer their students there. Students were more inclined to take advantage of the student-run and student-fee-funded one-on-one tutoring program (all Lewis & Clark students can have up to 2 hours per week per course of one-on-one peer tutoring). Despite its mission statement to help students with "any math problems they encounter, including those which arise in classes outside the Mathematics Department," only students in lower level math classes patronized the MSC. Likewise, Math 055, the course that justified the Math Skill Center's very existence, was not popular among students, as it did not offer any credit towards graduation. Nor was the course popular with faculty and administrators who felt that a selective college had no business offering a remedial math course.

The retirement of the MSC's longtime director in the spring of 2011 created an opportunity to completely rethink the need for and function of Math 055 and the MSC. A math department faculty member stepped in as interim director — an experiment in how much leadership the center required — and a new course was developed to replace Math 055: CS102 — Quantitative Reasoning, a course designed to improve quantitative reasoning skills rather than to remediate algebra. These events coincided with the college receiving a generous grant from the Teagle Foundation. The grant proposal, written in response to a growing realization that students were graduating from the college without having improved their quantitative reasoning abilities, had two main objectives: 1) Develop a pilot course (CS102) to address the issue, and 2) expand the role and function of the Math Skills Center to support such a course and to support strengthening QR skills in general.

The remedial math content of Math 055 was absorbed into the new course's curriculum, however, unlike Math 055, CS102 was structured more traditionally in that it would *not* be self-paced with MSC tutor support, nor taught in the MSC, nor taught by the [interim] MSC director. Without Math 055 to teach and support, and other tutoring options available on campus, it wasn't clear that the Math Skills Center would be utilized at levels justifying its expense. Under the interim director, and with reduced staffing, the usage and efficacy of the MSC were under close observation in 2011/12 by an administration looking to tighten its belt.

Fortunately, it was observed that the MSC was still just as necessary as ever, if even just to support CS102 through peer tutoring, so a new director was appointed for the 2012/13 year. At that time, the faculty piloting CS102 handed off the entirety of the teaching, assessment, and curriculum development duties to the new director, and thus a new relationship between the MSC and *quantitative reasoning* was born.

By virtue of this shift in curricular priorities, and with a great deal of financial help from the Teagle Foundation's grant, the MSC has undergone a major transformation. Moved to a new, more central space, and given a new name, the Symbolic and Quantitative Resource Center also has a new and expanded mission: to support and enhance symbolic and quantitative reasoning across the curriculum and campus community.

The SQRC is still in its early phases of this transition and is still navigating its evolving role in supporting the pursuit and development of quantitative reasoning skills at the college, both in and out of the mathematics department. Many new services that will appeal more broadly to the campus community are on the horizon. At present, the SQRC primarily accomplishes its new mission by simply expanding its drop-in tutoring services to students in all quantitative courses at all levels, not just lower level math classes. To this effect, the center has tutors qualified to assist students not just in mathematics but also in statistics, computer science, physics, chemistry, economics, logic, quantitative reasoning, and quantitative research methods courses. But more than just a homework help center, visitors to the SQRC can find additional resources for their courses such as course textbooks, computers with software such as R, Excel, SPSS, STATA, LaTeX, and Mathematica, as well as enthusiastic tutors who are trained to help students engage with the material in a way that will foster greater independence with it.

As previously alluded to, the center has just completed its relocation from an obscure basement location to a more central, accessible, and visible location in the main entry of the building where all math and social science courses are taught. This relocation has had a profound effect on the SQRC. Though exact figures aren't available, we know that more than double the number of students visit the SQRC in its new location. The new location also well positions the SQRC to become a greater part of the campus community, greatly expand its curricular reach and scope of services, and rebrand itself as a center that provides both academic support *and enrichment*.

The new SQRC occupies 655 square feet of an almost exactly square space (see Figure 1), right inside the main door of the newest and largest classroom building on campus. The space was formally used as a café (long since defunct) followed by an underutilized student lounge space. Through some summer remodeling and necessary funding by the college, the SQRC could open its doors in the fall of 2013. A wall-to-wall, floor-to-ceiling sliding glass door — the only barrier between the center and the noisy passing time traffic — was converted into a wall-to-wall, floor-to-ceiling white board, a favorite attribute of the new space. The sliding glass door was fixed in place at the edge of a new doorway that was installed, forming the interior entrance to the SQRC. The door is secured by a keypad lock to which only tutors have the code (they are able to program it to stay unlocked for student passage). The SQRC also has an exterior entrance, unlocked from the inside by sliders, or from the outside with a Math Department key (not available to tutors). Tutors open the center from the interior entrance, unlock the exterior doors, and then students using the center can (and do) access the room from both entrances. It is nice to have a location with exterior windows so that students can easily tell when it is open if it is dark outside and so that both tutors and students feel safer there at night.

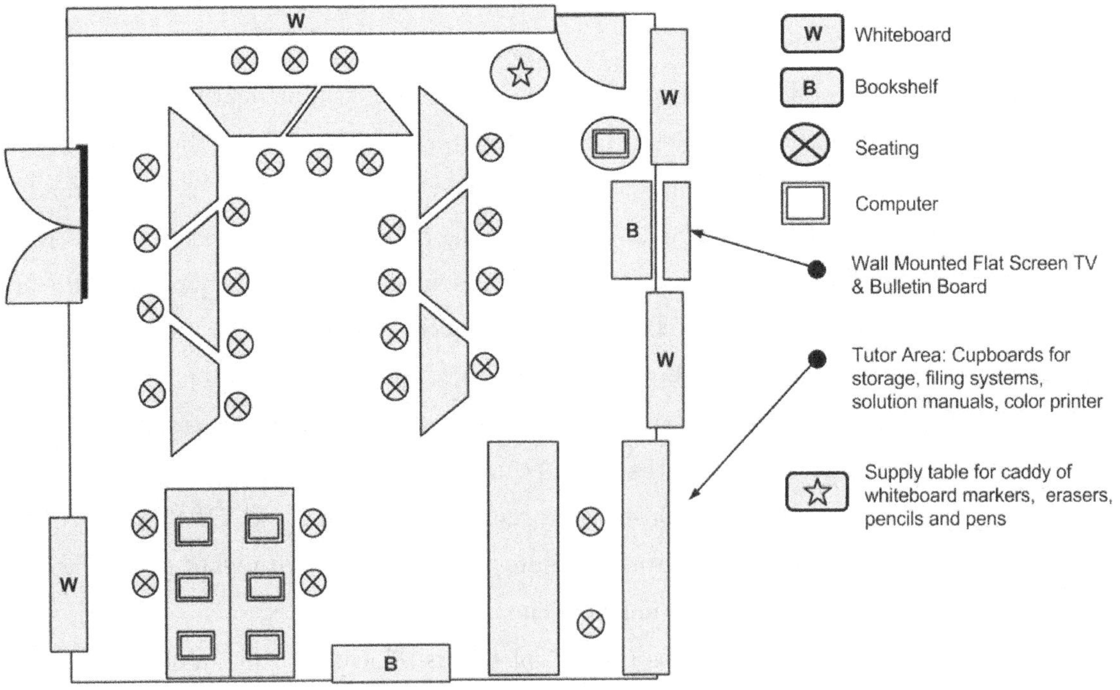

Figure 1: SQRC Floor Plan

Just inside the interior door is a computer sign-in station; all students visiting the center are required to sign in using a Google form (see Appendix). After signing in, the students can take a seat at one of 8 trapezoidal tables (which can be rearranged into many different formations), or at one of the six computers in the SQRC: 4 iMacs and 2 PCs. The computers are kept on tall tables that can be walked up to, but four stools are provided. Rarely are all 6 computers in use at one time. The SQRC has two bookshelves for student use. A small bookshelf holds supplemental course resources (such as old textbooks and laminated 'Quick Study' guides) as well as office supplies for student use such as whiteboard markers, scratch paper, graph paper, a stapler, 3-hole punch, pencils and pens, rulers, electric pencil sharpener, tape, etc. A larger bookshelf is the new 'textbook library'. It holds all current textbooks for SQRC supported courses, another very popular feature of the new space. Students can use the books only while they are in the center and many students do use them to work on their homework in between classes.

The room still has the back counter and front counter of the defunct café; these have been repurposed as tutor and file storage. The front counter provides a nice space for students to walk up to for quick questions. The back counter has a color printer that can be used by students to print out materials and assignments for quantitative courses. It also holds a countertop filing system that is used for students to turn in their homework, or for professors to leave supplemental materials, or student work to be handed back. (Files are made as needed for any professor that requests them.) Solution manuals are locked in one of the cupboards and are generally only utilized by tutors. Graphing calculators are also kept locked up and can be checked out by students with their student ID. We do allow students to leave the center with a calculator as they are often checked out for use on an exam. Thanks to the vestiges of the café, the back counter also has a sink, mini-fridge, and a coffee maker. The latter two are employee perks that the tutors greatly appreciate. Tutors can sit in one of two stools (soft seating with a chair back) that are generally positioned in between the two counters.

On Sundays and during the evening, when the center is busiest, two nearby classrooms are reserved for SQRC overflow. Tutors are instructed to move study groups, or other groups of students studying similar material (i.e., Chemistry and Physics) to the overflow rooms before the center becomes overwhelmed.

2 Center Organization and Services

Organizationally, the SQRC is an autonomous entity, independent of any other department. Its personnel is comprised of student tutors and the director. The director, officially reports directly to the Dean of the College, but generally consults with the Math Department chair for most matters. Historically, the Math Skills Center employed between 16 and 20 tutors a semester, but the SQRC has required an increase in personnel due to an increase in traffic and its expansion of services. It currently employs about 30 tutors, five of which exclusively tutor chemistry, and three of which are hired by the physics department to primarily tutor physics. The remaining can all tutor math as well as other subjects such as computer science and economics. The SQRC currently operates with an annual budget of $12,600, of which $11,000 is for payroll, and the remainder for supplies and other operational miscellany. However, given the greatly increased demand of the SQRC, and the end of the Teagle grant's financial support of the SQRC, a budget of $25,000 was approved for the following school year. At present, chemistry and dedicated physics tutors are paid out of their department's budgets.

Despite its departmental autonomy, the SQRC is unofficially a part of the close-knit Math Department family. The director, who is required to hold at least a Master's degree in Mathematics, and who has always been hired by a Math Department search committee, keeps an office in the math department, attends department meetings, and relies heavily on the department's administrative assistant.

This relationship with the math department actually led to some difficulty in determining a new name for the center. With the relocation of the center and planned expansion of services, and also in an effort to shed the reputation of being just a support center for those struggling in basic mathematics, it was generally agreed upon that a new name was in order. Determining the new name was left up to the director and the math department, though it ultimately needed approval from the Dean. A first and popular suggestion was 'Quantitative Resource Center'. However, math faculty felt that since the center would still *primarily* focus on math, they didn't want its new name to suggest otherwise. Also, though mathematics and quantitative reasoning are certainly related, one is not a subset of the other. Pure mathematics is more symbolic in nature than quantitative; hence, 'Symbolic' was appended to the name 'Quantitative Resource Center'.

The director, a full-time exempt staff member on a 9-month contract, has three primary responsibilities: running the center, teaching CS102, and administrating the Quantitative Reasoning Exam (QRE).

- The first entails submitting and managing the yearly budget, staffing the center — including interviewing, hiring, supervising, and training the tutors — advertising, marketing and assessing the center's services, communicating with the math department and campus community regarding matters pertinent to the center and tutors, and making sure that the center is in good working order with regard to supplies and educational resources.

- As for teaching, the college currently offers two or three sections a semester of CS102 and the Director of the SQRC is its sole instructor, responsible for the instruction, assessment, and curriculum development of the course as well as advising students in the class.

- Finally, with the help from the Math Department's administrative assistant, the director is in charge of administering the 'Quantitative Reasoning Proficiency Exam.' The test places students into CS102, or determines that they are proficient enough in intermediate algebra to move onto their college level math and science courses. Responsibilities of administering the test include maintaining information about the test on the college's website, providing review materials, assessing the testing instrument (currently using the commercial product ALEKS PPL), advertising and proctoring the exam at various times during the year, and advising students based on their results.

The SQRC is typically open everyday except Saturday, with hours during both daytime: mid-morning until 5 PM, and evening: 6 PM until 10 PM. The actual schedule varies from semester to semester and day to day depending on the number of tutors, their availability, and demand. The schedule is posted on the doors of the SQRC, announced in classes, and placed as a Google Calendar on the SQRC's website.

Usually one to two *math* tutors staff the center at all times with an additional chemistry or physics tutor available in the late afternoons and evenings. Many of the math majors are qualified to tutor in other subjects such as physics, computer science, and economics and those specialties are advertised accordingly. For special courses, such as logic and quantitative research methods, specific tutors and tutor hours are advertised to students in course syllabi.

When students arrive at the center they are prompted by the tutor on duty to sign in at the sign-in computer (right inside the interior entrance). After signing in, they can take a seat and ask for help immediately, work independently or in a study group, or make use of the new textbook library. Students may also come in to turn in homework, or to use or get help with Excel, Mathematica, R, STATA, SPSS, LaTeX, or other software.

When tutors are assisting students they are encouraged only to write on the whiteboards, rather than on a student's paper. More importantly, the SQRC has about a dozen mini whiteboards, 2' x 3' pieces of white board material purchased and cut to size at Home Depot, which students are encouraged to use while in the SQRC rather than a pencil and paper. These whiteboards better facilitate group work and collaboration, and encourage productive brainstorming. The whiteboards are very popular among both students and tutors. Indeed, even professors often borrow several for group work in their classes.

3 Staffing, Hiring and Training

The SQRC tutoring staff is comprised exclusively of student peer tutors. Historically, tutors were recruited through their math professors, who were asked to identify strong students with a good demeanor and helpful attitude. This year, in an effort to expand the areas of expertise of tutors, professors were also asked to recruit students with strong computing skills, experience with statistics, and double majors. Currently the chemistry and physics departments recruit and hire their own tutors but the SQRC director schedules and supervises them. Similarly, students that tutor for more specialized non-math and science courses, such as logic and research methods, are recruited directly by the courses' professors.

Many of the referred tutors are math majors, but since the center has historically only served lower division courses, the only requirement is successful completion of Calculus 2. A requirement that tutors have successfully completed Discrete Mathematics and Calculus 3 (multivariable calculus) is currently being considered as many students now come into the center for help with these classes.

The following is the process of hiring math tutors: each spring professors are invited to recommend strong students with helpful demeanors as potential new SQRC tutors. The director then emails the referred students an application and invitation to arrange an interview. Hiring is ultimately based on goodness-of-fit, how many positions there are to fill, and whether or not a student has work-study. The position does not require work-study but those with work-study are given preference in hiring. In addition, rising sophomores are given higher priority than juniors or seniors as the SQRC is looking to find students who will be on staff for 2 – 3 years of their college career. A tutor's position has always been automatically renewed if they express interest in continuing the position; they are not required to re-interview and they generally stay on from semester to semester, year to year, returning to the SQRC even after they have studied overseas for a semester. Tutors start at minimum wage (currently \$9.10/hour in Oregon) and get a \$0.25/hour raise per semester of service.

This hiring process has consistently resulted in hiring great tutors for the center; however, there always seem to be well-qualified and enthusiastic students that fall through the cracks of the professor referral system. Thus, this year the SQRC will have an open application process but still ultimately require candidates to be recommended by their professors. Only those students who are strongly recommended by their professors will be invited to an interview.

As previously discussed, one to three tutors staff the center at any given time. The earliest shifts usually have only one tutor, as they tend to have the fewest number of students visiting; by 1 PM there are two tutors on staff, and after dinner (6 PM – 10 PM) there are two to three math tutors as well as a physics or chemistry tutor. The evening is the busiest time for the center. A tutor will typically work two 1.5 – 2.5 hour shifts each week.

The MSC's budget is not currently adequate for extensive training. Tutors attend one paid 2-hour mandatory meeting each semester that includes both orientation (in the fall) and some tutor training. A representative from Student Support Services attends and discusses learning styles, learning differences, and multicultural sensitivity. To mitigate this lack of training, less experienced tutors usually only work with more experienced tutors, rarely by themselves on a solo shift.

The Writing Center at Lewis & Clark uses a 1-credit hour Practicum course as a means of training its tutors. Such a course is being piloted in the spring semester of 2014 at Lewis & Clark for the SQRC. If successful, it will be offered yearly in the fall and will be required of all new tutors. The course will meet for one hour, weekly, with the intent of training tutors in best practices with regards to effective tutoring, training tutors to use various types of software used in quantitative courses, and as an opportunity to talk about what is coming up in various courses so that tutors can be ready for challenging material.

4 Community Interactions

Interaction between the old Math Skills Center and the campus community was minimal outside of the Math Department, and somewhat scant even then. Indeed, its location made it difficult to find for all but the very determined! Although there was general acceptance that the MSC's existence was essential, its reputation as a remedial math hub and support center for underprepared students meant that non-math faculty and the administration kept their distance. However, due largely to the new and improved central location of the SQRC, as well as its new focus on quantitative reasoning rather than remedial mathematics, community interaction and appeal have increased considerably.

The SQRC is mainly a place where students go to meet with peer tutors, but given that the center is located in the same building where all of the math and social science classes are taught, professors often pop in on their way to or from class to drop off papers or supplemental materials and also to assist their students. Additionally, all math professors can use their key to access the center, and other professors can obtain the door code if they request it. This has enabled some professors to hold review sessions in the center when it is not otherwise open, which they are welcome and encouraged to do. Having professors hold office hours in the center has been considered but is not, at this time, on the horizon.

The entire faculty has generally expressed their support and enthusiasm for the SQRC and their desire to interact with it, thus further plans to increase its exposure are underway. Plans include faculty development, especially in supporting individual and departmental attempts to infuse non-quantitative courses with elements requiring quantitative reasoning. This may take the shape of faculty trainings, as well as individual mentorship and course support in the SQRC.

Many potential opportunities and ideas for the growth and trajectory of the SQRC, especially as it pertains to community exposure, will be subject to the soon-to-be formed SQRC Advisory Committee. Until now, neither the MSC, nor SQRC, have had an advisory committee. The proposed committee will comprise of 6 – 8 members of faculty across the curriculum and will focus on, among other things, ways that the SQRC can interact with and impact the entire campus community. One exciting possibility that will be discussed is how the SQRC can help create and support a scientific component of the Writing Center!

5 Assessment at the Center

The only formal assessment of the center to date occurs during the external evaluation of the math department. Internal efforts to rigorously assess the center are in their nascent stages. One item for assessment is who is using the center, and why. In the past the tutors would tally the number of students in specific classes in the center during their shift, with a large group of classes left for the 'other' category. The new sign-in system also collects this information, and more (See Appendix). This will enable the director to monitor unique versus return visits (student IDs are required) and also allow us to compare the grades of the students who visited the SQRC with those who did not, as a way of possibly assessing its efficacy. A student feedback form, on paper and on the computer, will also be instituted this year. Absent of this feedback, assessment has largely been anecdotal, or via tutor survey, but even these methods have motivated positive change. Indeed, the relocation of the center was the direct result of loud and frequent complaints that the old center was hard to find and felt 'dungeon-like' (a further punishment for those who struggled with mathematics).

At the end of a tutor's shift, he or she fills out a shift report (See Appendix). The shift report is also a Google Form, the responses of which populate a spreadsheet that is reviewed by the director daily. These shift reports are a valuable source of information for the director and a good way of keeping track of how the center is running on a day-to-day basis. Part of the shift report is called 'Top Topics' where tutors can enumerate specific questions or concepts that multiple students asked about. In the future this information will be sent to professors so that they know for what topics their students are using the center.

6 Acknowledgements

We would like to thank the National Science Foundation for its generous support of the QMaSC Workshop through NSF grant number 1255945. Any opinions, findings, and conclusions or recommendations expressed in this material are those of the author(s) and do not necessarily reflect the views of the National Science Foundation.

I would like to thank the Teagle Foundation for its generous collaborative Engaging Evidence grant to Lewis & Clark and Willamette College. Through this grant the director of the SQRC has been able to connect with other center directors and become fully engaged and immersed in the national community of center directors and quantitative reasoning. If not for this grant, the writer would not be writing right now.

7 Appendices

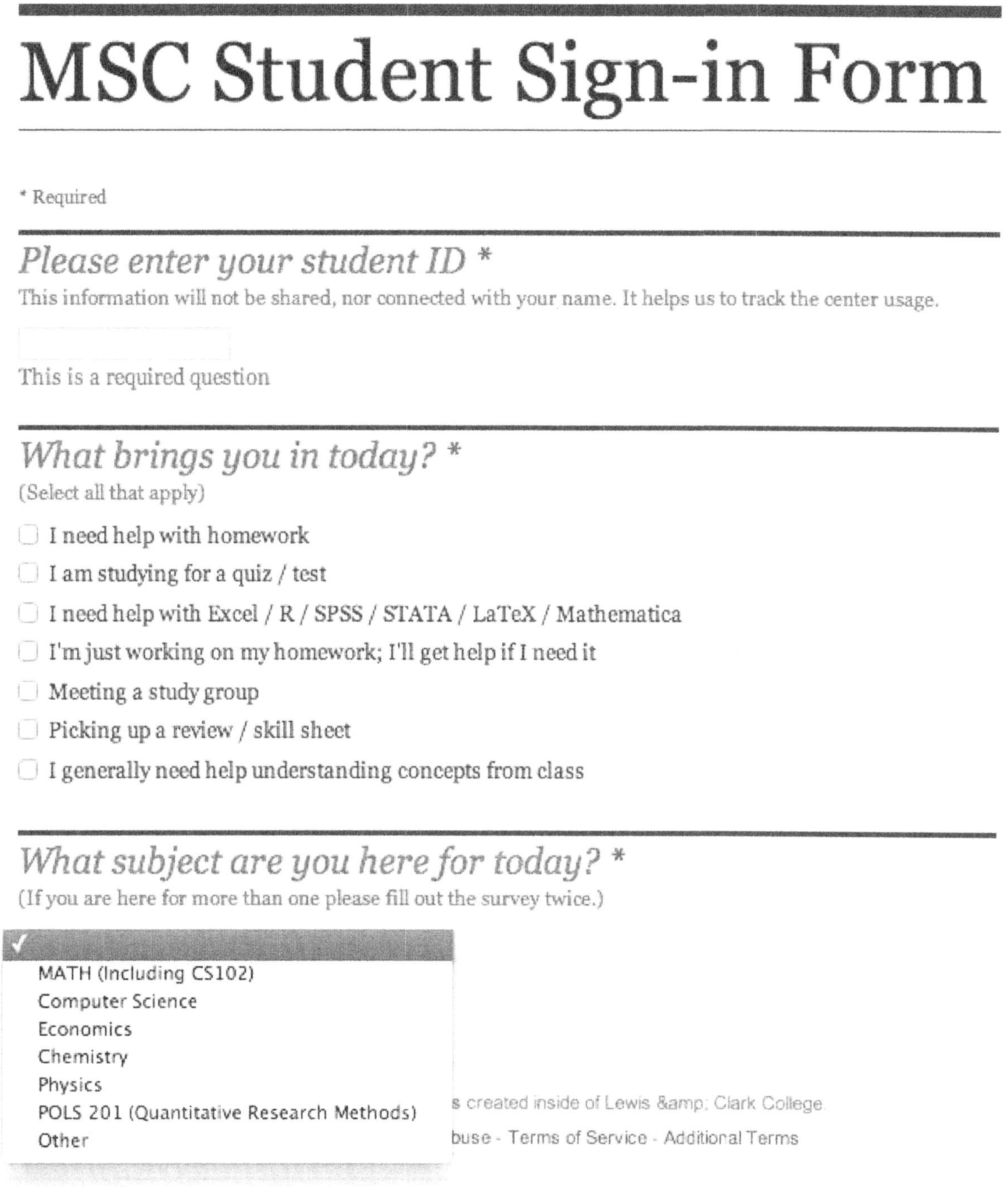

MSC Student Sign-in Form

* Required

Please enter your student ID *

This information will not be shared, nor connected with your name. It helps us to track the center usage.

This is a required question

What brings you in today? *

(Select all that apply)

- I need help with homework
- I am studying for a quiz / test
- I need help with Excel / R / SPSS / STATA / LaTeX / Mathematica
- I'm just working on my homework; I'll get help if I need it
- Meeting a study group
- Picking up a review / skill sheet
- I generally need help understanding concepts from class

What subject are you here for today? *

(If you are here for more than one please fill out the survey twice.)

✓
MATH (Including CS102)
Computer Science
Economics
Chemistry
Physics
POLS 201 (Quantitative Research Methods)
Other

s created inside of Lewis & Clark College

buse - Terms of Service - Additional Terms

Figure 2:

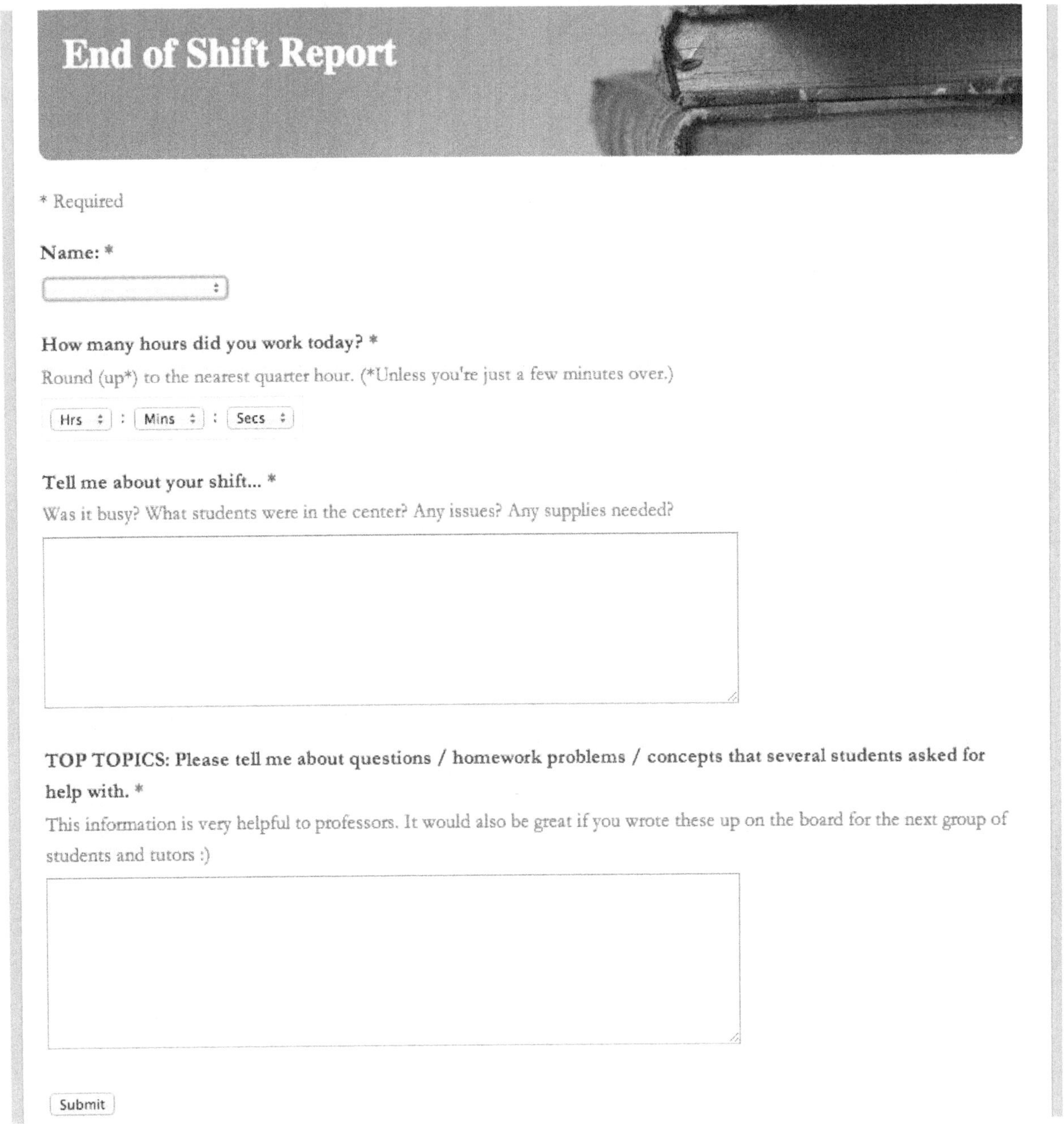
End of Shift Report

* Required

Name: *

How many hours did you work today? *

Round (up*) to the nearest quarter hour. (*Unless you're just a few minutes over.)

Hrs : Mins : Secs

Tell me about your shift... *

Was it busy? What students were in the center? Any issues? Any supplies needed?

TOP TOPICS: Please tell me about questions / homework problems / concepts that several students asked for help with. *

This information is very helpful to professors. It would also be great if you wrote these up on the board for the next group of students and tutors :)

Submit

Figure 3:

Case Study: Math and Computer Science Academic Support Services at Bethel University

© Patrice Conrath,
Bethel University

1 Introduction to the Center

Bethel University, in St. Paul, Minnesota, is a liberal arts institution serving approximately 6,000 students in both undergraduate and graduate programs [1]. This case study will provide information on academic support services for undergraduate students in the College of Arts and Sciences (CAS) which has a student population of about 2,700 students. The CAS offers 88 majors and many students choose to live on-campus all four years. Bethel's College of Adult and Professional Studies and Graduate School also has an emerging center called the Academic Resource Center, which offers weekly math and writing support. This math support is available for four hours on one evening per week (3 - 7 pm) and is staffed by a Master's level tutor. The rest of this document will focus on services offered to undergraduate students.

Math and computer science support services for CAS students include math placement testing, Math Lab, Computer Science (CS) Lab, and individual tutoring in addition to faculty/teaching assistant (TA) office hours. These programs exist under the umbrella of the Academic Enrichment and Support Center (AESC), which also coordinates resources including academic counseling, free peer tutoring, the Writing Center, Business Help Sessions, Science Tutor Help Sessions, Christian-

Suggested Citation: Patrice Conrath (2016), "Case Study: Math and Computer Science Academic Support Services at Bethel University ", http://dx.doi.org/10.5038/9780977674435.ch23 in G. Coulombe, M. O'Neill, M. Schuckers (Eds.) *A Handbook for Directors of Quantitative and Mathematical Support Centers*, Neck Quill Press, http://scholarcommons.usf.edu/qmasc_handbook.

This material is based upon work supported, in part, by the National Science Foundation under Grant DUE-1255945. Any opinions, findings, and conclusions or recommendations expressed in this material are those of the author(s) and do not necessarily reflect the views of the National Science Foundation

ity and Western Civilization TA Sessions, Nursing Exam Reviews, Multilingual Student Writing Support, and Biokinetics and Applied Health Science Help Sessions.

The AESC office physical space includes offices for the co-directors, the administrative support staff, the Writing Center, and a meeting space for group tutoring sessions and staff meetings. The Math Lab and Computer Science Lab meet in three classrooms near the Math and Computer Science Department. The main classroom used by the Math Lab is a high tech student-centered active learning environment with five pods and Writewalls. The pods each seat nine students and have independent displays. Materials for the labs, including answer keys, nametags, an attendance laptop, and laptops for student checkout, are housed in a large rolling cart that is transported to the labs for each session. Tutors hang signs, including course names for grouping students, at each session and a permanent bulletin board outside the main lab gives schedule and tutor information (e.g., which student tutors are available for specific courses). Two of the classroom spaces have stayed consistent for many years, so the areas are known as the Math Lab and Computer Science Lab spaces, even though classes meet in the same rooms throughout the week.

Bethel has philosophically chosen to incorporate full-time faculty to serve as AESC directors, as a portion of their load, to build faculty buy-in and use departmental communication and collaborative best practice support in their representative fields. This model was patterned after Calvin College's academic support model when AESC first began in 1988. Shortly thereafter, Math Lab became a part of the AESC services in 1992, and the Computer Science Lab was added in 2001. The faculty-driven model gives loads for managing AESC services. While a full-time load for Bethel faculty is seven courses (three per semester and one in January), loads for managing AESC services range from one to three courses, with two faculty serving in a larger capacity as co-directors for the office. The AESC staff, which includes a full-time, ten month office manager, meets biweekly to communicate about recruitment, assessment results, logistical needs and brainstorm as a "think tank" about arising needs and challenges. The co-directors of AESC report to the Associate Dean of Academic Affairs helping to provide communication bi-directionally. This model provides naturally strong and positive relationships and communication between departments, Academic Affairs, and the AESC office which enhances staffing, training, assessment, and other efforts.

The AESC mission is to:

> ...develop, administer and evaluate academic support and enrichment services intended to promote students' academic success. We encourage students to identify their strengths and weaknesses as learners, observe the connection between their study strategies and their results, and to develop and implement a plan for improvement.

By including both "support" and "enrichment" in the title, the AESC office intentionally sought to welcome both struggling and advanced students to utilize its services. AESC has been successful in serving the whole student population and many students naturally include AESC services as part of their weekly routine. During the 2014-2015 academic year, AESC services totaled more than 11,500 student contacts, averaging about 4.39 contacts per student in CAS. Of these contacts, 4156 were realized in the Math and Computer Science labs, so these services are a significant piece of the AESC resources.

The Math and Computer Science labs align with AESC's mission and seek to provide environments that will foster growth and support opportunities for students enrolled in introductory math and computer science courses. During a Math Lab session, it's normative to find both a struggling General Education math student discovering new insights from a peer tutor's fresh explanation of a concept and Discrete Math students enjoying a lively theoretical debate over the most concise way to state a proof. Being a tutor in the Math or Computer Science lab is seen as an honor. Tutors enjoy not only the content review that naturally occurs when explaining concepts to others, but also the camaraderie and friendships that organically develop while working together in the busy labs. Bethel's Christian heritage also promotes attitudes of service and encouragement, so the environment is generally positive and non-competitive. In addition, faculty embrace these resources, regularly interacting with the tutors and lab director (their colleague) and encouraging their students to utilize the labs and tutoring services.

2 Center Organization and Services

All math and computer science services are coordinated by the Director of Math and Computer Science Services (Math/CS Director) who has faculty status and reports to both the Math and Computer Science Department and AESC (see Appendix A for reporting structure).

The director is responsible for math placement and skills testing, hiring, and managing tutors who serve in the Math and Computer Science Labs or work individually with students. Recently another lab director for Computer Science was added to support the growing computer science and technology aspects of the position.

Math placement exams are currently used to test students entering Precalculus and Calculus 1. Placement testing utilizes in-house developed exams and is administered using in-house developed online testing and reporting tools. Tests can be taken at any time through an online format in a "self-enrollment Moodle course." Skills testing takes place at the end of the first week of classes, holding these students accountable for learning the prerequisite material. These tests are similar to the placement exam and are considered the first exam in the course. The Math/CS Director oversees testing procedures; hires proctors; updates communication on the testing web site; interacts with the Admissions and Registrar's offices; maintains and updates forms and follow-up materials; and informs and fields questions from advisors, faculty, and students.

Student requests for individual tutors originate in the AESC office, where the student meets with an AESC counselor for an initial assessment. The counselor consults online lab attendance records, the student's professor, and the Math/CS Director to determine the best plan for the student. Some students are referred back to Math Lab or CS Lab, while others are assigned peer tutors who meet individually with students for approximately two hours per week. Link-ups are tracked and managed by AESC support staff, and ongoing assessment for tutees and tutors is initiated by the AESC office using a Google form. The feedback from the student (see Appendix B) is reviewed by the Math/CS Director when the individual tutor meets to discuss progress with their student(s). This meeting is also recorded using a Google form (see Appendix C).

Bethel's lab schedules differ from most large universities in that Math Lab is open for only 13 hours per week, and CS Lab is open 2 hours per week in the fall and 4 hours per week in the spring. However, within these hours, tutors have approximately 1500–2500 visits per semester and about 200 visits in January (three week interim courses). Lab hours include mostly evening hours since Bethel is a residential campus where most students are able to attend these evening sessions. Two sessions per week are provided during daytime hours for commuting students.

The CS Lab serves students in Introduction to Programming, Computer Science 1, and Scientific Computing with the following schedule in the spring semester:

Sunday: 7:45 - 9 pm

Monday: 4:15 - 5:15 pm

Tuesday: 6 - 8 pm

Fall semester CS Lab hours are: Sun. 8 - 9 pm, Tues. 6:30 - 7:30 pm. Math Lab serves students in Math for the 21st Century and Creative Problem Solving (General Education math courses), Math for Elementary Education 1 and 2, Statistical Analysis, Business Calculus, Precalculus, Calculus 1 and 2, Multivariable Calculus, Differential Equations, Linear Algebra, and Discrete Math, along with the Computer Science courses listed above with the schedule below:

Fall and Spring Semester Hours

Sunday: 7:30–9:45 pm

Monday: 5:45–7:00 pm

Tuesday: 4:15–5:15 pm & 8:45–10:00 pm

Wednesday: 8:30–9:45 pm

Thursday: 10:15–11:00am & 6:00–8:15 pm

Interim Hours (January)

Sunday: 8:00–9:30 pm

Monday–Thursday: 6–7:30 pm

Attendance at each session varies from 2 - 70 students, with highest attendance on nights before homework is due (making Sunday, Tuesday, and Thursday nights busiest). See Appendix D for a sample tutor schedule.

3 Staffing, Hiring, and Training

As noted in the previous section, Math and Computer Science services are coordinated by the director who also serves as a faculty member in the Math and Computer Science Department. This director hires all math and computer science tutors, who then become employees of AESC. Some tutors engage in tutoring activities for a variety of disciplines. Tutoring is provided exclusively by undergraduate peer tutors who are often math or computer science majors and have been academically strong in the courses served by the labs. A tutor may be hired for a specific course or to work with various courses depending on their skill and interest.

To begin the process of hiring for a new semester, current tutors are surveyed using a Moodle survey to gather scheduling information and course interest (which is matched to faculty recommendation and references) (see Appendix E). Remaining positions are filled by first soliciting recommendations from Math and Computer Science faculty using a Google document. The Math/CS Director reviews tutor applications and interviews all tutor candidates (see Appendix F), contacts references, and hires the top students to serve as both individual and lab tutors.

Strong returning tutors who demonstrate the ability to multi-task are given "Lead Tutor" positions in which they not only tutor, but also manage lab sessions. These tutors have the following responsibilities: set up materials before each session, start and manage the online attendance tracking system, manage tutors to make sure students are being well served, and give feedback to the director after each session. Two student managers are also hired to help with the creation of the schedule, tutor communication, and logistical support (creating nametags, updating the tutor poster, keeping an eye on materials and signs, and other maintenance tasks). The student managers also serve as tutors, so they are intimately involved in all tutoring aspects and often bring creative ideas to improve our services.

All AESC tutors go through an initial two and a half hours of training and then continue to be trained within the discipline where they serve. General initial training topics include: understanding the AESC philosophy for creating successful independent learners, the role of tutors, tutoring multilingual learners, structuring a tutor session, learning styles, and logistical details such as timesheets and assessment. Tutors also receive continuing online training each new year, and have access to online Moodle modules on a variety of topics that may come up during tutor training (e.g., best practices for supporting students with learning disabilities, three sessions on supporting multilingual students, and using emotional intelligence in tutoring).

Math and Computer Science tutors take part in an additional two and a half introductory hours of training focused on working with math and computer science students and participate in monthly lab meetings, interspersed with periodic email communication as needs and issues arise. Lead Tutors complete an additional hour of initial training, meet biweekly with the director, and have ongoing weekly communication with the director, mainly through email and the online attendance system.

Modes of tutor training include face-to-face meetings facilitated by the director (often jointly led with Lead Tutors), 15-minute Lead Tutor directed training sessions after lab sessions, or written communication/interactive surveys that tutors are motivated to complete online as they are paid for this flexible meeting. Lead Tutors continue to rise to the challenge of leading tutoring services and often give very insightful comments during training sessions for meeting tutoring challenges and figuring out new ways to serve our students well. Topics discussed in training include lab policies and procedures, working with large groups, helping needy students, modeling good study skills, and handling comments about professors and courses.

4 Community Interactions

Our unique organizational structure, where center directors also have faculty status in their representative departments, fosters healthy relationships between the Math and Computer Science services and the Math and Computer Science department. Therefore, communication issues are minimized, and there are no political issues or turf wars. The services are fully embraced, supported, appreciated, and advertised by the Math and CS professors.

To advertise our tutoring services, all students in courses served by the labs are given bookmarks with the schedule and room information on the first day of classes. Professors also include hours and information in their course syllabi and on their course pages in Moodle. We often do a second round of advertisements in each class after the first exam. Tutors share a short video presentation on our services at this point of need (when students first realize they may be struggling) and students listen to the tutor, as their peer, who encourages their attendance.

Math Lab and CS Labs have been well received and are often a part of students' normal work-week schedule. Students come to the labs to work on homework, get homework checked, or ask questions. To help those who may be more hesitant to join a group, extra "welcome tutors" are utilized during the first week of labs, to help new students sign in and find groups. Course signs hung consistently in the same locations also help students to develop patterns so they can easily find tutors and students in their classes. Our latest advertising effort is a positive, short commercial video that will be shown in classes and linked from course Moodle pages and the AESC web site to help students see the lab before they attend (somewhat like students' use of Google earth to spy on a location before they go there!). In the coming school year, we plan to improve our AESC website and testing website to provide fresh, current information (e.g., daily or weekly study tips) so that students will visit our sites on a regular basis and begin to use the preexisting study skill modules which incorporate video, handouts, and PowerPoints with Adobe Presenter (put together by the co-director of AESC for a sabbatical project [2]).

As the Math and CS services provide assistance to students in multiple departments, we gain support from other academic departments such as Business, Physics, and other sciences. No off-campus tutoring occurs through the Math and CS services, however many tutors do take part in tutoring in the community, and Bethel helps to coordinate some of these efforts.

5 Assessment

The AESC office functions as an academic department, so it takes part in Bethel's accreditation process. Bethel is accredited by the North Central Association of Colleges and Schools of Higher Learning. Therefore the AESC office submits a report of activities, assessment results, and documentation of the use of data to improve services, which are spearheaded by AESC's student learning objectives. The AESC faculty team created an assessment outcomes map which specifies when and which assessments are used to assess student learning outcomes. This team recently engaged in creating new assessments in the area of tutor enrichment. This report is reviewed Bethel's administration annually and the accreditation team.

Internal assessments happen in several different formats. As stated above, tutors working individually with students must meet with the Math/CS Director periodically to discuss their tutoring link-up. Those receiving services are also required to submit feedback using a Google form (see Appendix B). The director is able to access these student comments during the meeting, and most students agree to share their comments with the tutor. These comments are fruitful for providing encouragement for the tutor and discussing issues and perceptions of the student. Strategies for improving the tutee's academic achievement (e.g. study skill changes perhaps directing students to the online study skills workshops [2]) and to build the relationship are discussed and meeting notes are recorded on a Google form (see Appendix C). The AESC office manager tracks these Google forms to ensure that appointments between the tutor and director happen. Pay is withheld until the assessment is complete.

Communication between tutors and the director happens at the end of each tutor lab session via an automated e-mail through the online attendance system. As each system closes, an email with attendance figures and tutor comments is sent to the director and any interested professors. This allows for immediate attention to any issues that arise in the session. Lead Tutors also give great feedback from each session and help to tackle any problems or issues that may arise in the sessions. Often, a week before a monthly training meeting, Lead Tutors are contacted to add comments to a Google docs brainstorming document. Lead Tutors share current issues and training needs, and strategies are discussed. Then, each Lead Tutor is assigned a topic to discuss at the next meeting. Lively discussions occur at these meetings as tutors usually have bright ideas for dealing with issues and have an intimate understanding of the student culture and courses, since many of them had the same professors. Also, we find that tutors who may have heard the same exact training topics at the initial training are usually more attentive to the same training content when delivered by a peer tutor using language and scenarios they now understand. Tutors also email professors with comments specific to their courses after each session. This active communication network improves our services and also lets students know that their specific needs are supported and handled with care.

More formal assessment of the labs and tutors happens at the end of each semester. All lab attendees are sent a link to an online assessment using a Qualtrics survey (see Appendix G). Tutors also complete an AESC Tutor Self-Assessment & Enrichment Survey through Qualtrics (see Appendix H). These tools provide valuable comments for encouraging tutors and improving our services and training. The director reviews these comments, implements changes, and often uses direct quotes from the surveys in training materials to add power to a training point.

6 Additional Topics

In this section, two special features of Bethel's Math and Computer Science services will be highlighted along with a current issue we are facing.

In-house Developed Attendance Systems: In 2003, we recognized the need for moving beyond paper sign-ins to an automated system. So, during the fall 2003 semester, students in the Computer Science course, Web Programming, focused their efforts on creating an attendance system. The professor used a competition to motivate student teams to produce a quality product. The winning team members were hired to complete the attendance and reporting application during the spring semester, and it was implemented in the Math Lab during the next academic year. Since then, students were employed to create similar attendance and reporting systems for other labs, providing useful reports and communication for AESC staff, instructors, tutors, and the directors.

The systems needed to be updated, so after reviewing commercial software and finding them deficient, we decided to once again create an in-house developed attendance tool with oversight from our Information and Technology Services department. The tool is now a unified AESC attendance application that provides attendance records, reports, and easy mechanisms for communication. Although the start-up investment is high in this type of application, we have found that it is very beneficial to have customized reports and other functions that fit our needs and unique schedules. Utilizing students to help create the application also makes this option cheaper and provides a wonderful real life application for their Web Programming course.

Strong Community Atmosphere: When discussing tutor recruitment with colleagues from other universities, some have noted that recruiting tutors is extremely difficult. At Bethel, we are usually able to develop a strong team of tutors. Tutoring is viewed in a positive light, and often as an act of service. Those who tutor upper division courses wear this like a badge of honor and majors strive to have all courses listed on the schedule by their name. The professors also encourage students to serve in the labs. If students do well in a class, they are often expected to tutor for that course. Students are assigned a session for the semester along with two to ten other tutors, so they enjoy working together and become close to this group of colleagues as they serve the students together. Several students have developed a love for teaching through their service in the labs. Community is also developed through activities such as student designed t-shirts and Math and CS Lab pizza parties where students create their own math and CS shaped pizzas.

Current Issues – Supporting Students with Skill Deficits: At Bethel, we have recently noticed a downward trend in the entry-level mathematics proficiency of students enrolling in General Education and foundational business classes that require math content. While our professors have raised questions about the quality of some of these incoming students, the admissions personnel and Vice President of Data Enrollment and Management contend that the ACT average has remained relatively stable over several years. They do, however, state that the variance among ACT sub-scores has changed. This seems to be the result of students excelling (perhaps choosing to specialize) in one content area to the detriment of another. Consequently, a student may have outstanding English or reading scores but post extremely low math scores, resulting in a reasonably acceptable ACT composite score. This could then support the observations of both math faculty and admissions personnel that increasing numbers of Bethel students have sub-par ACT math scores accompanied by relatively stable ACT composite scores. There are some interesting observations relating ACT scores to college success found in an "Inside Higher Ed" article, "ACT's Validity Questioned" [3] that relate to this difference between composite and sub-scores.

Bethel offers General Education math courses, but a course sequence for math and science majors begins with Precalculus. Thus, we do not offer any developmental math courses. In order to ensure that students can complete the courses successfully, we currently test for placement into Precalculus and Calculus 1. For these students, Bethel offers two ALEKS preparatory courses allowing them to complete a more in-depth review of Bethel's prerequisite course objectives. We have chosen ALEKS, because it develops a plan for remediation that is tailored to specific student needs (as opposed to more rigidly structured packaged courses). ALEKS also has learning modules that are not "template math" and seem to follow good pedagogical methodology, focusing on conceptual understanding versus procedural fluency. Students in Calculus 1 self-remediate quickly, if needed, in preparation for a "Skills Test" given early in the Calculus 1 course. This test holds them accountable for their preparation. Precalculus students are also given a plan for self-remediation prior to the "Skills Test"; some take the time to do the remediation and some drop out.

Prospective elementary teachers who wish to enroll in the first of two math content courses must also satisfy certain prerequisite skills criteria. Those with ACT or SAT scores must:

- have scored 24 or higher on the math portion of the ACT, or
- have scored 560 or higher on the math portion of the SAT score, or
- successfully complete Bethel's ALEKS Math for Elementary Education Prep course

Those without ACT or SAT scores must:

- successfully complete Bethel's ALEKS Math for Elementary Education Prep course (recommended), or
- score at or above the 80% level on the Math for El.Ed. Pretest

Given our initial success with ALEKS remediation in prerequisite skills for Math for El. Ed. 1, we have become interested in expanding our use of ALEKS to other courses. It is our hope to find ways to better facilitate success in our entry level math courses by channeling underprepared students through appropriate ALEKS courses prior to enrollment. This plan will give students, from a variety of courses, an additional options for shoring up their skills before taking a course.

7 Acknowledgments

I would like to thank Dr. Ruth Nelson, AESC Co-director and Associate Professor of Psychology and Jesse Cabak, middle school math teacher, and former Math Lab tutor for their editing help. Thank you also to Dr. Scott Brown, Professor of Mathematics, contributing editor for part 5, "Current Issues".

We would like to thank the National Science Foundation for its generous support of the QMaSC Workshop through NSF grant number 1255945. Any opinions, findings, and conclusions or recommendations expressed in this material are those of the author(s) and do not necessarily reflect the views of the National Science Foundation.

8 Bibliography

[1] "Bethel at a Glance." http://www.bethel.edu/about/bethel-glance/. Accessed: March 2, 2016.

[2] S. Mulberry, "On-line Study Skills Workshops." https://googledrive.com/host/0B-iVbRlqYrPLVmdLOWtMbkFSdOk/AESC-Study-Skills-Workshops.htm, 2010. Accessed: March 2, 2016.

[3] S. Jaschik, "ACTs Validity Questioned." Inside Higher Ed., June 21, 2011. http://www.insidehighered.com/news/2011/06/21/study_suggests_most_colleges_use_act_inappropriately.

9 Appendix A: Institutional Structure for Bethel Math and Computer Science Services Director

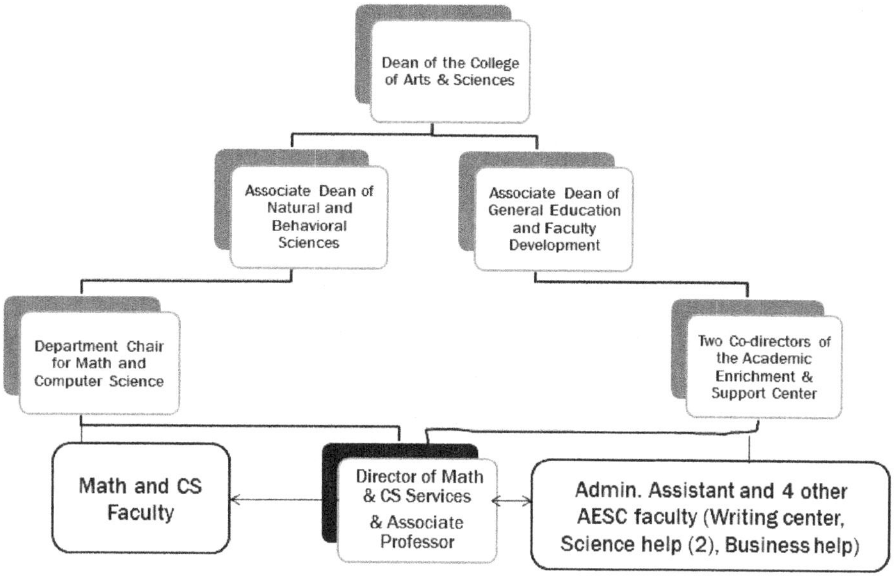

Figure 1: Sample Reporting Structure for the Director of Math and Computer Science Services

Communication and Meetings:

1. **Math and Computer Science Department Chair & Faculty:** regular, daily communication takes place through hallway conversations, meetings and email
2. **AESC Faculty:** bi-weekly meetings in the AESC office and email communication as needed
3. **Associate Dean of General Education and Faculty Development:** monthly reports are submitted by the Math and Computer Science Services Director to the Co-directors of the AESC office who bring the AESC office report to their monthly meeting with the Associate Dean.

10 Appendix B: Student Tutor Relationship Google Form Assessment

How are Things Going With Your AESC Tutor?

Please answer the questions below in regards to the individual AESC tutoring you've been receiving this semester.
* Required

Your Full Name (First and Last): * StudentName Name of tutor: * TutorName

Course in which you are tutored: * Calc. 1

Number of hours you spent with your tutor last week? * 2 hours

Number of hours you spent with your tutor this week? * 2 hours

Do you have a regular schedule for tutoring? * Yes

Did you contact your prof. for help this week? * Yes

Did you attend all classes and labs this week? * Yes

In the last week, how much time did you spend reading your textbook(s) for this class?* 1 hour

In the last week, how much time did you spend reviewing class notes for this class? * 3 hours

In the last week, how much time did you spend completing homework/assignments for this class? * 3 1/2 hours

Are you attending the help session or tutoring lab that applies to your course? Yes

In the last week, how many TOTAL hours did you spend studying? * 8 hours

How do you rate your current understanding of the course material? *

	1	2	3	4	5	
Very confused	○	○	○	●	○	Understand well

How do you rate your ability to keep up in this course? *

	1	2	3	4	5	
Impossible to keep up	○	○	●	○	○	Very confident I can keep up

What are you doing to improve your tutoring sessions and experiences in the course? * {text box}
What teaching techniques did your tutor use that really helped? * {text box}
Have you tried more effective ways to study for your course based on your tutor's suggestions? {yes, no} . If "yes", how are you studying differently? {text box}
What additional support would be helpful for your success in this class? {text box}

Is it OK if we share your responses with your tutor? * Yes

11 Appendix C: Tutor Interaction Assessment Google Form

AESC Tutor Link-up Evaluation Form

This form is used to evaluate the tutor's link-up with their tutee. It is filled out during the tutor's meeting with one of the AESC lab directors or counselors.

* Required

Tutor Name (Your Name): * TutorName Tutee Name: * StudentName

Course in which the student is being tutored: * Precalc. Professor ProfName

Are you regularly meeting with your tutee? *

(•) Yes ( ) No

In the last two weeks, how many hours have you spent meeting with the student? * 2 hrs

How well prepared is the student when they come to meet with you? *

	1	2	3	
Not Well Prepared	( )	( )	(•)	Well Prepared

Please comment on your sessions with this student. How do you generally feel the sessions are going? What does a typical session look like (i.e. what do you do in meetings with the student)? *
{text box}

Do you feel the student has improved in their understanding of course material? *

(•) Yes ( ) No

Do you feel the student has improved their study skills or study methods in the time you have worked with them? *

(•) Yes ( ) No

Do you have any problems or concerns relating to this tutor link-up? *

( ) Yes (•) No

If you answered "YES" to the question above, please explain your problems or concerns.

[✓] Send me a copy of my responses.

12 Appendix D: Sample Math and CS Lab Schedules

Math Lab

Bethel Math Lab, Spring 2016 Schedule

Serving Students in Introductory Math Courses in HC113 & HC 114

Sunday , HC 113 7:30 - 9:45 pm	Tuesday, HC 113 4:15 - 5:15 pm	Wednesday , HC 113 & HC 114 8:30 - 9:45 pm
Jordan Shawn 7:30 - 8:45 pm Andra 7:30 - 8:45 pm Claire Chris Iesha (9 - 9:45 pm) John Zack Carrie (8:00 - 9:00 pm) Ben	**Nathan** Ruth Claire Kirsten Karl	**Kayla** Greta Dan Elise Zack
Sun. 7:30 - 9:45 pm, HC114, Stats.	Tuesday, 4:15-5:15 pm, HC 114, Stats.	Thursday, HC 113 & HC 114 10:15-11 am
Lydia Sarah Jensen	**Lydia** Eli	**Annie** Ella Ryan Dan
Monday, HC 113 5:45 - 7:00pm	Tuesday, HC 113 8:45 - 10 pm	Thursday, HC 113, 6 - 8:15 pm
Alissa Ella Chris	**Kayla** Jazmin Ryan Ben Alyssabeth Jakob	**Annie** Eamonn 6 - 7:15 pm Ben Seth Kirsten JT Ben Anna
	Tuesday, 8:45 - 10 pm, CC 325 Stats.	Thurs., 6 - 8:15 pm, HC114, Stats.
	Lydia Andrew Amy	**Amy** Jensen

Math Lab Director Information:
Prof. Conrath, office phone: ***, email: conrath@bethel.edu

Subs: Holly, Jenna

***Bold Font = LeadTutor
(and first person in each session)

Courses	Precalc./	Multi.	Math for	Math for	Statistical	Linear	Differentia	Computer

{CSLab}

Bethel CS Lab, Spring 2016 Schedule

Serving Students from Computer Science 1 in CC230

Sunday 8-9 pm	Ben Ieasha	Monday 5:15-6:15 pm	Peter Ryan	Tuesday 6-8 pm	Annie Nadia on call: Ben	Sub List:	Kayla Jordan Ben

CS Lab Director Information: Dr. Thomas, office phone: ****, email: ****

13 Appendix E: Moodle Returning Tutor Survey

Current Math Lab & CS Lab Tutor Survey For Fall 2013

Thank you for filling out the Math Lab & CS Lab survey. We will be recruiting new tutors soon, so we want to know if you are returning. Please fill this survey out by **Tues. April 9, 2013, 10:15 am.**

Question #1, Response Required

Please check all that apply:

[x] I would like to work in Math Lab this Fall(courses & times are similar to the spring schedule)

[x] I would like to be an individual math tutor for Fall 2013 semester (this means that I am willing to work individually with at least one student, about 2 hours/week)

[ ] I would like to work in CS Lab this Fall(the course will be Scientific Computing, times will be similar to spring)

[ ] I would like to be an individual C.S. tutor for Fall 2013 semester (this means that I am willing to work individually with at least one student, about 2 hours/week)

[ ] I will not be able to work in Math Lab or CS Lab this next fall (go on to question #2).

Appendix E: (continued)

Question #2, Response Required

If you are NOT returning as a tutor, please state why (for example, "graduating"). Then, you are done with the survey and can submit it after this question. Thank you!

Question #3, Response Required

If you will be returning, please mark all courses you feel confident tutoring:

[ ] Math for the 21st C [ ] Math for El. Ed. 1 [ ] Math for El. Ed. 2

[x] Precalc., Calc. I & Bus. Calc. [x] Calc. II [ ] Multi. Calc.

[ ] Stat. Anal. [ ] Discrete (only fall) [ ] Diff. EQ (only spring)

[x] Linear (only spring) [ ] Intro. to Computing (fall & interim)

[x] Scientific Computing (only fall) [ ] C.S. 1 (only spring)

The following **two** questions relate to **Fall '13 Math Lab**, skip if you won't be working in the Fall

Question #4, Response Required

I would like to work this many times per week in the lab this fall in Math Lab:

[ ] 1 [ ] 2 [x] 3 [ ] 4

Question #5, Response Required

I am available to work the following times for fall Math Lab (we will only schedule you a maximum of the number of times you listed above, please list *all* times you are available):

[ ] Sundays: 7:30-9:45 pm [x] Mondays: 5:45-7 pm [ ] Tuesdays: 4:15-5:15 pm

[x] Tuesdays: 8:45-10:00 pm [x] Wednesday: 8:30-9:45 pm [ ] Thursdays: 10:15-11am

[x] Thursdays: 6:00-8:15 pm (Similar questions follow related to CS Lab)

14 Appendix F: Initial Tutor Interview Form

Bethel Math/CS Tutor Initial Interview - Day/Date/Time:__________________________

5/8/12

Potential Tutor____

1. Which courses are you interested in tutoring? Where did you take these courses and who was/were your profs.? Classes to tutor: Math for the 21st Creative Prob. Solving Math for El. Ed. 1 or/and 2 PreCalc.

Bus. Calc. Calc. I Calc. II Multi. Stats. Discrete C.S. 1 Intro. to Programming Linear Diff EQ

2. Why are you interested in tutoring?

3. What are some of your past experiences that may be useful for a tutoring experience?

4. Why do you think the area you will be tutoring in is difficult for some students? How would you deal with someone who is very anxious?

5. Your job as a tutor involves helping the person you are working with to also become a better student. What kinds of study habits, specific to math, do you have that would be helpful to communicate with other students? (...discuss test preparation strategies too)

6. What do (did) you like about the course(s) you would like to tutor?

7. Describe the tutor training, hook-up, and check-up process. References. Questions?

Indiv? Y N	Ap sent to Vicki?___
M.L.? Y N Hrs that DONT work:	Updated google doc___
	Rxn:
Emailed References?___ Heard back?___ ok? ___	
Tutor notified?___	

15 Appendix G: Qualtrics Bethel Lab Attendee Survey

1. How often do you visit the Math Lab? {responses: I've only been to Math Lab once, Once/month, Every other week, Once/week, 2-3 times/week, 4-5 times/week}
2. Which math course(s) have you received math support for: {responses: all courses served are listed}
3. If you have gone to Math Lab once or twice, please explain why you dont attend the lab more regularly. {response: text box}
4. If you regularly attend the Math Lab, what do you find to be the most helpful for you? {response: text box}
5. If you regularly attend Math Lab, what could we do to improve our services? {response: text box}
6. If your instructor gave extra credit for attending Math Lab, how much did the extra credit incentive influence your attendance? {responses: Not at all. I would have come if extra credit was not offered, Somewhat, A lot, The only reason I attended was for the extra credit, The only reason I attended at first was the extra credit but then I kept coming because it was helpful.}
7. Any other comments or specific feedback? Suggestions? Thank you for taking the time to complete this survey. {response: text box}

16 Appendix H: Tutor Self-Assessment Survey

Bethel AESC Tutor Self-Assessment & Enrichment Survey

A. **How many semesters have you tutored?** Answers: 1 8

B. **What area(s) do you tutor in?** Answers: Math, Computer Science, Sciences (Biology & Chemistry), Business & Economics, Writing Center, Nursing, CWC, Physics, Other

C. **What type of tutoring do you engage in?** Answers: Help Session/Lab, Individual tutoring

D. **Please answer each of these questions as it relates to your tutoring experiences. Thank you for your honest feedback!** Answers for the following questions are: Strongly Agree, Agree, Neither Agree nor Disagree, Disagree, Strongly Disagree, N/A

1. Tutoring increased knowledge in content & study strategies.
2. Tutoring helped me to review and better understand course material.
3. Tutoring helped me make connections from prior courses to current courses.
4. Tutoring has increased my confidence in the subject matter.
5. Tutoring has increased my understanding of effective study strategies.
6. Tutoring has increased my own time management skills.
7. Tutoring increased teaching skills.
8. Tutoring has increased my ability to facilitate discussion or group problem solving.
9. Tutoring has helped me to learn multiple ways of explaining and approaching problems.
10. Tutoring has increased my listening skills.
11. Tutoring has increased my ability to problem-solve, especially in the moment.
12. Tutoring has helped me to ask better leading questions.
13. Tutoring has helped improve my communication skills.
14. Tutoring helped with future vocational preparation.
15. Tutoring has helped me develop and further refine skills for my future profession.
16. Tutoring has helped with my acceptance to graduate school.
17. Tutoring has made me more employable (e.g., internships, summer jobs, post school jobs).
18. Tutoring has further developed leadership skills (e.g., managing an individual link-up or organizing small or large group sessions).
19. Tutoring helped with interpersonal skills & character development.

20. Tutoring has increased my interpersonal skills such as reading students' nonverbals and deescalating students' anxiety and frustration.
21. Tutoring has helped me to develop deeper empathy for students.
22. Tutoring has increased my patience.
23. Tutoring has brought humility to my learning.
24. Tutoring has helped me to work more effectively with other students.
25. Tutoring has positively impacted my overall ability to relate to other students.
26. Tutoring has provided intrinsic benefits.
27. Tutoring allows me to use the gifts God has given me to serve others.
    a) Tutoring has shown me that I enjoy helping students learn.
28. Tutor training (AESC tutor training and training sessions with tutor coordinators) prepared me well for tutoring students.

E. **What additional training would have been beneficial?**

F. **Give any ideas you have for improving the effectiveness of our services.**

G. **If you are discontinuing your work with AESC, explain why. If you are continuing to tutor next semester, explain why.**

H. **Are there any additional benefits of tutoring you have experienced?**

Case Study: The Mathematics & Statistics Workshop at Bates College

© Grace Coulombe,
Bates College

1 Introduction

Bates College is a private, four-year liberal arts college with approximately 1700 students, located in Lewiston, Maine. In the early 1990s, several faculty members recognized the increasing importance of quantitative approaches as an essential part of the intellectual life of the college. Furthermore, they acknowledged the need to provide support for students underprepared to succeed in the natural sciences. As a consequence, the Mathematics and Statistics Workshop (MSW) was created in 2000. Long before the creation of the Workshop, students at Bates were required to satisfy a one course quantitative requirement—a "q-course." It was only with the creation of the MSW that the college endeavored to fully support students in completing that requirement. Therefore, from the outset, the intent was to create a center that would "enable all Bates students, starting from their individual backgrounds, to develop, build on, and take responsibility for their own quantitative skills and understanding." [1] The Math Workshop was aptly named to mirror that of the pre-existing Writing Workshop, which had enjoyed nearly 20 years of successful service to the college prior to the creation of the MSW.

The continuing mission of the Math Workshop is to increase learning in mathematics and quantitative literacy by teaching and tutoring Bates students, as well as supporting the faculty

Suggested Citation: Grace Coulombe (2016), "Case Study: The Mathematics & Statistics Workshop at Bates College", http://dx.doi.org/10.5038/9780977674435.ch24 in G. Coulombe, M. O'Neill, M. Schuckers (Eds.) *A Handbook for Directors of Quantitative and Mathematical Support Centers*, Neck Quill Press, http://scholarcommons.usf.edu/qmasc_handbook.

This material is based upon work supported, in part, by the National Science Foundation under Grant DUE-1255945. Any opinions, findings, and conclusions or recommendations expressed in this material are those of the author(s) and do not necessarily reflect the views of the National Science Foundation

across several disciplines. Although it is clear that the MSW seeks to provide enrichment for all q-courses, few professors currently seek assistance in designing q-specific modules for their courses.

Initially the Workshop was centrally located in the old library building. The director's office occupied a room with a two-way mirror overlooking a seminar room formerly used for the observation of psychology experiments. During that first year, the Workshop room housed only one seminar-style table and six chairs and therefore could not support assisting more than 6 simultaneous drop-in visitors. As such, the MSW was plagued with many problems. Because there was insufficient space for both tutors and students to sit, tutors often had to stand while assisting students. The two-way mirror created an awkward dynamic, as students in the Workshop room often felt they were surreptitiously observed by the director from her office. Furthermore, there was no way to control the temperature in the center, which sometimes averaged in excess of 85 degrees.

In 2001, the Workshop moved to a house on the edge of campus, the former home of the history department. The living room of this house, at 319 square feet, became the Workshop room, filled with one large conference table, two small tables each accommodating 1–2 students, one computer, and one chalkboard. The director's office was adjacent to the Workshop room. Gradually, a few small improvements were made through the addition of another chalkboard, a computer upgrade, and the replacement of two small tables with nicer, but still small tables. Because it was separated from the center of campus by a walk of 3–5 minutes, this Workshop space provided a lovely distraction-free zone for students to focus on their work.

However, the space quickly proved too small for the number of students who needed its services. The Workshop room could comfortably accommodate 12 students, but at peak times it was often packed with 18–20 students. At a few times during the semester the MSW might see in excess of 30 students during a 2–3 hour block for drop-in visits, so that it had students sitting on the floor or in the back hallway near the bathroom. The MSW occupied this house for 11 years.

In August of 2012, as part of a wider college initiative to centralize resources, the Math Workshop relocated to the ground floor of the library. This new space, of approximately 1700 square feet, is configured with 4 large tables, 2 medium tables, two small tables, two semi-private cubicles, two computer stations, a projector and screen, a private conference room, a portable chalkboard, and the director's office. In short, the MSW can now comfortably accommodate at least 50 students and tutors simultaneously. The perimeter walls and support pillars are covered with a combination of chalkboards, bulletin boards, and chalkboard paint to allow for most vertical surfaces to double as writing surfaces.

An iPad is used as a dedicated sign-in registry to track visitor usage. Unfortunately, the MSW cannot afford to hire a tutor to sit at a sign-in desk to welcome visitors, ask them to sign-in, and then direct them to a suitable tutor or working group. Consequently, the Workshop occasionally suffers from inconsistency in sign-in practices. Tutors must be reminded regularly to ensure that student visitors sign in on the iPad.

Figure 1: The Mathematics & Statistics Workshop, 2013

The MSW space is open, with nothing more than a two-foot high bookshelf to delineate the math center from the remaining stacks and group study space on the ground floor of the library. Despite the lack of an enclosing wall, students respect the space as reserved for MSW visitors during open drop-in hours. Outside of drop-in hours, it is common to find a variety of math and science students using the MSW space for group study. Fortunately, the Workshop is not located in or near a designated quiet zone of the library, so the MSW does not receive complaints about the noise level during peak times. The openness and visibility of the Workshop help to promote the idea that the center is available for everyone to use and, most importantly, that there is no shame in seeking assistance.

2 Center organization and services

The Math Workshop is led by a full-time director. The director reports to the Dean of the Faculty and submits an annual report each summer detailing the work of the center. The director also holds an appointment as a lecturer and teaches one course per year. The MSW does not have an assistant director or an administrative assistant. All of the work at the center is completed by the director and, on average, 35 peer tutors per academic year.

Support for students comes in a variety of forms and is provided by peer tutors and the director.

- Drop-in Help Sessions: The MSW has morning (Mon.–Fri. 10:30am-12pm), afternoon (Mon.–Thurs. 1–4pm, Fri. 1–2:30pm), and evening (Sun.–Thurs. 7–9pm) drop-in hours for students seeking help without an appointment. Students meet with tutors or use the dedicated computers pre-loaded with mathematical and statistical software. At this time, budget constraints prevent the Workshop from having an increased number of drop-in hours. Early mornings are reserved for students to schedule one-on-one private appointments with the director in order to accommodate students who do not thrive in a group tutoring environment.
- Attached Tutors for Selected Courses: In addition to the availability of drop-in hours, each

section of pre-selected courses has a tutor specifically assigned to work with the students in that class. The list of such courses includes Calculus I, Calculus II, Linear Algebra, Multivariable Calculus, Probability, Mathematical Statistics, and q-courses as needed — such as Working with Data and Great Ideas in Mathematics. Attached tutors attend class at least once per week alongside enrolled students. Depending upon the amount of group work that takes place in the classroom, the attached tutor may also serve in the capacity of a teaching assistant roaming the room to help answer questions. Students schedule appointments, individually or in small groups, to meet with the attached tutor outside of class. Attached tutors also work at least one evening per week in the drop-in center.

- Private Tutoring: Students may request to work individually with a peer tutor in supported math and q-courses. Most private tutoring appointments take place outside of regularly scheduled drop-in hours. (Formerly, subject specific tutoring was administered and funded by the Dean of Students for non-math and non-q-courses. Starting in the fall of 2016, all non-math tutoring will be offered through the newly formed Academic Resource Commons.)

- Support with Technology: Depending upon the expertise of tutors in a given semester, the Workshop provides tutorial sessions to assist students in the use of graphing calculators and software programs including MATLAB, Mathematica, Maple, Prism, SPSS, and the mathematical typesetting program LaTeX.

Support for Faculty — provided solely by the director.

- Class Visits — Classes may either visit the center or the director will visit classes in order to introduce students to the location and services of the MSW. In each case, students are provided with a bookmark detailing MSW services, location, contact information, and the URL for web resources. For some courses, the reverse side of the bookmark is customized to include useful formulas.

- Diagnostic Quizzes and Subsequent Reviews – Upon request, the director works closely with departments and individual instructors to design non-credit-bearing diagnostic quizzes to assess student proficiency with pre-requisite algebra, pre-calculus, or q-skills necessary for a given course. Each department or instructor identifies a threshold for determining sufficient proficiency of the requisite skills. While all students are welcome to attend the subsequent "just-in-time" review sessions, those students scoring below the predetermined threshold are strongly encouraged, and occasionally required, to attend the associated reviews. The number and duration of review sessions is determined by the amount of material the department wishes to be covered. These diagnostic quizzes serve a vital and proactive role in helping to identify at-risk students early in the semester so that they may be connected with appropriate resources.

For example, the diagnostic quiz for students of Scientific Approaches to Environmental Issues, the methods course for environmental studies majors, focuses on skills using exponents, logarithms, significant figures, dimensional analysis, and linear and exponential models. After the quizzes are scored, the Workshop director and the course instructor discuss the results with the students as well as the resources available, including a targeted review session, for students to improve their proficiency with such skills.

- In-Class Workshops and Guest Lectures — The director works closely with classroom instructors to design and lead in-class workshops or guest lectures as a one-off primer for a given set of topics on quantitative skills to accompany a specific course. Such workshops may also include information on writing with numbers.

  For example, the MSW annually visits a class meeting of Interdisciplinary Studies: Methods and Modes of Inquiry, the methods course for students majoring in an interdisciplinary program. Students read, "Size: It's Personal" Chapter 2 of Blastland and Dilnot's The Numbers Game. This reading jumpstarts the conversation about writing with numbers. The authors charge readers to clarify the significance of numbers by asking, "Is that big?" [2] This discussion, coupled with a short primer on basic statistics, provides the starting point of continued class discussions on constructing arguments making use of quantitative data.

3 Staffing, Hiring and Training

The hiring of peer tutors is a multi-step procedure consisting of an application, an interview, and training. Interested students start by completing an online application to be a tutor, which asks for information about relevant experience, previous coursework, references, an unofficial transcript, and the types of tutoring roles in which the student is interested. There are multiple tutoring roles within the MSW: drop-in tutors, private tutors, attached tutors, and community tutors. Community tutors are those who wish to work with local high school and community college students, such tutors are not paid by the MSW.

The second stage of the application process consists of a two-part interview, which usually lasts about 30 minutes. Prior to the interview, applicants are provided with a list of math problems they may be asked to solve or discuss during the interview. The applicant first meets with the director to discuss details of relevant experience, as well as the reasons the applicant wishes to be a tutor. In addition, the tutor is asked to discuss the solution to a common math problem with the director. Next, the applicant is interviewed by at least 2 experienced tutors. During this part of the interview, the applicant and tutors engage in a mock tutoring scenario to gauge how the applicant will respond to a variety of common issues that might arise in a tutoring session. After the completion of interviews, the director meets with tutor interviewers to discuss each candidate and the associated hiring decision.

The problem-solving and mock tutoring session portions of the interview are crucial in helping to identify potential problem areas for the director to monitor during the first few weeks of work for the new hire. In the cases where an applicant displays considerable nervousness or stumbles through the interview, the applicant may be hired for a probationary period. This tutor will work under close supervision for at least one semester; if acceptable progress is made, then the tutor will be rehired for future semesters. If the tutor does not show improvement, they will not be re-hired.

In short, this two-stage application process coupled with the professorial references arms the director with considerable evidence and subsequent justification on which either to hire or not to hire a given applicant.

Tutors are required to participate in training sessions. The bulk of the training is conducted in a 4.5-hour session. Depending upon scheduling and coordination with other peer-to-peer services on campus, additional 1-hour (or 90-minute) training sessions are offered combining tutors/mentors from the various services. For example, a group training session on ethics is conducted for math, writing, peer science leaders, and non-math subject tutors.

Prior to the initial training session, tutors receive an electronic copy of a tutoring manual addressing topics related to payroll paperwork, submitting time sheets, basic procedures, tutoring goals and responsibilities, general tips, and expectations. Upon arrival at the training session, tutors take an informal quiz to verify that they have read the manual. The 4.5-hour block of training starts with a quick review of selected elements of the training manual, followed by an in-depth series of activities and discussions focused on problem solving, learning styles, strategies for effective sessions, common tutoring scenarios, working one-on-one, working with small groups, and professional behavior.

4 Community Interactions

As a member of the faculty, the director is required to serve on a variety of committees, some for a limited amount of time and others indefinitely. The director has served on the following: the Curriculum and Calendar Committee (4 years), the Learning Commons Advisory Group, Faculty Commons Advisory Board, the Peer Writing Coordinator search committee (2007), and the Writing Specialist for the Sciences and Quantitative Writing search committee (2014).

In 2004, when Bates was considering changing distribution requirements, the director participated in researching quantitative reasoning requirements at peer institutions. A primary focus of this research involved taking a team of 3 faculty members to a MAA-PREP Workshop on Quantitative Literacy Across the Curriculum. After returning from the workshop, the director reported to the general education design teams regarding possible directions to take in strengthening the existing q-requirement. One such suggestion included a two-course q-requirement, entitled Q1 and

Q2, intended to mirror the three-course (W1, W2, W3) writing requirement. This model allowed for Q1 to be completed by means of an introductory quantitative reasoning course (or by passing a q-skills proficiency test). The completion of Q2 would be satisfied by completing any one of several q-courses across the curriculum, but ideally, students would take their Q2-course within with their major. In the end, the design team did not adopt this suggestion, but opted to focus more on other (non-q) features of the general education requirements. As such, the q-requirement at Bates continues to be satisfied through the completion of one course or course equivalent from a pre-approved list. The only change implemented by the design team regarding q-courses was the process by which such courses gain the q-classification. Following the recommendation of the MSW director, the list of criteria for classification of q-courses is based upon readings from *Mathematics and Democracy* [3].

In 2008, the director spearheaded a Dean of Faculty sponsored workshop on "Expanding Q courses: Beyond the Natural Sciences." Guest speakers were brought in to discuss exciting work done in a selection of humanities or social science based q-courses at peer institutions. Despite an attempt to publicize this event widely, attendance was severely limited due in part to the arrival of a major blizzard coupled with what is considered to be a lack of faculty "buy-in" on non-traditional q-courses.

In early 2009, the Dean of Faculty's office charged a committee, including the MSW director, to revisit the q-requirement and its associated assessment. Bates and Bowdoin Colleges were awarded funds by the Teagle Foundation for a 3-year joint project to design a q-skills assessment test administered to incoming first-year students and a strategy for assessing the college quantitative requirement. The Bates-Bowdoin Quantitative Reasoning (BBQR) collaborative project began with a careful study of Bowdoin's existing q-skills test, redesigning, and piloting the new exam. As a consequence, the q-skills test is now used at both Bates and Bowdoin Colleges. The project also resulted in multiple information sessions and workshops (on both campuses) focused on quantitative reasoning across the curriculum, reasoning from evidence, and how faculty advisors might use the q-skills test results to best advise first-year students. While the BBQR project has ended, work continues at each college regarding next steps.

Despite this seeming endorsement from the college administrators to highlight the need for quantitative reasoning across the curriculum, the wider faculty at Bates has yet to fully embrace that goal. The director's attempts to promote quantitative literacy have not succeeded in reaching beyond the faculty in the natural sciences and those social sciences that frequently utilize statistical tools. Consequently, the majority of services emanating from the Math Workshop are focused on traditional mathematics, scientific, and statistical topics. Rarely does the Workshop get to focus on subtle and nuanced instances of quantitative reasoning embedded in contextual environments in the humanities.

Lastly, the director intermittently collaborates with writing specialists, from the Writing at

Bates program, to design and conduct a variety of workshops on writing with numbers or presentation feedback for students in courses with a significant quantitative component. Thus far, the workshops on writing with numbers have been only lightly attended, whereas the workshops providing timely feedback for students a few days before they give a public presentation have found more success.

5 Assessment at the Center

Every five years the Dean of Faculty's office conducts an internal assessment of the Workshop by soliciting letters and surveys from students and faculty who have used the services of the MSW. The responses, combined with past annual reports and MSW visitor data, are used to evaluate the director and MSW services.

Within the MSW, assessment takes a variety of forms. At the end of each semester, online surveys are used to gather information about the quality of services, the proficiency of the both drop-in and private tutors, and whether or not the physical layout of the MSW is conducive to learning. This information is used to improve the services provided and to determine whether new services may needed.

Statistical summaries regarding the usage of the MSW are generated using the visitor registry data gathered from the iPad sign-in app, private tutoring reports, attached tutoring reports, and sign-in data from specialized help sessions and workshops. This data is included as part of the annual report submitted to the Dean of the Faculty each August.

The following is a sub-collection of a data summary provided to the Dean of the Faculty regarding the usage statistics of the MSW in 2012-2013:

- 408 students used at least one service of the Math Workshop.
- 2005 drop-in tutoring sessions were provided to 398 students.
- 517 private tutoring sessions totaling 741 hours were provided to 47 students.
- 91 students scheduled 337 private tutoring sessions with their *attached tutor* totaling 307.75 hours. (In 2012-2013, each section of Calculus I, Calculus II, and Probability was assigned an attached tutor for the semester due to anticipated high demand for private tutoring.)
- 10 students used private tutoring services only, no drop-in hours.
- 3031 combined interactions with students: private tutoring, drop-in help, and formal review sessions. (This number doesn't include informal advising meetings.)
- 76% of *all* drop-in tutoring sessions were devoted to Calculus.
- 54% of the afternoon drop-in sessions were devoted to Calculus.

- 50% of students seeking assistance were first years, 28% were sophomores, 13% were juniors, and 9% were seniors.
- 47 students received *private* tutoring in the following topics: GRE Prep, Pre-calculus review, Calculus I and II, Linear Algebra, Multivariable Calculus, Abstract Algebra, Introduction to Abstraction, Intermediate Microeconomics, Econometrics, and Statistics for Economics.
- 3 consultations provided for Senior Thesis students majoring in geology, biology, and environmental studies.
- The 54 peer tutors employed by the Workshop had majors in Mathematics, Physics, Chemistry, History, French, Spanish, Music, Art and Visual Culture, Biology, Biological Chemistry, Engineering (3-2 program), Economics, and Politics.

6 Additional Remarks

Many of the obstacles encountered in the life the Math Workshop are similar to those of most QMaSCs at similar institutions. Finding a suitable space, both sufficiently large and comfortable, was certainly a challenge. However, patience won out, and the MSW currently occupies a space large enough to accommodate the future growth of program offerings as well as any potential increase in the number of daily visitors.

Likewise, the MSW continues to face challenges in securing sufficient funding. In the mid-2000s, the director rightly argued that math/q-tutors deserved a higher wage compared with other campus jobs that do not require specialized knowledge or formal training. The Office of Student Employment (SEO) agreed to a wage increase to $10 per hour for all trained MSW tutors. Unfortunately, the economic crisis of 2008, coupled with an increased demand for tutors, caused the MSW to exceed its budget for several years in a row. Consequently, the director struck a deal with the Dean of the Faculty (DOF) and SEO to reduce the wage of trained tutors to $8.50 per hour in exchange for an additional $7000 per year in the budget allocated to the MSW for student wages. This budget increase is a testament to the valuable services provided by the MSW and the DOF's commitment to providing academic resources; however, the MSW tutors continue to be underpaid compared to tutors and mentors at peer institutions.

The MSW continues its never-ending campaign of fostering a reputation as a location providing valuable services that are not limited to remedial mathematics. By far, the most effective means of promotion and public relations comes from word-of-mouth recommendations by the students who repeatedly use the Workshop's services. Class visits by the director alone can never compete with a strong testimonial from a friend or classmate when students are considering whether or not to visit the Workshop.

The director regularly posts a "Problem of the Week" on the central support pillar of the Workshop. The first three students to solve the puzzle earn their choice of a puzzle or a lightweight messenger bag bearing the MSW logo. Repeatedly, the problem of the week captures the attention of students who are merely wandering through the ground floor of the library. Most often, students who regularly visit the workshop are among the winners, but the list of previous winners also includes those not enrolled in any math, science, or q-course. In some cases, the winners have inquired about the possibility of becoming a math tutor.

Directors who are new to running a QMaSC or who are creating a QMaSC from the ground up should expect to encounter their fair share of obstacles. A healthy dose of patience is required as a director works to gain faculty buy-in for activities and programs surpassing remedial support, building a reputation for providing valuable services, bargaining for a space that allows for expansion, and advocating for staff and student workers.

7 Bibliography

[1] D. Haines, L. Malloy, C. Schwinn, and A. Scott, "Proposal for the Establishment of the Bates College Mathematics Center." submitted to the Bates College Administration, July 1998.

[2] M. Blastland and A. W. Dilnot, *The Numbers Game: The Commonsense Guide to Understanding Numbers in the News, in Politics, and in Life,*. New York: Gotham Books, 2010.

[3] L. A. Steen, *Mathematics and Democracy: The Case for Quantitative Literacy.* Princeton, NJ: National Council on Education and the Disciplines, 2001.

8 Acknowledgements

I would like to extend my sincerest appreciation to Misty Beck, Writing Specialist in the Social Sciences and Interdisciplinary Programs at Bates for her thoughtful comments on drafts of this case study. Likewise, I am thankful for the support and advice of Hillory Oakes, the former director of Writing at Bates. My heartfelt thanks are offered to Michael Schuckers and Mary ONeill for their comments on the early drafts of this document. Lastly, I would like to thank the National Science Foundation for its generous support of the QMaSC Handbook project through NSF grant number 1255945.

9 Appendix

Links to web forms:

Tutor Job Application

http://www.bates.edu/math-stat-workshop/for-tutors/tutor-job-application/

Tutor Request Form

http://www.bates.edu/math-stat-workshop/for-students/private-tutor-request-form/

Bi-Weekly Private Tutoring Report Form

http://www.bates.edu/math-stat-workshop/for-tutors/private-tutoring-report-form/

Bi-Weekly Attached Tutoring Report Form

http://www.bates.edu/math-stat-workshop/for-tutors/attached-tutor-activities-report-form/

Private Tutor Evaluation Form

http://www.bates.edu/math-stat-workshop/private-tutor-eval/

Drop-in Hours Evaluation Form

http://www.bates.edu/math-stat-workshop/drop-in-eval/

Sample Bookmarks for Students of Calculus

Mathematics & Statistics Workshop

Location: Ground Floor Ladd Library
Director: Grace Coulombe

FALL AND WINTER SEMESTERS

Drop-in Hours: 1–4 P.M. Mon. – Thurs.
Calculus Help: 7–9 P.M. Sun. – Thurs.
Mornings & Fridays by appointment only

SHORT TERM

By appointment only

TO MAKE AN APPOINTMENT

call Grace at 786-8374 or
email gcoulomb@bates.edu

MATH WORKSHOP SERVICES

✦ Drop-In – students are welcome to drop in *without an appointment*, from 1 – 4 P.M. Monday through Thursday to review math topics as needed or to use the computer resources.

✦ Private Tutoring – one-on-one help with a peer tutor at your convenience.

✦ Calculus Help Sessions – nightly drop-in help sessions for students in Calculus I and II staffed by peer tutors which meet Sunday – Thursday from 7 – 9 P.M. throughout the fall and winter semesters.

✦ Attached Tutors for Math 105 and 106 – students may make appointments to work one-on-one with the tutor specifically assigned to their section of Math 105 or 106.

✦ Support with Technology – help with graphing calculators and software programs such as MATLAB, Minitab, Maple, SPSS, and the mathematical typesetting program LaTeX.

For online resources visit our website:
http://www.bates.edu/math-stat-workshop/

Back of Calculus Bookmark

Derivative Formulas

$\frac{d}{dx}x^n = nx^{n-1}$ $\frac{d}{dx}c = 0$

$\frac{d}{dx}e^x = e^x$ $\frac{d}{dx}b^x = b^x \ln b$

$\frac{d}{dx}\ln x = \frac{1}{x}$ $\frac{d}{dx}\log_b x = \frac{1}{\ln b}\cdot\frac{1}{x}$

$\frac{d}{dx}\sin x = \cos x$ $\frac{d}{dx}\cos x = -\sin x$

$\frac{d}{dx}\tan x = \sec^2 x$ $\frac{d}{dx}\sec x = \sec x \tan x$

$\frac{d}{dx}\arcsin x = \frac{1}{\sqrt{1-x^2}}$ $\frac{d}{dx}\arctan x = \frac{1}{1+x^2}$

Derivative Rules

Sum Rule $\frac{d}{dx}(f(x) - g(x)) = f'(x) - g'(x)$

Product Rule $\frac{d}{dx}(f(x)g(x)) = f'(x)g(x) + f(x)g'(x)$

Quotient Rule $\frac{d}{dx}\left(\frac{f(x)}{g(x)}\right) = \frac{f'(x)g(x) - f(x)g'(x)}{(g(x))^2}$

Chain Rule $\frac{d}{dx}(f(g(x))) = f'(g(x))\,g'(x)$

Integral Formulas

$\int x^n dx = \frac{x^{n+1}}{n+1} + C,\ n \neq -1$ $\int \frac{1}{x}dx = \ln|x| + C$

$\int e^x dx = e^x + C$ $\int b^x dx = \frac{1}{\ln b}b^x + C$

$\int \sin x\,dx = -\cos x + C$ $\int \cos x\,dx = \sin x + C$

$\int \sec^2 x\,dx = \tan x + C$ $\int \sec x \tan x\,dx = \sec x + C$

$\int \frac{1}{1+x^2}dx = \arctan x + C$ $\int \frac{1}{\sqrt{1-x^2}}dx = \arcsin x + C$

Natural Logarithm Properties

$\ln(ab) = \ln a + \ln b$

$\ln\frac{a}{b} = \ln a - \ln b$

$\ln a^r = r \ln a$

Useful Trigonometric Identities

$\sin^2 x + \cos^2 x = 1$ $\sec^2 x = 1 + \tan^2 x$

$\sin(2x) = 2\sin x \cos x$ $\cos(2x) = \cos^2 x - \sin^2 x$

$\sin^2 x = \frac{1-\cos(2x)}{2}$ $\cos^2 x = \frac{1+\cos(2x)}{2}$

The Unit Circle

For each point (x, y):
$x = \cos\theta$, $y = \sin\theta$

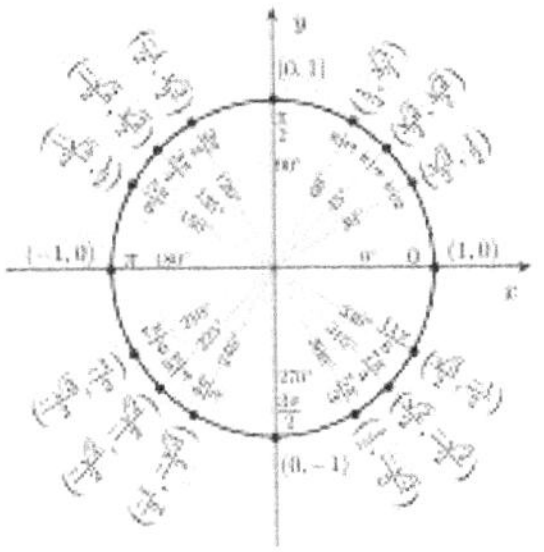

Back of Env. Studies Bookmark

Exponent Properties

- $a^m a^n = a^{m+n}$
- $(a^m)^n = a^{mn}$
- $(ab)^m = a^m b^m$
- $\frac{a^m}{a^n} = a^{m-n},\ a \neq 0$
- $\left(\frac{a}{b}\right)^m = \frac{a^m}{b^m},\ b \neq 0$
- $a^{-m} = \frac{1}{a^m},\ a \neq 0$
- $a^{\frac{1}{n}} = \sqrt[n]{a}$
- $a^0 = 1,\ a \neq 0$

Logarithms

$\log_b x = a$ *means* $b^a = x$,
with $b, x > 0$.

- $\log x$ refers to log base 10, i.e. $\log x = \log_{10} x$
- $\ln x$ is the natural log and represents $\log_e x$, where $e \approx 2.71828$, i.e. $\ln x = y$ means $e^y = x$

Common Models

- Linear: $y = mx + b$
- Slope of line: $m = \frac{\Delta y}{\Delta x}$
- Exponential: $y = ae^{bt}$; a, b constant

Geometric Formulas

A = area, C = circumference,
V = volume, SA = surface area,
P = perimeter = sum of side lengths

- Rectangle: $A = lw$, $P = 2l + 2w$
- Triangle: $A = \frac{1}{2}bh$
- Circle: $A = \pi r^2$, $C = 2\pi r$
- Trapezoid: $A = \frac{1}{2}(b_1 + b_2)h$
- Sphere: $V = \frac{4}{3}\pi r^3$, $SA = 4\pi r^2$
- Cone: $V = \frac{1}{3}\pi r^2 h$, $SA = \pi r(l + r)$

Metric Conversion Prefixes

10^1	deka (da)	10^{-1}	deci (d)
10^2	hecto (h)	10^{-2}	centi (c)
10^3	kilo (k)	10^{-3}	milli (m)
10^6	mega (M)	10^{-6}	micro (μ)
10^9	giga (G)	10^{-9}	nano (n)
10^{12}	tera (T)	10^{-12}	pico (p)
10^{15}	peta (P)		

Case Study: Math Learning Center at Oregon State University

© Thomas Dick,
Oregon State University

1 Introduction

Oregon State University is a land grant research university with approximately 25,000 students. Science and engineering are academic strengths and virtually all undergraduate students take at least three mathematics courses. Hence, the mathematics department has enormous responsibility for lower division (freshman/sophomore) mathematics service courses, as well as upper division courses serving its own undergraduate majors and graduate courses serving its Masters and PhD programs.

The Math Learning Center (MLC) has been in existence for over 30 years and serves as an important academic support for all undergraduate students taking lower division mathematics courses, from College Algebra through Differential Equations taken by STEM majors. The mission of the MLC is simply stated: to support student learning of mathematics. Professor Emeritus Gary Musser was the driving force behind creating the original Mathematical Sciences Learning Center (serving both Mathematics and Statistics) after he had visited several other such centers across the country.

The original center included an extensive assessment service, providing for testing (and retesting) for a number of modular algebra courses as well as the foundational courses for students

Suggested Citation: Thomas Dick (2016), "Case Study: Math Learning Center at Oregon State University", `http://dx.doi.org/10.5038/9780977674435.ch25` in G. Coulombe, M. O'Neill, M. Schuckers (Eds.) *A Handbook for Directors of Quantitative and Mathematical Support Centers*, Neck Quill Press, `http://scholarcommons.usf.edu/qmasc_handbook`.

This material is based upon work supported, in part, by the National Science Foundation under Grant DUE-1255945. Any opinions, findings, and conclusions or recommendations expressed in this material are those of the author(s) and do not necessarily reflect the views of the National Science Foundation

preparing to become elementary teachers. After budget cuts in the 1990s, this function of the center was discontinued. The MLC does provide for proctored make-up midterm and final exams.

The Math Learning Center is in the same building as the Mathematics Department. The close proximity of the MLC to faculty and graduate teaching assistant offices has always been considered to be extremely important. It serves to emphasize the connection between the department and the MLC to both students and departmental faculty and staff.

2 Center Organization and Services

The hours of the Math Learning Center (MLC) during the academic year are 9 am – 5 pm on Monday through Thursday, and 9 am – 4 pm on Friday from the second week of the term through the last week of classes. The MLC also supplies tutors for an evening study area in the university library that is shared with tutors from the learning centers for Chemistry and Writing. (The library tutoring is available on Sunday through Thursday evenings during the term). For each of the nine large residence halls on campus, one tutor is made available one night a week in a designated common study area in the hall.

The services offered by the MLC are:

- a well-lit study area with tables and chairs,
- free drop-in tutoring (provided by GTA's, faculty volunteers, and a few hired undergraduate tutors).
- resources (textbooks, solution manuals, a few graphing calculators, etc.) to check out for use in the MLC.
- make-up testing (faculty or TA's can leave make-up tests in a secure file cabinet at the MLC reception desk; clerks administer and refile the completed test to be picked up at the instructor's convenience),
- a mathematics education resource classroom - used by mathematics classes for prospective elementary/middle school mathematics teachers as well as our Math Excel (Emerging Scholars Workshop) program,
- a computer lab that is available to undergraduate math majors, math graduate students, and faculty teaching classes using computers.

At the evening "satellite" sites (library and residence halls), only tutoring is provided.

3 Staffing, Hiring, and Training

A tenured faculty member serves as the faculty director for the MLC. The faculty director oversees the modest budget, coordinates scheduling of the Emerging Scholars workshops (Math Excel program, described in more detail below), and serves as communication liaison with other units on campus (library, registrar, first-year student programs, and other academic units). Day-to-day oversight responsibilities of the MLC are an explicit part of the position descriptions of three fixed term (year-to-year contract) senior instructors, representing 25 – 50% of their full-time load, the remainder of their responsibilities being regular classroom teaching. Their duties include hiring/scheduling tutors and helpers.

The front reception area desk of the MLC is staffed by work-study undergraduates. The clerks check out resource materials to students, monitor the small (25 seat) instructional computing laboratory in the MLC, and proctor make-up examinations in a special partitioned area of the MLC.

The primary function of the MLC is to provide drop-in tutoring assistance across all of our lower division mathematics courses. Usually 4–5 tutors are available each hour. Approximately 55 graduate students having teaching assistantships are required to put in three hours per week tutoring in the MLC as part of their regular responsibilities. Faculty tutors are volunteers, and 10–12 faculty tutors volunteer one hour a week in the MLC each term. Faculty are encouraged to view an hour in the MLC as an "office hour away from the office," where it is perfectly acceptable for them to let their own students go to the "front of their line" during their MLC hour.

There is a small budget to hire a few undergraduate tutors (5–10 per term), and these students are expected to have completed at least first-year calculus. If undergraduate tutors are interested in earning general course credit instead of pay, they may register for an upper division variable credit "projects" course, with 1 credit awarded for every 3 regularly scheduled weekly hours worked in the MLC throughout the term. For example, if an undergraduate works 6 hours per week as a tutor in the MLC, he or she would earn 2 credits through the projects course.

Students are limited to 10 minutes with a tutor if there is a line of students waiting, but they can simply go back to the end of the line. There are lines of students waiting to see tutors, especially during weeks when courses are having midterms.

There is no special training for tutors. As part of our regular teaching orientation for new Graduate Teaching Assistants (GTAs), we do provide some instructions on interacting with students during their scheduled MLC tutoring hours.

MLC helpers are instructed to:

1) start out by asking what course the student is taking (so the tutor will have some sense of what mathematical techniques are appropriate — for example, a student in College Algebra should not be shown how to use calculus to find the vertex of a parabola);

2) ask the students to talk through their thinking and to describe their progress on a mathematical problem up to the point where they feel they are "stuck" — this talking process gives the tutor not only a chance to diagnose the student's difficulty, but also helps the student to articulate the mathematics (remarkably, this "talk through your thinking" instruction often

results in students recognizing an error themselves);

3) refrain from the urge to simply work through a problem for the student, but rather to talk them through it ("don't use your pencil – let them use theirs!").

The MLC front desk clerks (undergraduate work study students) do gather hourly statistics on usage of the center by noting the number of tutors available as well as the number of students in the MLC at that time. Normally, 20–40 students are in the MLC every hour with 4–6 tutors available. This provides us with a rough measure of usage that we might call "student-hours" in the MLC, and we estimate our usage at 12,000 student-hours per quarter term, or 36,000 student-hours per academic year.

4 Special program: Math Excel (Emerging Scholars Workshop)

The Math Excel program at Oregon State University began in 1998 with initial seed funding from OCEPT (Oregon Collaborative for Excellence in the Preparation of Teachers). It is based on Uri Treisman's Emerging Scholars Workshop model, which has been implemented on many college campuses over the last thirty years. Treisman's Emerging Scholars model is based on groups of students meeting regularly in a social atmosphere to work collaboratively in solving mathematics problems related to their introductory coursework. Oregon State University's implementation of the Emerging Scholars Programs was patterned closely on (and named after) the University of Kentucky implementation - Math Excel. The Math Excel program has been sustained long past its initial seed funding, and the assignment of graduate teaching assistants as Math Excel workshop leaders is simply considered a particular type of GTA assignment.

The primary responsibilities of a workshop leader include the design of a session's worksheet, as well as the facilitation of students' problem solving efforts during the workshop session itself. Math Excel workshop sessions at Oregon State University are offered twice a week for selected lower division courses, including College Algebra, Precalculus, Differential Calculus, and Integral Calculus, and on occasion, Multivariable Calculus and Differential Equations. Duncan and Dick [1] documented the success of the program relative to student achievement, averaging approximately half a grade point higher than predicted (by mathematics SAT scores) over 19 different sections of Math Excel during the first two years of the program (see [2] and [3] for additional discussion of the experiences of our Math Excel program).

5 Bibliography

[1] H. Duncan and T. Dick, "Collaborative workshops and student academic performance in introductory college mathematics courses: A study of a treisman model math excel program," *School Science and Mathematics*, vol. 100, pp. 365–373, November 2000.

[2] T. P. Dick, "Lessons learned in a math excel workshop: The importance of maintaining high cognitive demands," *Journal of Mathematics and Science: Collaborative Explorations*, vol. 6, pp. 65–74, Fall 2003.

[3] C. M. Hake, M. L. Little Crow, and T. P. Dick, "Persistence in mathematics by underrepresented studentsexperiences of a math excel program," *Journal of Mathematics and Science: Collaborative Explorations*, vol. 6, pp. 87–102, Fall 2003.

[4] Oregon St. University, "Math Learning Center." `http://www.math.oregonstate.edu/mlc`.

[5] Oregon St. University, "Math Excel Program." `http://www.math.oregonstate.edu/mlc_excel`.

Case Study: The William B. Law Learning Commons at Tallahassee Community College

© Vera L. Mayes,
Tallahassee Community College

1 Introduction to the Center

Program History

Tallahassee Community College, located in Tallahassee, Florida, serves approximately 14,000 students. TCC offers A.A. and A.S. degrees and certificate programs. The college has partnerships with five universities, making it possible to obtain a four-year degree on the TCC campus.

Until 2008, Tallahassee Community College had four separate learning centers: The Reading Center, the English Skills Center, the Math Center, and the Writing Center. The Reading and English Skills Centers were designed to serve only students enrolled in Academic Support (Developmental) classes. The Math Center and the Writing Center were designed to serve the entire campus. Five years ago, the centers were joined in a centrally located building that had been redesigned to house all four of them. Because the Learning Commons was designed to serve the entire campus, other services were added, including academic computing (open access computers), as well as student success advising. The Library is co-housed in the same facility.

Program Location and Facility

The Learning Commons adjoins the Library and is centrally located on campus next to the Student

Suggested Citation: Vera L. Mayes (2016), "Case Study: The William B. Law Learning Commons at Tallahassee Community College", http://dx.doi.org/10.5038/9780977674435.ch26 in G. Coulombe, M. O'Neill, M. Schuckers (Eds.) *A Handbook for Directors of Quantitative and Mathematical Support Centers*, Neck Quill Press, http://scholarcommons.usf.edu/qmasc_handbook.

This material is based upon work supported, in part, by the National Science Foundation under Grant DUE-1255945. Any opinions, findings, and conclusions or recommendations expressed in this material are those of the author(s) and do not necessarily reflect the views of the National Science Foundation

Union and the main administration building. It is divided into two floors, with the content areas of math, science, accounting, and computer programming served on the first floor and the communications areas of reading, English, writing, and ESOL (English as a Second Language) served on the second floor.

A portion of an existing building on campus was redesigned specifically as a one-stop learning assistance facility. It is a well-lit, thirty-five thousand square foot, comfortable space with a variety of types of seating, flexible table arrangements, six study rooms, three classrooms, staff offices, and a multitude of computers for student use. The Writing Area is located in the Library section of the building next to the Reference Department to facilitate collaboration between the two areas.

The Specialists' (content area) offices are centrally located and designed in an open format to foster collaboration among the staff.

Program Objectives

The TCC Learning Commons strives to be a comprehensive, integrated learning center that provides assistance and resources to all TCC students. The Learning Commons offers students and faculty a broad range of services, including diagnostic assessments, learning materials, electronic resources, individual conferences, one-on-one and small group tutoring, whole-class support, workshops and seminars, success strategies, technology support, and individual learning plans. Throughout the Learning Commons, students have open access to computers, computer applications, and technology support. Tutoring is available in a range of content areas, including math, science, accounting, economics, reading, writing, language skills and support for non-native speakers of English. Distance delivery of learning resources is provided by the Learning Commons web site. A searchable portal site, the new Virtual Learning Commons, launched a beta version in the fall of 2013.

Services and resources are available in a variety of modes to meet different learning styles and demands of students. The aim of the Learning Commons is to be proactive in helping students identify their learning needs and then use the resources and services that are available to improve their academic performance. The Learning Commons fulfills its mission by its commitment to high academic standards; effective use of technology; integration of resources and support services; just-in-time delivery of services; a diverse, trained and professional staff; and faculty participation and confidence.

2 Center Organization and Services

Reporting Lines

The Learning Commons is a separate entity on campus as part of a recent reorganization plan. In 2013, the school created a new position, Director of the Learning Commons, who reports to the Provost of the College and is part of the executive team of the college. The two Learning

Commons Coordinators (one for Math and Science, one for Communications) report to the Director. The Provost, in turn, reports directly to the college president. The College president reports to the Board of Trustees, which is responsible for establishing college-wide positions and funding. Each floor of the Learning Commons has a coordinator and an assistant coordinator. Content area specialists bring curriculum knowledge, best practices in education, technology expertise and the ability to work across campus with faculty and staff. Other Personal Services (OPS) staff members are hourly, temporary employees hired to flesh out content areas as well as meet peak demand time frames, and serve as receptionists. All OPS staff and content area specialists report to the coordinator or assistant coordinator on their floor. The specialists assist the coordinator in supervising and training the work-study student staff (mostly receptionists) and OPS tutors. In addition, a specialist for technology is located on each floor, and both supervise the OPS people responsible for providing technology assistance.

Sources of Funding

The College General Fund provides most of the funds for tutoring and technology. Some of the tutors are paid through Perkins funds for specific content areas. Some of the student staff members qualify for the Federal Work-Study Program.

Services and Students Served

Learning Commons' services include content area tutoring in math, science, accounting, economics, computer programming, computer software assistance, reading, English skills, writing, ESOL skills, and a variety of other areas depending upon faculty volunteers from various departments on campus. Assistance includes self-help, brief staff-assisted, and intensive staff-assisted services to assist students in improving their academic performance. Drop-in services come with support provided by staff members, if needed. Students can have half-hour writing conferences, and several of the gateway math courses can have one-on-one or small group conferences available to them, too. Developmental courses at the first level in math, reading, and English skills have weekly labs that provide faculty-guided use of learning resources in a classroom setting, as well as staff-guided use of learning resources.

The Learning Commons was designed to encourage student collaboration. Students frequently use the commons to meet in small, collaborative working groups to learn and to apply specific academic skills, with staff support provided when needed. Furniture can be easily rearranged, computers are placed in pods to better facilitate project collaboration, and moveable whiteboards are available throughout the space. Study rooms are also available for student study groups or for quiet independent study.

Workshops are offered that involve presentations by staff or faculty members on content knowledge or on topics related to improving student academic performance; these workshops may include guided group use of learning resources.

The Learning Commons is open to all TCC students. In addition, students enrolled in other programs on campus, such as High School Equivalency Diploma Program (GED), Adult Basic Education (ABE), and the University Partners (Barry University, Flagler University, Embry-Riddle Aeronautical University and Saint Leo University), are also permitted to use the services that are provided by the Library/Learning Commons.

3 Staffing, Hiring, and Training

Staffing

Permanent staff of the Learning Commons includes a Director, two Coordinators (one for each floor), an Assistant Coordinator on each floor, and Learning Specialists whose content areas match those of the content tutoring provided. Temporary staff includes OPS tutors that cover content and technology needs, receptionists, and work-study students. There are approximately 100 employees during a typical semester.

All professional staff members have degrees that reflect their content area and are chosen because they demonstrate the ability to mentor and train the OPS staff. They are also required to have a minimum of one year of teaching experience or experience in a learning center.

Some of the developmental classes have labs that meet weekly in the Commons, and Learning Commons' staff assists during these sessions. In addition, faculty members are encouraged to host some of their office hours in the Commons so students can meet them on a common ground. Faculty members also help by presenting review workshops, especially for math and science courses, just prior to final exams. Staffing levels are based on usage and demand. Although writing conferences are available at all times when the Commons is open, tutoring for other content areas is not and the staff for those areas have specific availability schedules, which are posted in-house and on our website.

Hiring

Permanent staff members in the Learning Commons are hired after being recommended by a selection team (led by the appropriate coordinator) and finalized by the recommendation of the Director. The Learning Specialists take a content test in their chosen area to ascertain their current level of ability. These are board-approved positions, and they are evaluated on a six-month, and then a yearly, basis. OPS applicants take tests in their chosen content fields, must have two recommendation letters, and must pass an interview that examines their philosophy and ability to tutor their content area. Their ability to communicate clearly is also evaluated during the interview. The screening tests that the potential OPS tutors take in the qualitative areas have been designed in coordination with the faculty, so they have a high degree of influence on the quality of the tutors hired.

Training Guidelines

Training in the Learning Commons is based on the CRLA (College Reading and Learning Association) model. We are currently certified at Levels I, II, and III. This means that the topics must meet CRLA criteria as a minimum, but we can still add other topics as we deem them appropriate. Training is mandatory in the Commons. After tutors complete the required Levels I and II, they are then put in what we call the 2+ group. This level continues their professional development, and allows them to develop a project each semester that will benefit the Commons. Level III is by application and acceptance only. The first semester of Level III is based on the CRLA topics, and the second semester is an internship.

A training day is scheduled at the beginning of each semester involving issues of concern for all tutors. It includes cross-training between floors to better serve students and faculty. Training day includes information about the specific content areas, as well as activities designed to foster relationships throughout the Commons. External consultants are sometimes needed and hired to provide specialized training for resources or services. Staff training continues throughout the semester on a regular basis, either weekly or bi-weekly, depending upon the content area. Since there are so many staff members to train, each floor divides its staff into small groups or cohorts. Cohorts meet weekly or bi-weekly, depending upon the level and the number of topics to be covered. Time is tracked and entered into a database designed specifically to track topics, length of time, mode, and completion of tasks, to determine if requirements have been met. [1]

Learning Specialists or coordinators are designated as training facilitators. Presentations are done by a variety of individuals including the facilitators, faculty, other campus staff and the tutors themselves. Some training assignments are done online, or as specific writing assignments. The presentations are based either on best practices in training, leadership or mentoring or on specific content areas as guided by CRLA requirements. The Learning Commons developed a WIKI where staff can define CRLA topics, develop lesson plans, define learning outcomes for each session, and develop ways to assess those outcomes. The training facilitators for each cohort can easily access the lesson plans and materials for each training session. In this way, we are assured that the same material is presented to all cohorts. In addition, this practice stores information that is readily accessible for tutors to make up missed sessions. Tutors and staff are given the opportunity to evaluate sessions and to provide input for future sessions.

4 Community Interactions

The Commons' permanent staff members serve on a variety of committees and initiatives across campus, including grant initiatives and leadership committees. The staff participates in campus trainings along with faculty and other staff on campus, and they are sometimes called upon to do presentations across campus. Each staff member serves as a liaison with both full-time and part-time faculty. They meet with faculty members to discuss resources, the needs of their students, and what is needed in the Commons to support what is done in the classroom. Because the Commons/library is a hub for students, the campus police make a point of routinely patrolling the facility to keep it a safe place. Support from all units on campus is critical to keep the facility functional and attractive. As mentioned before, the Learning Commons is open to all TCC students, including students enrolled in other programs on campus.

5 Assessment at the Center

The Learning Commons was developed in part because the data collected showed the need for more space and resources. A powerful part of the data showed the effect that the old centers had on success rates.

Data are collected on number of visits, length of visits, time of day, and classes. The data are fed into the college's mainframe so that reports can be run to meet a variety of needs. The IT team has developed some basic reports that pull student data from the mainframe together with attendance data from Accutrack. Table 1 gives a snapshot of Usage of the Learning Commons and Its Growth [2]

Table 1: Snapshot of Usage of the Learning Commons and Its Growth, [2]

	Fall 2008	Fall 2012
Weekly average number of visits	3,995	6,482
Courses supported	220	383
Unique students	7,180	8,454
Total Visits	62,388	*100,475

* *Prior to combining in one facility, this was the level of usage for all four centers combined in one year.*

When the cohort data does not show a positive correlation between success and visits, the Commons' staff will investigate resources still needed. They will also collaborate with faculty to look at continued student needs. In addition, other data such as usage of resources and attendance at workshops is collected and analyzed. Yearly reports track interactions with faculty, types of technology available, divisions that are collaborated with, along with modes of resources for each division and course.

Table 2: Comparison of Success Rates, [3]

	Fall 2012	
	Pass Rate no conferences	Pass Rate 3+ conferences
Core English classes	68%	90%

	Pass Rate no visits	Pass Rate 3+ visits
Core Math/Science classes	57%	70%

Table 3: Fall 2011 Success Data Based on Use of Anatomy Models in the LC,[3]

	No Checkouts	1–3 Checkouts	4+ Checkouts
BSC2085L	54.4%	72.2%	81.8%
BSC2086L	81.7%	82.1%	94.4%

Both student and faculty surveys are used to determine satisfaction with services and to determine future needs. The surveys allow the staff to see if students feel that they are given strategies, not just answers, in their interactions with staff. The surveys ask faculty members if the Learning Commons has sufficient resources for their courses. An official program review just implemented on campus will also include the Learning Commons as a unit. It will track the success of the Commons in meeting goals, and will be repeated every three years. In addition, one of the college's main QEP (Quality Enhancement Plan) goals has included the Learning Commons and the support for students that it provides, so that the Southern Association of Colleges and Schools (SACS) reports include data from the Learning Commons. The Learning Commons retains an advisory team that consists of staff, faculty and administrators. This group meets once a semester and considers feedback from the different divisions and across campus. It also allows the staff to highlight new services and talk about needs on an ongoing basis.

6 What Makes the Learning Commons Unique?

The faculty and students on campus embrace the Learning Commons. Since opening five years ago, usage jumped astronomically (in the math area alone: 242%) and the faculty presence has increased. Faculty work with the staff to determine ways that the commons can help students get access to resources such as textbooks, handouts, anatomy models, and calculators. Since the goal is to help students develop into more independent learners, it is a great collaboration. The faculty also host review sessions or bring classes over for orientation or tours. The campus feels that this is a good resource and is receptive when asked to help support the Commons.

Another goal of the Learning Commons is to have a diverse staff, this includes diversity in educational background as well as ethnicity. This goal stems from matching the college's mission of diversity. The college has students from 86 different countries and actively recruits globally. It recently set up an International Students Services office to assist these students with their unique challenges. The administration also understands how important this academic support is to the students and to the college's efforts at retention. Even when budgets have been cut and positions have not been refilled in other areas across campus, this has not been the case in the Commons.

Representatives often visit the Commons from other colleges and universities to see how all the services and resources have been pulled together, and the Commons is used as a model. Elected representatives have visited to see how the monies are being used, and they are delighted to see students using the facility and the resources. Data show that since the Commons started five years ago, the retention and graduation rate across campus has increased by two percent. Students and their usage continue to be tracked to help justify requests for new space, increased budgets, and new positions. The Learning Commons will need to adjust to a new house bill that takes effect in the fall of 2014 and restructures developmental instruction in Florida. Data will be used to build a case and to ask for new money to support new services.

One of the last things that we feel makes us unique is that we are never done. We add new resources and services every year. We try to be responsive to students and their requests for services. We are always searching for excellence.

7 Bibliography

[1] V. Mayes, G. Schaberg, and M. Hamilton, "CLRA application, level 2," 2013.

[2] V. Mayes, S. Sampson, G. Schaberg, and M. Hamilton, "Learning commons end of year report 2012/2013," 2013.

[3] V. Mayes, S. Sampson, G. Schaberg, and M. Hamilton, "Learning commons end of year report 2011/2012," 2012.

Case Study: The Spinelli Center for Quantitative Learning at Smith College

© Catherine McCune,
Smith College

1 Introduction

The Spinelli Center for Quantitative Learning is located in a bright, open space on the second level of Neilson Library, the main humanities library of Smith College. It was originally called the Quantitative Learning Center (QLC), initiated in 2007 through a matching grant from Unbound Philanthropy which helped fund the center's first two and half years. Smith College is a small, liberal arts, college for women, located in the heart of a lively town in Western Massachusetts. Smith has an open curriculum in which there are no distribution requirements, instead almost all requirements are determined by the major. Approximately 2,500 students are on campus in Northampton, while 150 study elsewhere (junior year exchange either abroad or at another college in the USA). The Spinelli Center's mission is to support quantitative material across the curriculum, with the informal motto "We *want* to help you!"

The space occupied by the Spinelli Center was previously home to narrow stacks of documents and books, which means the space has low concrete ceilings, a surplus of good lighting, and some interesting logistical challenges. The center has two tall cubicle offices, five large group-work spaces, and six computer stations. Four of the group-work spaces have whiteboards and the fifth has a SMART board. The six computers are Macs that dual boot Windows and are equipped with the

Suggested Citation: Catherine McCune (2016), "Case Study: The Spinelli Center for Quantitative Learning at Smith College", `http://dx.doi.org/10.5038/9780977674435.ch27` in G. Coulombe, M. O'Neill, M. Schuckers (Eds.) *A Handbook for Directors of Quantitative and Mathematical Support Centers*, Neck Quill Press, `http://scholarcommons.usf.edu/qmasc_handbook`.

This material is based upon work supported, in part, by the National Science Foundation under Grant DUE-1255945. Any opinions, findings, and conclusions or recommendations expressed in this material are those of the author(s) and do not necessarily reflect the views of the National Science Foundation

same software found in the science computer classrooms, with the addition of statistical analysis programs used in the social sciences. The center has no printer, but students can print downstairs in the library's central computer room.

The Spinelli Center directly supports the departments of Chemistry, Economics, Engineering, Physics, Psychology, and Sociology, by employing tutors attached to particular courses or sections of courses. The center indirectly supports courses in the Department of Mathematics and Statistics through general quantitative tutors and through review workshops. Support to other parts of the curriculum is provided by counseling students working on seminar or senior projects that involve data and data analysis and by collaborating with faculty who wish to include quantitative research methods or data-driven class projects in their courses. Students in the Department of Education and Child Study are supported in their licensure process partly through the Math 101 and Math 103 courses, partly through the workshop series "Basic Math Review for Standardized Tests," and partly through individual tutoring sessions. Spinelli Center staff teach Math 101: Algebra, a developmental math course for students who are not prepared to take traditional college level mathematics courses, and Math 103: Math Skills Studio, a two-week intensive review of pre-calculus and calculus offered during the January Term. Refer to "Review Sessions and Classes" in the next section for descriptions of these courses.

2 Organization and Services

Students primarily access services provided by the Spinelli Center by dropping in during scheduled tutoring hours or during staff office hours. Some students may also contact tutors via email to arrange an individual or small-group tutoring session. Names of qualified tutors are listed on a bulletin board in the center. Instructors post the email addresses of tutors associated with their courses on their course web pages. Tutoring services are publicized on the Spinelli Center website and on some individual course web pages, and tutors make class visits at the start of the semester to introduce themselves. Tutors are also observers on course Moodle web pages, so the tutors can email students directly (and vice-versa). Additionally, review workshops are coordinated with relevant departments who assist in publicizing them. Services of the Statistics Counselor are advertised through email, class visits, and collaboration with instructors. More details related to these services are given below.

Tutoring (http://www.smith.edu/qlc/tutoring.html)

- Drop-in hours for assistance with general quantitative questions during the day (Monday – Thursday, 9am–10:30am and 12pm–5:30pm; Friday 9am – 1pm) are staffed by Q-tutors (general quantitative tutors). Students who attend these hours tend to seek out help with algebra, pre-calculus, calculus, general physics, general chemistry, introductory economics, and introductory statistics.

- Drop-in hours for assistance from a Master Tutor depend on the instructor, course, and tutor. Schedules are regularly updated on the website, with most sessions during the hours 6–10pm, Sunday – Thursday. A student selected as a Master Tutor is attached to each of the following courses: the basic four chemistry courses, the two general physics courses, mathematical methods in physics, the core engineering courses, some elective courses in engineering, the five core courses in economics, statistics for psychology, statistics for sociology, and statistics for government.
- Individual appointments or small-group meetings are available with a Master Tutor, Q-tutor, or with one of the professional staff.

Statistics Counseling

Students can see the Statistics Counselor for assistance on senior projects and seminar projects that require the collection, management, and/or analysis of data. Instructors can ask the Statistics Counselor for assistance on their own work or obtain training for a student assistant on a data-driven project. Upon request, the Statistics Counselor can give guest lectures such as "An Introduction to Quantitative Methods." Assignments and, if needed, grading rubrics can be provided. Typically, the material for the data assignments is taken directly from the course material and from the students taking the course. The Statistics Counselor can assist in planning statistical projects or assignments and guided statistics lab sessions for life sciences courses in which students collect their own data. Examples of departments where the Statistics Counselor has collaborated are: Afro-American Studies, Anthropology, Biology, Economics, Engineering, Environmental Science and Policy, Exercise and Sports Studies, Government, History, Psychology, Sociology, and Studies of Women and Gender.

Review Workshops and Classes (http://www.smith.edu/qlc/workshops.html)

Two-hour review sessions on a single topic are the most popular workshop offered. These include the presentation of content and supervised problem practice. The current topics offered each semester are Lines, Logarithms and Exponents, Trigonometric Functions, Differentiation, and Integration. Topical Workshops are scheduled in coordination with instructors in physics, chemistry, and mathematics. For example, the review on integration is scheduled to coincide with the physics topic "Center of Mass."

A series of nine two-hour sessions reviewing basic math topics that are commonly needed in studying for standardized tests (GRE, Massachusetts Teachers Exam for Licensure, etc.) is regularly offered. Seniors and students majoring in education or students wishing to become licensed to teach in Massachusetts comprise the target audience for basic mathematics reviews. The sessions are fast-paced, densely packed with material, and no time for solving problems is given. The hope is for students to use the opportunity to do a rough sweep of mental cobwebs and to realize that they need to spend quality time with a study guide and one of the Q-tutors.

Instruction on developmental math topics is given in the four-credit course MTH 101, generically labeled Algebra. The Spinelli Center's Director teaches this course each semester, spending deep, quality time with a few topics (arithmetic, ratios, linear equations, lines, simultaneous equations, triangle trigonometry) and focusing on changing student attitudes around mathematics through group work and coaching. The course attracts a growing number of education majors and social science majors, in addition to the non-traditional aged students for whom the course was originally intended, and also students from a nearby non-traditional college. Math Skills Studio, Math 103, a "boot camp" for review of pre-calculus and some calculus topics offered during the January Term, is co-taught by the Director and the Quantitative Skills Counselor. Math 103 was originally intended for engineering students needing review of pre-calculus and some calculus topics in order to succeed in the general physics courses. The course also welcomes students needing review of pre-calculus in order to succeed in calculus, students needing review of calculus to succeed in intermediate economics courses or in calculus II, or students needing a mathematics course for teacher licensure. The course includes 32 contact hours over two to three weeks. Each session consists of content review and supervised problem solving. Instructors in courses from other departments (for example, Environmental Science and Policy or Sociology) occasionally request a class visit from the Director to give "An Introduction to Excel" lecture.

Online Resources

The Spinelli Center resources webpage, http://www.smith.edu/qlc/resources.html, contains resources for quantitative content review, for information and inventories related to learning styles, and for math anxiety evaluation and resources.

Q-Skills Assessment

The Spinelli Center worked with a small group of faculty to develop an online Quantitative Skills Assessment Survey that is given to incoming first-year students and to exiting seniors. The survey has become an annual event, and the office of Institutional Research now collaborates with the Spinelli Center in administrating this survey. The results are intended to eventually help departments assess whether students graduate with adequate quantitative skills appropriate to their major and to determine whether there is a problem with the acquisition of quantitative skills during the students' time at Smith College.

3 Staffing, Hiring, and Training

The professional staff of the Spinelli Center consists of the Director, the Quantitative Skills Counselor, and the Statistics Counselor:

- *The Director* is a 10-month position that is purely administrative (not faculty). The director reports to the Associate Dean of Faculty, and the position is currently held by someone with a PhD in pure mathematics. The Director's duties can be roughly grouped into center administration (50%), faculty/student support and tutoring (20%), and teaching (30%).

- *The Quantitative Skills Counselor* (QS Counselor) is a 9-month position of 23 1/3 hours per week (0.5 full-time equivalent for 12 months). The QS Counselor reports to the Director, and the position is currently held by someone who already holds a masters degree in geography, holds a bachelors degree in mathematics/environmental science, and is pursuing a masters in biology. The QS Counselor's duties can be roughly grouped into tutor administration (45%), curriculum support (25%), teaching (20%), and center administration (10%).

- *The Statistics Counselor* is an hourly position of approximately 10-12 hours/week for about 35 weeks. The Statistics Counselor reports to the Director, and the position is currently held by a graduate student from a nearby university. The Statistics Counselor duties can be grouped as curricular support (65%), administrative (15%), outreach (15%), faculty development and support (5%).

The remaining employees of the center are student tutors (approximately 30 per year):

- *Q-tutors* (general quantitative tutors) are student tutors with a broad background in quantitative courses, usually including at least the first semester of calculus. In addition, Q-tutors often have experience in multivariable calculus, discrete mathematics, or linear algebra, and usually have introductory knowledge in two of chemistry, economics, engineering, physics, and statistics.

- *Master Tutors* are student tutors attached to a particular course or section. Usually the tutor has taken that particular course.

Hiring

The student tutors are hired through a process designed to select students who understand content, who have good communication skills, and who will be good employees. The Director and QS Counselor meet with departmental liaisons in late March to discuss plans and needs for the following academic year. The liaisons then solicit names of potential tutors, vet those students within their departments, and return their lists to the Spinelli Center by the end of the first week of April. Often a strong student appears on lists from multiple departments. The Spinelli Center staff solicits applications from those students, holds interviews, and offers positions in a formal email after allocating students in a grid of "needed/available" positions. The minimum expectations for tutors are stated in the solicitation letter, on the application, and again in the formal hiring letter. At two of those stages the student acknowledges understanding of the requirements by inserting her initials next to statements such as:

"___ I will attend tutor training on Saturday September 7th from 1-4pm."

The goal is to be done with the hiring process on the last day of classes, usually at the end of April. A few students are asked if they are willing to be back-up tutors in case someone cannot keep their commitment to tutor come the fall. Usually, two tutors need to be replaced over the summer.

Training

Some informal, subliminal training starts during the interview process. For example, applicants are asked, "What do you do when a student asks you a question beyond the scope of the course, to which you don't know the answer?" If the reply does not include acknowledging that she doesn't know the answer or if the reply does not include referring the student to the instructor for the course, the interviewer points out that this situation could be an excellent way for the tutee to begin an interesting conversation with the instructor and to show the instructor the scope of her thinking.

Training formally starts on the first Saturday of the fall semester from 1–4pm. Approximately 65 student tutors and assistants are trained in this session, around 30 of those are Spinelli Center tutors, and the other 35 are tutors who work for the Mathematics, Statistics, Computer Science, Physics departments, or for the Spatial Analysis Lab. For the first training session of the academic year, the tutors are expected to read the Tutor Training Manual and complete an associated online quiz before coming to the first training session (some tutors use the quiz as a guide for reading the manual, which is consistent with our goal). In this way, the tutors have already covered some administrative topics and basic information around tutoring practices before the training, and the training session time can be devoted to more engaging topics. The training continues in the form of monthly meetings over both semesters, and another large session is held on the first Saturday of the spring semester from 1–4pm.

At the fall training we discuss professionalism, interpersonal boundaries, the expectation that tutors should be "aggressively enthusiastic" about offering help and identifying themselves, the importance of administrative tasks (filling out tutor logs and time sheets), being the face of the center and the academic department that they represent during their shift, FERPA and confidentiality, good and bad non-verbal cues, and scheduling. Waiting time, micro-aggression, how to read a quantitative text, and study skills coaching are additional topics covered during lunch meetings and the spring semester training. The longer training sessions at the beginning of each semester typically involve role-play and discussions led by the returning tutors.

4 Assessment

When a student attends a tutoring session or a workshop, the tutor records the student's name and identification number and then enters it into an online Tutor Log (see the Q-Tutor log http://tinyurl.com/q6w7yqz or the Master Tutor log http://tinyurl.com/qcne62y). These logs help in assessing the usage of the Spinelli Center's various services, identifying scheduling problems, and planning the budget for the following year.

A service evaluation survey is given at the end of each semester to assess the Spinelli Center's tutors and services. The timing of the survey is not ideal because it is given during the last two weeks of the semester. A low response rate is expected from a significant part of the targeted population, resulting in selection bias and other issues. Service is not assessed immediately following a tutoring session because many tutoring sessions (engineering, physics, and introduction to organic chemistry) have 20–30 students in attendance during the 7–9pm session, who are intent on finishing homework that is due the following day. It is unlikely the tutees would take time to fill out a survey, even if it is on paper. Additionally, the cost of the labor to convert paper responses to computerized data would be prohibitive.

The Center plans to analyze possible platforms for more timely ways to administer service surveys, during the academic year 2013-2014.

5 Community Interactions

The Spinelli Center has supported students, courses, or instructors in the following departments: Afro-American Studies, Anthropology, Biology, Chemistry, Economics, Education and Child Study, Engineering, Environmental Science and Policy, Exercise and Sports Studies, Government, History, Mathematics and Statistics, Physics, Psychology, Sociology, and Studies of Women and Gender.

The Department of Mathematics and Statistics has its own staff of student tutors called Math TA's and Stats TA's who staff drop-in hours Sunday – Thursday from 7–9pm. The Spinelli Center offers support of mathematics and statistics courses during the day through the Q-tutors and through appointments with the QS Counselor, the Director, and the Statistics Counselor. In this way, the Spinelli Center services complement that of the Math TA's and Stats TA's. The single-topic review sessions offered by the Center are designed to support the mathematics instructors so that less time is spent in class reviewing prerequisite material.

The Spinelli Center has invited the Math TA's and the Stats TA's to the fall tutor training sessions during the past three years, and this has become a good point of collaboration and support. Most recently the invitation has been expanded to include department TA's and student staff from the Department of Computer Science, the Geographic Information Systems (GIS) lab, and the Department of Physics.

The Spinelli Center can be viewed as the academic twin of the Jacobson Center for Writing, Teaching, and Learning, which used to be the sole academic support center. The Spinelli Center already collaborates with the Jacobson Center (especially with the scientific writing counselor), the Wurtele Center for Work and Life (focused on work-life balance and practical aspects of how to be a student), and the Office for Institutional Research. The Director intends to collaborate with the Sherrard Center for Teaching and Learning (faculty development) and some of the interdisciplinary centers recently established at Smith. The Director of the Spinelli Center is a member of the Statistics Program (a group of instructors from across the curriculum who instruct, supervise or use statistics professionally) and is a member of the advisory board for the academic program Translation Concentration (a concentration is similar to an academic minor but is different because it is interdisciplinary; this concentration stretches across many language and cultural studies departments).

The Director of the Spinelli Center has been a regular participant in the Northeast Consortium on Quantitative Literacy (NECQL) since 2008, and the Spinelli Center (then QLC) hosted the NECQL annual meeting in April 2009. Smith is an institutional member of the National Numeracy Network (NNN), the Director has attended two annual meetings of the NNN, and the Director serves as a reviewer for the NNN's online journal *Numeracy.* A team from Smith College attended Carleton's PKAL-QuIRK (Project Kaleidoscope-Quantitative Inquiry, Reasoning, and Knowledge) initiative in October 2009, during which they developed short and long range plans, most of which have come to fruition. The Director has helped or mentored other founding directors of new quantitative centers and has served as an external reviewer for established centers.

6 Acknowledgements

I thank the Smith College Dean of Faculty, Danielle Ramdath for her support and mentoring, my children for their patience with my working on evenings and weekends, and Viola J. Spinelli for all of the ways that she has supported my growth as an administrator and the center's growth as an integral support service for Smith students.

Case Study: Quantitative and Symbolic Reasoning Center at Hamilton College

© Mary B. O'Neill,
Hamilton College

1 Introduction

Hamilton College is a small, highly selective four-year liberal arts college, located in central New York. The college offers students an open curriculum, with no distribution requirements. There are, however, four graduation requirements: 1) complete a major, 2) pass three writing-intensive courses, 3) fulfill the Qualitative and Symbolic Reasoning (QSR) Requirement (one course), and 4) pass the physical education requirement. The college serves approximately 1800 undergraduate students from all over the country, but primarily the Northeast. According to information from the Office of Admission, students come from 46 states and 40 countries; more than one-quarter of the student body consists of diverse students from the U.S. or are international students. The most popular majors are economics, mathematics, government, biology, psychology, history, English/creative writing, and world politics. Hamilton College has a strong reputation of supporting excellent writing, quantitative literacy, and communications skills. Tutoring, collaborative learning, and academic support are looked on favorably and are encouraged as an integral part of student success. Other programs that offer academic support are the Nesbitt-Johnston Writing Center, the Oral Communication Center, Language Center, HEOP (Higher Education Opportunity Program), and ESOL (English for Speakers of Other Languages).

Suggested Citation: Mary B. O'Neill (2016), "Case Study: Quantitative and Symbolic Reasoning Center at Hamilton College", http://dx.doi.org/10.5038/9780977674435.ch28 in G. Coulombe, M. O'Neill, M. Schuckers (Eds.) *A Handbook for Directors of Quantitative and Mathematical Support Centers*, Neck Quill Press, http://scholarcommons.usf.edu/qmasc_handbook.

This material is based upon work supported, in part, by the National Science Foundation under Grant DUE-1255945. Any opinions, findings, and conclusions or recommendations expressed in this material are those of the author(s) and do not necessarily reflect the views of the National Science Foundation

In the late 1970's, Hamilton received an IBM grant to, "investigate possible deficiencies in quantitative skills and to make recommendations, if appropriate, on how the college might remedy any such deficiencies." In 1984 The Quantitative Literacy Committee designed a written examination to be given to all incoming first-year students to test quantitative skills necessary to make informed judgments about problems that have an essential numerical content, skills that should be part of the "intellectual equipment of any educated citizen." This exam was intended as an advising tool. The department of mathematics also gave a placement exam to students who wished to take calculus, directing them to the appropriate calculus course. Ironically, students would often pass the calculus exam, but not the QSkills Exam.

Following its investigation and development of the quantitative skills exam, the committee recommended that the college establish a Quantitative Skills Tutorial Center. Students scoring below a certain cut-off would be advised to pursue studies, including tutoring at the center, to improve their quantitative skills. Student tutors were majors in biology, chemistry, economics, mathematics, and psychology. All tutors had a firm mathematical background.

From 1990 to 1993, the center was located in a classroom building separated from the Department of Mathematics. Because of classroom needs, it was moved to the building that housed the Department of Mathematics. This closeness to the math department helped, as students came in readily from the math classes held on the same floor. The name of the center was changed to the Quantitative Literacy Center (QLit. Center), and tutoring took place after 2:00 pm, when classes were no longer held in the space. An adjacent storage closet was set up as an office for the director. It was generally agreed that the location of the QLit. Center and of the office was a good one, as it was in the middle of the campus, and in a building that saw much student traffic; however, the facilities left much to be desired, most notably the size of the center and the director's office.

In 1995 the faculty voted to make quantitative literacy a requirement. To fulfill this requirement, by the end of their second year, students could either 1) pass the QSkills Exam (taking the exam is required of all students), 2) take a "Q" course, or 3) take a non-credit tutorial facilitated through the QLit. Center. Students who missed the exam and students who matriculated in the second semester were given the option of going to the center to take the exam individually. This was a time-consuming task for the director of the center. Students who did not pass the exam were not given the option of a retake.

The QLit. Requirement was replaced in 2009 when the faculty passed a new Quantitative and Symbolic Reasoning (QSR) Requirement. The guidelines for students in the class of 2014 and later state:

Hamilton expects that every student will demonstrate facility in quantitative and symbolic reasoning by completing one or more courses in at least one of the following three categories:

1. **Statistical Analysis**. The use of statistical analysis to describe data and to make inferences.

2. **Mathematical Representation**. The use of mathematical models such as those based on graphs, equations and geometric objects to represent patterns, relationships and forms.
3. **Logic and Symbolic Reasoning**. The use of formal logic or symbolic reasoning such as in the following examples: the proper construction of a computer program or a formal proof; the analysis of language in linguistics; or the study of music theory.

This requirement should be fulfilled by the end of the second year at Hamilton College. `https://my.hamilton.edu/catalogue/academic-regulations#StandardsforQuantitativeWork`

In 2011, the center's name was changed to the Quantitative and Symbolic Reasoning Center (QSR Center) to reflect its support for the new QSR Requirement.

The current QSR Center is located in the Christian A. Johnson building, which also houses the Department of Mathematics, Study Abroad Offices, the Digital Humanities Center, and classrooms and offices for the departments of mathematics and languages. Information about the mission and role of the QSR Center can be found at `https://my.hamilton.edu/qsr`.

2 Organization and Services

Services provided by the QSR Center:

- Support for the QSR Requirement
- Drop-in and group peer tutoring in quantitative courses
- Space for study groups
- Self-paced modules for skills review
- Computer lab and printer
- Library of quantitative resources
- Topical quantitative workshops and review for post-graduate exams
- Help beyond professors' office hours

The QSR Center supports the QSR Requirement and the faculty by offering drop-in peer tutoring in introductory level courses containing a math/quantitative/symbolic component. Students are encouraged to form study groups and to meet in the center at a regular time when there is a tutor in their subject available. The center offers review sessions for various topics throughout the semester. Materials are available for review and for "how to study" in the various disciplines.

In addition to the original disciplines of chemistry, economics, mathematics, and psychology, help with biology, physics, philosophy (symbolic logic), and departments' stats courses have been added. When a tutor is qualified, other QSR courses are supported, such as Elements of Music.

Statistics regarding QSR Center usage are calculated using student sign-in data. A student could be in for a short period of time or may stay for hours. During 2012-13, the courses served included Calculus (70%), Chemistry (11%), Economics (10%), Physics (4%), Philosophy (symbolic logic) (2%), Psychology/Music (1%). Biology and Computer Science take up the remainder. Total visits were 737 in the fall, and 446 in the spring, for a total of 1,183.

Students who need to go beyond the help at the center, or who need tutoring in subjects that are not offered at the center, may meet with the QSR Center director to request a one-on-one tutor through the Peer Tutoring Program. Information about that program is on the website `https://my.hamilton.edu/peertutoring`.

Center tutoring times have also changed since the beginning, from being open four hours a day Monday through Thursday, to almost 50 hours a week, closing on Saturday. Beyond the tutoring available, students have the opportunity to work on their own, work on the computers, use the printer, avail themselves of the library and resource materials, play out ideas and problems on the many white boards, sit in the lounge chairs and read, or to discuss their needs with the director.

The center is also a resource for departments and for faculty. Faculty may request that a special topic be covered; or request that a tutor meet with a class for a review session. The director may consult with faculty about the content of QSR courses that they teach or may meet with faculty to develop QSR aspects of their courses, whether they count towards the requirement or not. The center director, along with the QSR Committee, sponsors topical workshops, and brings in speakers on various QSR topics.

Periodic topical sessions are offered through the center, facilitated by the center tutors. During the 2012-2013 academic year, tutors planned and presented the "Math Commons." Students were invited to sit in on sessions covering the following topics: Personal Health and Fitness, Graphs and Charts, Personal Finance, and Algebra Review.

On a White Board, tutors can record "Top Topics" that have been covered during the day. Tutors and students can look at the board when they come in and see what is being worked on. A summary of top topics can be compiled and e-mailed to the appropriate faculty.

3 Staffing, Hiring, and Training

The staff of the QSR Center consists of a Director and peer tutors. Each year the director selects two or three upper-class experienced tutors as Senior Tutors, who help with tutor training, special planning, and data collecting.

Peer tutors are majors who have been recommended by the departments the center supports. Sometimes a student who is not a major, but who has done well in a particular course, is recommended. Tutors should have excellent academic and personal qualifications. All tutors must have taken calculus. Following a meeting and interview in the spring, the director examines the applicants' transcripts and references. The tutors are among the top students in the college who have a desire to work with their fellow students and to practice the academic and personal skills necessary to be good tutors and communicators. Approximately 24 tutors are hired for a year, but only about 18 or 19 of these tutors are active at any given time because many Hamilton students study abroad.

Tutors are trained in the early fall in the administrative and personal aspects of tutoring. The tutor guidebook details responsibilities and issues that may come up during a tutoring session. Topics include confidentiality, dealing with personal issues, using interpersonal skills to put students at ease, tutoring in a multi-cultural setting, dealing with students who "just want the answer," and what to do when you can't help a student. Tutors discuss what makes a good learner—they talk about the learning process and about the importance of study skills. There are periodic meetings with representatives from the departments that the center supports, as well as meetings in small groups to go over problems and to work on the challenges of tutoring several disciplines during a short period of time. Tutors also learn how to fill out report forms and time sheets properly. At end of each shift, the tutor fills out a form telling the date, time worked, topics covered, and number of students the tutor worked with. The tutors are asked to elaborate about what worked and what didn't, and how they chose to solve the problems. A group like this, working together for many hours and weeks, becomes like a family, and the atmosphere is not just serious and hard work — it is good for students to be part of this, too.

Much is said about the benefits of tutoring for tutees, but tutors gain much from the experience besides earning the top hourly wage on campus. Tutors get to review material, they develop administrative experience, they develop their interpersonal skills, build up their resumes, expand professional relationships with professors, develop new friendships, and help the college. QSR Center tutors are often valedictorians and salutatorians, and are selected for department honors honor societies. The director is often called upon for references for graduate school, medical school, fellowships, and more.

Prior to the 2013–2014 academic year, the position of director was a $\frac{3}{4}$-time, ten-month appointment. Now the center director holds a full-time ten-month, administrative position. The director reports to the Associate Dean of the Faculty and is an ex officio member of the QSR Committee. The director's responsibilities involve all aspects of running the center, soliciting and training of tutors, monitoring the tutorials, consulting with faculty, maintaining the center web site, and coordinating the one-on-one Peer Tutoring Program. The director will teach a "Q" course each year. The increase in responsibility is an indication of the commitment of the college to this academic requirement and support.

4 Community Interactions

The Director of the QSR Center, along with other directors of Academic Support Centers at Hamilton who report to the Academic Dean, meet monthly to discuss issues that pertain to running academic support programs, such as budgets, reports, web site information, and other topics as they come up.

The director also works closely with departments that the center supports. Information about other departments' review sessions and support is kept up-to-date by the director; this information is posted on the center bulletin board. Course instructors are asked to put information about the QSR Center on their syllabi. The director meets with new faculty during New Faculty Orientation. The center also disseminates information to matriculating students during New Student Orientation in the fall and at the beginning of the spring semester for January admits and transfer students. The center is represented at Family Weekend, Fallcoming, Commencement, and at campus service days.

Center services are communicated to the college community via the web site, e-mail, fliers, class visits by center tutors, articles in the campus newspaper, Hamilton College web news, an annual open house, the Peer Tutoring Newsletter, and on Facebook.

The center director is an ex officio member of the QSR Advisory Committee, composed of faculty who teach QSR courses. This committee reports to the Dean of Faculty. Changes in the QSR Requirement and the QSR Program are made by the QSR Advisory Committee and are presented to the faculty via the Committee on Academic Policy.

The director attends meetings and conferences regarding quantitative support, such as NECQL (Northeast Consortium for Quantitative Literacy), JMM (Joint Math Meetings), and others.

5 Assessment

Student data is tracked by a sign-in program on a center computer located near the entrance. Students enter their names, class year, the date, course number, professor, and purpose of the visit. The list generated from this program is used to send out a questionnaire near the end of each semester.

Several years ago the Quantitative Literacy Program was part of an external assessment of the Department of Mathematics. Most of the comments and recommendations pertained to the Quantitative Literacy Requirement and the exam, both of which were perceived as weak and have since been strengthened. An external assessment specific to the QSR Center has not been done.

6 Additional Topics: A New Space

In the spring of 2011 it was decided that the building housing the QSR Center would undergo major changes. The math department would move to the first floor to gain visibility and to have more study and work space for students. Three programs would take over the space on the first floor vacated by the museum of art. A space on the third floor that had been the Media Library was designated for the QSR Center.

The QSR Center director met with the associate dean, the head of physical plant, and the QSR committee, to discuss center needs. Most important was a larger space, more tables, and more computers, along with an area with a conference table for group work and presentations. The center needed a larger office for the director, including more room to move around in, to meet with students, and with adequate work surfaces. The challenge was to turn a windowless space into one where students could work comfortably. Outside walls on the third floor are three feet thick, so it would be impossible to break through to put in windows. However, architects' plans included breaking through inside walls to the hall in order to bring in light from the building's existing skylight and to create a spacious feeling to the center.

At a final tutor meeting at the end of the spring 2011 term, tutors were asked to envision their "ideal center," and to draw plans on large sheets of paper. They came up with some very practical ideas: more tables, a conference table, more computers, and white boards. They had fun adding extras, like a beanbag pit and a coffee shop. Below is a copy of one of the plans:

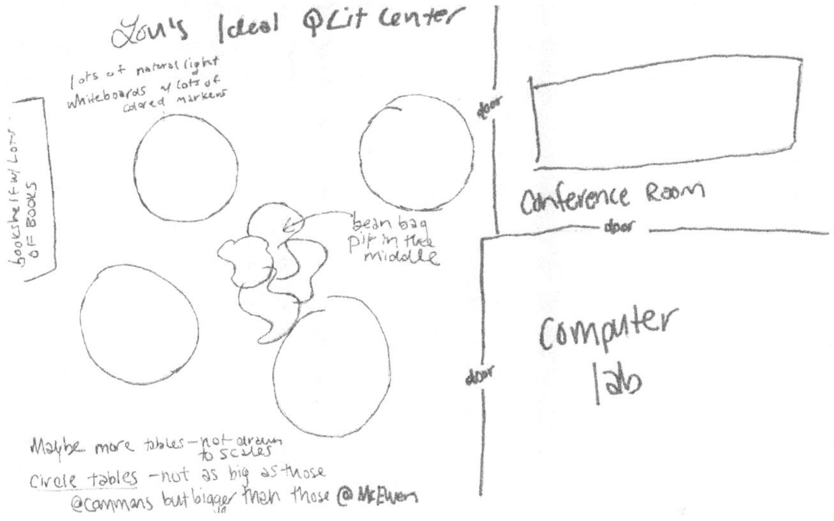

Figure 1: A student's ideal plan for the new center

Tutors especially liked the idea of many tables where they could quickly move from one student or group, to another. They also favored white boards over chalkboards.

Below is a copy of one of the architect's renderings, with windows looking out to the hall.

Figure 2: Rendering by Chiang/O'Brien Architects

Since the new center was completed, students have exclaimed that the space is so bright, open, and inviting that they don't even notice the lack of windows to the outside. At the end of our first year of operation in the new space, tutors named things that surprised them about the center.

Surprises in the New Center:

- The whiteboards, round tables, rolling chairs
- The added exercise climbing up the two flights of stairs
- The brightness
- The large conference table is a great study space, where students can spread out their material

Figure 3: Photo by Max Schnidman, Hamilton '14, September 2013

Contrary to the fear that the climb to the third floor would keep students away, we served a steady stream of students throughout the 2012-2013 academic year. The new QSR Center is a valuable asset to the QSR Program and to Hamilton College's academic community.

7 Acknowledgements

I would like to thank the National Science Foundation for its generous support of the QMaSC Workshop through NSF grant number 1255945. Any opinions, findings, and conclusions or recommendations expressed in this material are those of the author and do not necessarily reflect the views of the National Science Foundation. I would like to thank my co-PI's Grace Coulombe and Michael Schuckers, whose humor and intelligence make it a pleasure to work on this project. I thank McKayla Nuffer, Assistant Director of the Peterson Quantitative Resource Center at St. Lawrence University, whose attention to detail and whose technical skills have provided significant support for this project. Thanks to Hamilton College and the Dean of the Faculty for their generous support and to the QSR Advisory Committee, especially Professor Tim Kelly. Finally, I would like to thank my husband, Professor John H. O'Neill, who was a member of the original Quantitative Literacy Committee, and whose support and encouragement have been overwhelming, not to mention his proof-reading abilities (yes, he is an English Professor!).

Case Study: The Math Study Center at the University of Washington

© Patrick T. Perkins,
University of Washington

1 Introduction and History

The Math Study Center, a large tutoring facility at the University of Washington in Seattle, is run and funded by the Department of Mathematics. Its mission is to serve students taking large freshman mathematics classes at the UW. The Center serves students of precalculus as well as the three quarters of freshman calculus. The MSC was started in 1992 by Ken Plochinski, who had previously run a similar facility at the University of Michigan that was the source of the main ideas behind the MSC. The author took over as Director from 2000 to 2013, inheriting a smoothly run organization requiring very few changes. The Math Study Center, while being a great place for students to work on math regardless of whether or not they need help, is not a tutoring center in the traditional sense. It provides a comfortable place and a supportive atmosphere for students to come together and study, in groups or individually. The MSC is staffed by a number of tutors, both graduate student TAs and advanced undergraduates. The mathematics graduate students work at the center as part of their TA job and professors hold their office hours in the MSC. Tutors sit with students to help them get unstuck or to answer questions. The center does not provide long blocks of uninterrupted one-on-one tutoring. Textbooks, calculators, and other study materials may to

Suggested Citation: Patrick T. Perkins (2016), "Case Study: The Math Study Center at the University of Washington", `http://dx.doi.org/10.5038/9780977674435.ch29` in G. Coulombe, M. O'Neill, M. Schuckers (Eds.) *A Handbook for Directors of Quantitative and Mathematical Support Centers*, Neck Quill Press, `http://scholarcommons.usf.edu/qmasc_handbook`.

This material is based upon work supported, in part, by the National Science Foundation under Grant DUE-1255945. Any opinions, findings, and conclusions or recommendations expressed in this material are those of the author(s) and do not necessarily reflect the views of the National Science Foundation

be checked out for in-room use. A unique feature of the MSC is the variety of kinds of help the students can receive.

The MSC occupies a large room in the basement of the Communications Building, across the street from the Department of Mathematics. It is open Monday through Thursday from 9:30am to 9:30pm. It is also open Friday from 9:30am to 1:30pm and Sunday from 2:00pm to 6:00pm. When students enter the MSC, a desk assistant greets them at the front desk. The director's office and a small conference room are to the right. The main room, left of the entrance, is a square about 40 feet on each side. A variety of math related posters cover the walls. Many colorful cardboard polyhedrons, made by a math class years ago, hang from the ceiling. The room contains 8 large tables, each seating 10 students. Five smaller tables are scattered around the room. The large tables are arranged in 4 groups, two tables for each of the four classes that are served by the MSC. There are signs over the tables indicating which classes they are for. Ideally, students in the same class should sit together and help each other or perhaps form study groups. Seven cubicles with tables, chairs, and whiteboards are located around the outside of the room near the walls. The cubicles can be used by students who form study groups. They are also used by professors and teaching assistants to hold their office hours.

2 Center Management and Services

The author of this case study, who served as director of the Math Study Center for 13 years, is a Senior Lecturer in Mathematics at the University of Washington. The director taught four classes per year because the duties of Director count for half of the usual teaching load of eight classes. During that time, the director's office was in the MSC, across the street from the rest of the math department. Students frequently dropped in to ask questions. The director's duties included hiring, handling payroll, staffing, and scheduling. Scheduling, which is a significant task in a facility the size of the MSC, will be discussed later.

When the author started, he had no managerial experience beyond dealing with the TAs and graders for his classes. Learning to be a good manager was a significant part of the job. Dealing with more than 20 employees in their twenties often involved small weekly dramas. As such, the director had to be flexible and responsive to illnesses, as well as exam and homework jitters. One tutor had a sex change which among other things required the director to deal with the medical issues, as well as less evolved attitudes on the part of some of the other workers. It was important to build a sense of being a team among the workers. That way if one has a crisis, another is willing to cover for them. The director spoke to each of the workers at least once every two weeks. Often it was just a short conversation, involving details about previous chats such as how they did on a big midterm or how their first research talk went. This method fostered some trust and goodwill setting the groundwork for eventual talks with them about repeated lateness or poor tutoring.

The Math Department has a Chair, a position that rotates every few years, and a Department Administrator, a staff position. The MSC director reports to both of these people. At the beginning of each quarter, the director computes and submits an estimated budget to the Administrator for approval since the MSC is funded by the Math Department budget. While the Chair was always included in this email, the director deals mainly with this Administrator. Occasionally, the Chair stops by the MSC to see how things are going, but most of the time the director is left to run things as he sees fit. The director meets annually with the Chair and to give an update on the MSC. Because the graduate students who work as tutors are assigned to the MSC by the head of the Student Services office in the Math Department, the director works closely with her on staffing matters.

Originally, the director spent the various breaks creating work schedules for the next quarter, juggling about 25 student workers, each of whom was only available certain hours each day. Historical information about when the MSC was busy enabled the director to estimate how many tutors might be needed each hour. The director would spend about a week moving names into little boxes, trying to come up with a feasible schedule. Then he would meet with all of the workers on the first day of the quarter so that they could trade hours with each other until things were reasonably satisfactory. Tired of spending breaks in this way, the director wrote some heuristic software that proved fairly helpful. Things really improved when the Math Department hired Rekha Thomas, an expert in discrete optimization. She assigned, to a team of undergraduate optimization students, the project of solving the MSC scheduling problem. This team, led by Caleb White, came up with a very sophisticated solution to the problem, which did much more than just give a feasible solution used as a base for trading. The team implemented some very sophisticated constraints. For example, tutors do not like to enter and leave the MSC more than twice a day. Or, if they have a free hour between two classes, they often like to work at that time. The numerous features of the project are provided in a published paper [1]. Given that the system was written by mathematicians for mathematicians, it is not terribly user friendly and could use some work in this area. But it does an excellent job of making good schedules and takes about 6 seconds to run, rather than a week by hand.

3 Staffing, Hiring and Training

In Autumn 2012, a typical quarter, the MSC employed 15 undergraduate tutors and 6 math graduate students assigned to the MSC as part of their TA duties. The undergraduates worked between 8 and 12 hours a week. The grad students worked 6 or 12 hours a week, depending on their assignment. There were up to 100 students using the MSC at peak times during that quarter. The average number was about 50. We do not want students waiting 45 minutes for a tutor. Therefore, tutors are required to limit the time spent with each student to no more than 10 minutes. The ideal session is for the tutor to quickly get the student unstuck and back to work. Students are encouraged to talk to each other and get help in this way. There is a sign-up sheet on the wall that the students use to get help from a tutor. If the number of people on the list gets large, the director helps with the tutoring.

The bulk of the hiring for the year is completed during August. The undergraduate tutors are mostly students recruited from classes recently taught by the director. When serving as Director, the author often interviewed 25 people and hired about 5 each year. In interviews, he sought to get an impression of the candidate's character in addition to looking for people with good social skills, as well as the required math ability. The director also gave candidates a test on basic calculus. This open-book, untimed test was a list of problems from the calculus text used at the UW. The author admits that his hiring procedures could use some improvement, but they served him well enough for many years. No training was offered to the new tutors. Instead, the large number of returning tutors each year served as mentors to the new people. The director kept an eye on the new tutors during the first few weeks of their first quarter and answered many questions. It could be a little rough at first, but after a couple of weeks they always settled in and became comfortable with the job.

Each year the director also hired 4 or 5 friendly, helpful students to work as desk assistants. The desk assistant is the first person that a new student greets upon arrival. Because the director is not able to be in the MSC for all of the open hours, the desk assistants serve to keep an eye on the Center. They answer many student questions, particularly during the Autumn quarter. Desk assistants also check the tutors in and out on a time sheet and inform the director if a tutor is regularly late. There are books and calculators available for use in the MSC and the desk assistants took care of checking these in and out.

4 Community Interactions

At the first math faculty meeting of the school year, the director makes an announcement to describe the MSC, give the open hours, and encourage the instructors to refer their students. An email announcement is sent to the Math Department at the beginning of each quarter. Calculus instructors are given an overhead slide with information about the MSC. But most students already know about the MSC services and have no trouble finding their way to the Center. Word of mouth is the most effective advertisement.

The director maintains contact with the several other facilities on UW campus that offer math tutoring. Because the MSC usually has more job applicants each year than can be hired, the directors provides useful references to the other centers regarding those candidates that were interviewed and tested but not hired by the MSC.

On a number of occasions, people from outside the UW work as volunteer tutors in the MSC. These people are mostly recent immigrants who have not yet received a work permit. They seek to gain work experience and practice speaking English. Often, the director writes them a job letter later on.

The MSC occasionally receives phone calls from people in the Seattle area seeking help with various math-related issues such as fielding inquiries from parents seeking tutors for their children. The Math Department maintains a list of tutors for hire and the director is familiar with many of the people on the list. The director makes suggestions and asks for feedback on how things go. Sometimes people call with general math questions. When unable to give a good answer, the director gets them in touch with someone who is more knowledgeable on the topic.

5 Assessment at the Center

The director maintains an anonymous email account for student comments about the MSC. Despite getting many comments over the years, the author received very few complaints about the quality of tutoring provided. Several times a quarter the desk assistants take a census of the MSC. Every half an hour, for a week, they count the number of students using the center. This information is very useful when scheduling tutors for future quarters. As previously mentioned, when it gets busy in the MSC, the director helps with the tutoring. This helps to build the sense of being a team, and gave the director the chance to show the tutors how to serve students effectively in a short amount of time. Working directly with students enables the director to get informal feedback on how the MSC is doing. A few years ago, the Math Department ran a large program aimed at getting student feedback on the calculus sequence. The clear and strong message is that students valued the MSC highly and were very satisfied with the service they received.

6 Additional Topics

This year, academic year 2013-14, was my last running the Math Study Center. After 13 years I am returning to teaching full time and another faculty member is taking over. I look back with some satisfaction on the experience. I have learned that I am a pretty good boss. Hiring friendly, competent people and letting them get on with the job seems to be the secret. I had a very diverse group but wove them into a team by regularly interacting with them in a friendly way. I set clear goals and then let the workers figure out how best to fill them. I worked alongside them to set an example. Many workers came back year after year, in spite of the low wages forced on us by the state budget cuts. I also had fun with the scheduling problem. I learned a lot about discrete optimization and gained a new appreciation of how math can really be used. The MSC is a successful organization that provides a useful service at the UW. I can see some changes coming, mostly related to technology and computers. But this is for my successor to think about.

7 Bibliography

[1] C. Z. White, Y. Lee, Y. Kim, R. Thomas, and P. Perkins, "Creating weekly timetables for maximizing employee preferences," *The UMAP Journal*, vol. 25, no. 1, pp. 5–24, 2004. `http://www.math.washington.edu/~thomas/papers/MSCRevised.pdf`.

Case Study: Quantitative Learning Center at UConn

© Tom Roby,
University of Connecticut

1 Introduction

The University of Connecticut's Quantitative Learning Center (Q Center) is a resource to elevate the proficiency of students taking quantitative intensive (Q) courses across the undergraduate curriculum. The Q Center provides direct assistance to students via peer tutoring in most lower-level Q courses in math, chemistry, physics and statistics. It also assists faculty in incorporating best practices in their teaching and serves as a clearinghouse for information about Q resources and discussion of Q issues within the university. Additional information that supplements this case study is available at the Q Center's website: `http://qcenter.uconn.edu/`.

UConn is the state's flagship public research university, enrolling a total of about 30,000 students in a wide variety of programs across several campuses. In 2013, the main campus in Storrs had about 18,000 undergraduates; the number of entering freshmen was 3755, after many years of being around 3200, with occasional surges. The nature of the institution (large and research-focused in contrast to smaller or teaching-focused) leads to some noteworthy differences in the challenges faced by those running the Q Center, to be described in Section 6.

The need for a Q Center at UConn came out of general education requirements (from 2004) that all undergraduates pass at least 2 writing-intensive (W) courses and 2 quantitatively-intensive

Suggested Citation: Tom Roby (2016), "Case Study: Quantitative Learning Center at UConn", `http://dx.doi.org/10.5038/9780977674435.ch30` in G. Coulombe, M. O'Neill, M. Schuckers (Eds.) *A Handbook for Directors of Quantitative and Mathematical Support Centers*, Neck Quill Press, `http://scholarcommons.usf.edu/qmasc_handbook`.

This material is based upon work supported, in part, by the National Science Foundation under Grant DUE-1255945. Any opinions, findings, and conclusions or recommendations expressed in this material are those of the author(s) and do not necessarily reflect the views of the National Science Foundation

(Q) courses. University committees determine which courses qualify for the W and Q flags based on standard criteria. W courses must require students to produce and rewrite at least 15 pages and the course size is capped at 19 to allow instructors to give students more personal attention.

Q courses have no such cap and are frequently taught in large lectures of up to 350, sometimes (but not always) with smaller breakout recitations. The main litmus test for such courses is that (high-school) algebra is an integral part of the course and used throughout. Technology is explicitly mentioned as an allowable tool but not one that should substitute for understanding since merely feeding data into a canned program and reading the results is considered insufficient. More details are available at `http://geoc.uconn.edu/quantitative-competency`.

The new Q requirement expressly calls for the existence of a university-wide Q Center, which had not previously existed. A writing center *had* already existed as a unit of the College of Liberal Arts and Sciences (CLAS) (as at many colleges around the country). Starting in 2004, the Writing Center was organizationally relocated to become a university-wide unit, and the Q Center was created.

2 Center Organization & Services

The Q Center is part of UConn's Institute for Teaching and Learning (ITL), which provides various types of support for students and faculty. In the past, the head of ITL reported directly to the vice-provost for undergraduate education and instruction, who reported to the provost. More recently, an organizational change, merging ITL with the college of continuing studies and some other student support services, resulted in the creation of a Center for Excellence in Teaching and Learning thereby adding one more layer of administration. For simplicity, this report describes the Q Center as it existed from its inception in 2004 through 2012, during most of which time the author served as its director.

After the initial build-up, the core Q Center staff consisted of a director and associate director (each half-time), a part-time assistant director who mostly developed and maintained the IT infrastructure, and a program assistant who varied between half and 80% time. The director and associate director were also half-time tenure-track faculty members in the department of mathematics.

Q Center History

In its first year, 2004–05, the Q Center was run by interim-director Chuck Vinsonhaler, former math department head, who started a small peer-tutoring program. The author was hired as the first director in Fall 2005 and served until stepping down after the Fall 2012 semester. In Fall 2005, Christy Hilliard started as the first halftime program assistant. From the beginning, building a new unit, essentially from scratch, within a large research university posed many challenges. Finding suitable spaces for activities, advocating for appropriate funding levels, and developing a strongly positive reputation each took a great deal of time and effort.

Early efforts focused on providing services to students, and the Center's peer-tutoring program turned out to be extremely popular, as shown in Figure 1. The number of visits to peer tutoring greatly increased each fall for the first five Fall semesters, forcing directors to scramble for more resources and to hire more tutors at the last minute. Eventually volume leveled off, but *actual* demand for services seems never to have declined. The decrease in volume in Fall 2011 is directly attributable to budget cuts that necessitated a decrease in the number of open hours and in the number of tutors on staff.

Figure 1: Q Center Visits & Staffing Over Time

Term	# visits	# open hours	# tutors	# half GAs	# Entering Freshmen
Fall 05	600	20	10	6	3260
Spring 06	1200	28	14	13	
Fall 06	3906	40	30–40	12	3240
Spring 07	3866	50	45	11	
Fall 07	7000	58	40–55	13	3179
Spring 08	6550	58	55–60	17	
Fall 08	9279	58	60–80	13	3606
Spring 09	7895	58	80	13	
Fall 09	11,334	58	80–90	12	3221
Spring 10	8012	58	80–90	12	
Fall 10	10,457	58	70-75	12	3339
Spring 11	7949	53	70–75	12	
Fall 11	8638	44	65	12	3327
Spring 12	8494	44	65	12	
Fall 12	8772	44	55–60	10	3114

The term "GAs" refers to Graduate Assistants working as supervisors in the Q Center. GAs are generally hired from various Q departments, sometimes as half of their appointment (10 hours/week), sometimes as their full appointment (20 hours/week). The term "TA" refers to Teaching Assistants working within their own departments, not for the Q Center.

Peer Tutoring

The main service provided by the Q Center is *free, on-demand, drop-in* peer tutoring for many Q courses, particularly those taken by the majority of students to satisfy Q requirements. The center is typically staffed Monday–Thursday from 11a.m.–11.p.m. and Sunday 1–11p.m.. Occasionally, in lean budget years, the number of hours on Sundays and mornings might be reduced. At any given time there are typically 7–14 undergraduate tutors on duty, supervised by two graduate students (GAs) in Q departments. A glance at the schedule in Figure 2 helps to orient the reader to the

Figure 2: Q Center Tutoring Schedule, Fall 2010

Q Center Tutoring Schedule - Fall 2010 - Standard

Subject Codes: M = Mathematics, P = Physics, C = Chemistry, S = Statistics, I = Computer Science, E = Economics
Show All, or show only Q Center Supervisor, Q Center Tutor, *Receptionist*, or go to the archives.

Sunday

Time	Name	Subjects
1-3	**Marianne Larosa**	MP
	Lance Williams	MS
	Michelle B.	C
	Pat F.	CMP
	Derek H.	PMC
	Bob J.	MSP
	Aaron N.	MP
	Scott N.	MS
	Dave P.	MS
	Nick S.	MC
	Mingfeng Z.	M
	Aaron M.	
3-5	**Marianne Larosa**	MP
	Lance Williams	MS
	Craig H.	MC
	Derek H.	PMC
	Matt L.	M
	Maureen M.	MPS
	Britteny M.	M
	Aaron N.	MP
	Scott N.	MS
	Nick S.	MC
	Mingfeng Z.	M
	Aaron M.	
5-7	**David Cox**	PMC
	Keegan Soncha	C
	Sara E.	MSC
	Craig H.	MC
	Derek H.	PMC
	Bob J.	MSP
	Maureen M.	MPS
	Steve M.	CM
	Aaron N.	MP
	Atif R.	MCPSI
	David T.	CPM
	Aaron M.	
7-9	**Faith Hollingshead**	MSE
	Keegan Soncha	C
	Jon B.	MS
	Sara E.	MSC
	Craig H.	MC
	Mike I.	PMCSIE
	Bob J.	MSP
	Steve M.	CM
	Dave P.	MS
	Nick S.	MC
	Bensen T.	C
	Nathan W.	CM
	Caitlin W.	
9-11	**Faith Hollingshead**	MSE
	Keegan Soncha	C
	Igor A.	C
	Bryan B.	PMC
	Victor C.	MPI
	Marisia F.	CM
	Leslie K.	MS
	Dave P.	MS
	Nick S.	MC
	Bensen T.	C
	Nathan W.	CM
	Caitlin W.	

Monday

Time	Name	Subjects
11-1	**Beziel Shambamuto**	MPS
	Keegan Soncha	C
	Sean A.	CMP
	Matt B.	MSP
	Craig H.	MC
	Chad J.	PMC
	Britteny M.	M
	Michael P.	MCPSIE
	Ashley R.	MS
	Todd S.	MES
	Mingfeng Z.	M
	Mark F.	
1-3	**David Cox**	PMC
	Beziel Shambamuto	MPS
	Keegan Soncha	C
	Chris A.	MS
	Sean A.	CMP
	Gil D.	CPM
	Tyler E.	MSE
	Briana H.	MS
	Britteny M.	M
	Joe P.	MCI
	Steve R.	CM
	Nathan W.	CM
	Mike Y.	PMI
	Kellie Z.	MS
	Katie H.	
	Burton B. (2-3)	MPS
3-5	Burton B. (3-4)	MPS
	David Cox	PMC
	Keegan Soncha	C
	Jon B.	MS
	Tyler E.	MSE
	Sara E.	MSC
	Mike G.	PM
	Maureen M.	MPS
	Aaron O.	MI
	Steve R.	CM
	Courtney W.	CM
	Dave W.	MS
	Elaine W.	MS
	Mike Y.	PMI
	Katie H.	
5-7	**Hugo Panzo**	MSP
	Sara E.	MSC
	Alex F.	PMC
	Maureen M.	MPS
	Aaron O.	MI
	Josh R.	PM
	David T.	CPM
	Courtney W.	CM
	Dave W.	MS
	Mike Y.	PMI
	Aaron M.	
7-9	**Marianne Larosa**	MP
	Chris A.	MS
	Michelle B.	C
	Jon B.	MS
	Patt B.	MS
	Alex F.	PMC
	Jess G.	MS
	Josh R.	PM
	Ian S.	M
	Nathan T.	C
	Aaron M.	
9-11	**Lou Bachenheimer**	PM
	Chris A.	MS
	Bryan B.	PMC
	Patt B.	MS
	Jess G.	MS
	Chad J.	PMC
	Michael P.	MCPSIE
	Ian S.	M
	Pete T.	MSE
	Nathan T.	C
	Aaron M.	

Tuesday

Time	Name	Subjects
11-1	**Lou Bachenheimer**	PM
	David Cox	PMC
	Matt B.	MSP
	Michelle B.	C
	Jen F.	CS
	Jessica G.	MS
	Jess G.	MS
	Xuan G.	PMS
	John H.	C
	Aaron N.	MP
	Peter Y.	MES
	Kellie Z.	MS
	Nicole L.	
1-3	**Lou Bachenheimer**	PM
	Sankha Perera	MS
	Matt B.	SM
	Zack C.	C
	Lindsay C.	MS
	Jen F.	CS
	Craig H.	MC
	Aaron N.	MP
	Nathan W.	CM
	Peter Y.	MES
	Mark F.	
	Burton B. (2-3)	MPS
3-5	Burton B. (3-4)	MPS
	David Cox	PMC
	Sankha Perera	MS
	Omkar B.	PCMS
	Zack C.	C
	Leslie K.	MS
	Priyantha K. (3-4)	MS
	Aaron N.	MP
	Ashley R.	MS
	Todd S.	MES
	Courtney W.	CM
	Dave W.	MS
	Nicole L.	
	Lou Bachenheimer (4-5)	PM
5-7	**Lou Bachenheimer (5-6)**	PM
	Hugo Panzo	MSP
	Sean A.	CMP
	Michelle B.	C
	Omkar B.	PCMS
	Rudy C.	PMSI
	Marisia F.	CM
	Alex F.	PMC
	Mike G.	PM
	Craig H.	MC
	John H.	C
	Khrystyna N.	M
	Ashley R.	MS
	Colleen V.	MS
	Caitlin M.	
7-9	**Faith Hollingshead**	MSE
	Sankha Perera	MS
	Sean A.	CMP
	Rudy C. (7-8)	PMSI
	Gil D.	CPM
	Jessica G.	MS
	John H.	C
	Mike I.	PMCSIE
	Brittony M.	M
	Khrystyna N.	M
	Atif R.	MCPSI
	Caitlin M.	
9-11	**Faith Hollingshead**	MSE
	Sankha Perera	MS
	Brian B.	SM
	Gil D.	CPM
	John H.	C
	Leslie K.	MS
	Dave P.	MS
	Ian S.	M
	Nick S.	MC
	Aaron M.	

Wednesday

Time	Name	Subjects
11-1	**David Cox**	PMC
	Beziel Shambamuto	MPS
	Matt B.	MSP
	Mike C.	MPI
	Xuan G.	PMS
	Briana H.	MS
	Matt L.	M
	Michael P.	MCPSIE
	Mingfeng Z.	M
	Mark F.	
1-3	**David Cox**	PMC
	Beziel Shambamuto	MPS
	Brian B.	SM
	Garrett B.	PM
	Tyler E.	MSE
	John H.	C
	Briana H.	MS
	Khrystyna N.	M
	David T.	CPM
	Kellie Z.	MS
	Caitlin W.	
	Burton B. (2-3)	MPS
3-5	Burton B. (3-4)	MPS
	David Cox	PMC
	Chris A.	MS
	Garrett B.	PM
	Lindsay C.	MS
	Tyler E.	MSE
	Mike G.	PM
	Aaron O.	MI
	Colleen V.	MS
	Dave W.	MS
	Elaine W.	MS
	Mike Y.	PMI
	Caitlin W.	
	Sara E. (4-5)	MSC
5-7	Sara E. (5-6)	MSC
	Lou Bachenheimer	PM
	Al Calderon	M
	Sankha Perera	MS
	Chris A.	MS
	Omkar B.	PCMS
	Armand B.	MCS
	Lindsay C.	MS
	Divya K.	CMPS
	Dave L.	MCP
	Khrystyna N.	M
	Michael P.	MCPSIE
	Nick S.	MC
	Mike Y.	PMI
	Danny G.	
	Caitlin W.	
7-9	**Sankha Perera**	MS
	Chris A.	MS
	Josh A.	PCM
	Victor C.	MPI
	Jessica G.	MS
	Jess G.	MS
	Divya K.	CMPS
	Dave L.	MCP
	Steve M.	CM
	Khrystyna N.	M
	Michael P.	MCPSIE
	Nick W.	MPC
	Caitlin M.	
9-11	**Lou Bachenheimer**	PM
	Josh A.	PCM
	Bryan B.	PMC
	Victor C.	MPI
	Jessica G.	MS
	Jess G.	MS
	Steve M.	CM
	Ian S.	M
	Nick S.	MC
	Pete T.	MSE
	Caitlin M.	

Thursday

Time	Name	Subjects
11-1	**Sankha Perera**	MS
	Keegan Soncha	C
	Matt B.	MSP
	Jon B.	MS
	Matt B.	SM
	Zack C.	C
	Steve R.	CM
	Bensen T.	C
	Kellie Z.	MS
	Nicole L.	
	Burton B. (12-1)	MPS
1-3	**Lou Bachenheimer**	PM
	Keegan Soncha	C
	Josh A.	PCM
	Burton B.	MPS
	Matt B.	SM
	Patt B.	MS
	Craig H.	MC
	Steve R.	CM
	Nathan W.	CM
	Nick W.	MPC
	Mark F.	
3-5	Burton B. (3-4)	MPS
	Lou Bachenheimer	PM
	Lance Williams	MS
	Omkar B.	PCMS
	Patt B.	MS
	Zack C.	C
	Mike C.	MPI
	Alex F.	PMC
	Joe P.	MCI
	Dave P.	MS
	Ashley R.	MS
	Caitlin M.	
	Alex F. (3:30-5)	PMC
	Mike I. (4-5)	PMCSIE
5-7	Mike I. (5-6)	PMCSIE
	Marianne Larosa	MP
	Lance Williams	MS
	Sean A.	CMP
	Michelle B.	C
	Omkar B.	PCMS
	Zack C.	C
	Rudy C.	PMSI
	Marisia F.	CM
	Jess G.	MS
	Craig H.	MC
	Matt L.	M
	Ashley R.	MS
	Nick S.	MC
	Katie H.	
7-9	**Hugo Panzo**	MSP
	Josh A.	PCM
	Patt B.	MS
	Rudy C. (7-8)	PMSI
	Gil D.	CPM
	Marisia F.	CM
	Alex F.	PMC
	Jess G.	MS
	Dave L.	MCP
	Brittany M.	M
	Nick S.	MC
	Katie H.	
9-11	**Hugo Panzo**	MSP
	Brian B.	SM
	Patt B.	MS
	Gil D.	CPM
	Jess G.	MS
	Dave L.	MCP
	Ian S.	M
	Nick S.	MC
	Katie H.	

scope of the operation. GAs, "Q Center Supervisors", are listed by their full names at the top, tutors by first name and last initial, followed by the receptionist on duty. Color versions of this schedule for various semesters are available on the website (in the "archives" section) and display snapshots of the overall operation as it changes over time.

Although frequently requested, the center never offers one-on-one tutoring. Instead, tutors roam from student to student, provide bursts of assistance, and allow tutees to ponder the problem in between. Tutors are trained to avoid holding a pencil, as much as possible, and to provide encouragement with a minimum of hints, enabling students to reach a solution with as much independence as possible. It is made clear to tutors and students that this method will generally facilitate the greatest learning for the students, who do not have tutors to help them during exams.

The exemplary practices of UConn's Writing Center provide a useful comparison and contrast.

Students seeking help with their writing, typically a term paper, schedule a one-on-one appointment and bring a draft for discussion with the tutor. A student might do this once or twice a semester, and the atmosphere tends to be rather relaxed, like meeting a friend in a coffee shop.

By contrast, Q tutoring has an atmosphere more like an emergency room, with often-nervous students arriving for last-minute help on homework or on imminent exams. The supervisors and receptionists triage students to appropriate tutors in an attempt to balance the load. Most tutors can tutor for multiple subjects, so fewer tutors are needed to staff the Center at any one time while meeting the variety of demand distributions that arise during the semester. A fuller comparison of the different tones of the Q and Writing Centers is presented in [1].

Many visitors return, with the average tutee making about five visits to the Center. Some students sit in or near the Center, work on assignments, and request help as needed. Sometimes tutors encourage students to form study groups, and when a large number of students arrive for the same course, the individual tutoring sessions often transform into an impromptu small recitation.

Physical Locations & Learning Commons

As the Center grew, peer-tutoring happened in a variety of shared spaces, including the director's office, a small classroom, and after hours in a dining hall. Locating dedicated space in a large university with many competing demands was extremely difficult, so the Center made do with what was available. But then opportunity knocked.

Like its peers, the UConn library observed a rapidly diminishing demand for its onsite services. As print reference materials and journals moved online, the library faced a surplus of space but a shortage of patrons. In an effort to attract more students, the librarians created a *Learning Commons* on a floor formerly occupied by printed journals and reference materials. Ideally students would find the new space inviting for study alongside academic resources.

The recently-renamed *Homer Commons* now houses a large number of workstations and small "studios" for group work, as well as centers to help students with Q and W subjects, foreign languages, and computer issues. The Q Center began moving into the Commons in Fall 2006 and, by Fall 2009, all of its tutoring operations were consolidated in that one location. Because there were no physical walls in the assigned area, the Center could grow somewhat as tutoring volume increased. This was mutually beneficial because the visitors for Q tutoring, the "anchor" for the learning commons, give a large boost to the library's headcount. A map and more details on Homer Commons is available at: `http://homercommons.uconn.edu/`.

The Center's furniture supports its model of tutoring. Although some of the tables are of normal height, most are at a comfortable height for people to stand around. The task chairs that surround the tables have adjustable heights within a range from low to high. This makes it easier for the tutors to move around between the students whom they are assisting without the extra burden of sitting down and standing up or the strain of bending over students sitting at lower desks. All of the tables are small and move easily in contrast to most of the older, massive tables common in the library.

In the early years, the layout of tables defined the Center's space, and some people preferred a more flowing floor plan between different areas in the Commons. A significant number of students started arriving for tutoring help without signing in; this was problematic both for the accuracy of assessment and for providing correct numbers to the upper administration for justifying budget requests. Consequently, the Center's borders are defined by retractable tape barriers of the sort used to regulate queues. When the tutoring center is closed, the tape is retracted so that students may use the space freely for other purposes.

Workshops for Students

Different models for workshops aimed at students in Q courses were tried over the years, but attendance was generally poor. Students appear very fond of the individualized, on-demand attention they get from peer tutors and far less interested in sessions scheduled at particular times, which may strike them as similar to regular classes. However, the Center runs many successful workshops, given by a GA or instructor with a strong reputation, right before an exam.

IT infrastructure

As the Center grew so did the need to create IT solutions to help manage and assess center operations. GA Marc Corluy, hired in Fall 2006 to do extensive work on the public website, dramatically improved the reliability and flexibility of its infrastructure. Stylistic considerations are now relegated to one file in which one simple change modifies all of the pages on the site. Information that needs to appear in multiple places is stored in a MySQL database; as information changes, it only needs to be updated in one place, eliminating the common problem of conflicting information on different pages within one website.

Over several years, Marc and his successor, John Baber, created a number of tools that improved the quality of the public website including:

- various displays of tutor availability and expertise;
- the "tutoring by topic" function, which allows students to see the availability of tutoring on specific topics, e.g., "optics" is a topic within physics that few tutors know well; students looking for help on optics can filter the tutoring schedule by this topic; and
- lists of all Q courses, organized in various ways (e.g., by department), and notations as to whether tutoring is available "always", "never", or "sometimes" (when knowledgeable tutors are on duty and the center is not too busy).

The public site also includes an FAQ page, links to online resources, and a list of private tutors for hire.

However, the lion's share of improvements, created to streamline the Center's administrative and data gathering procedures, are housed on a private, secure section of the website only available to Q Center administrators. The online features that make it possible to run such a large tutoring operation with only a part-time program assistant include the following.

- A timecard system for tutors, who log in at the start of and log out at the end of their shifts. The online system replaced an error prone and labor-intensive paper system. All calculations are done through web-programming, reducing the human operations until the unavoidable manual transfer of data into arcane university systems.

  Corollaries of the above system include specific tools to track tutor absences, to track number of hours worked by a specific tutor within the academic year (primarily to manage work-study limitations), and to compare the average number of tutor-hours per week across different semesters.

- Tools for tutors to input their availability and for Center administrators to easily create a balanced schedule, covering all core subject areas with sufficient tutors to meet expected demand. Previously, creating this schedule by hand used to take a week, now it only takes a couple of days.
- Tools to track and analyze tutor evaluations and the number of visits, described in Section 5.

3 Staffing, Hiring, and Training

In a center of this size, the bulk of Q Center tutoring is done by undergraduates, while the administrators spend a good deal of time hiring, training, and evaluating workers.

Hiring & Recruitment

Students interested in becoming peer tutors complete an online application asking them to submit an (unofficial) transcript, some basic personal information, details of any previous tutoring experience, a list of courses they feel comfortable tutoring, and the name and email of a faculty reference. Tutors are required to have a minimum GPA of 3.4 overall and 3.5 in Q subjects. The application is available at `http://qcenter.uconn.edu/apply-to-be-a-q-center-tutor/`.

Directors call in for an interview those students with strong transcripts in areas of current need to assess their interpersonal skills and their ability to explain something technical. The interview is typically conducted by the associate director or an experienced GA. All of these filters are useful in maintaining a high level of quality among the tutors.

In the first few years, the directors asked faculty in Q disciplines to recommend that their best students consider applying to work as a Q Center tutor. Once the Q Center developed a reputation as a good place to work, word-of-mouth became the most effective recruitment tool. Retention levels are high, though some tutors are lost to other desirable positions such as ones involving research.

Good GAs are retained as many semesters as possible, but typically there are openings to fill each spring semester for the following academic year. Positions are filled through an open search but also in consultation with liaisons in each of the four most important Q departments, Mathematics, Chemistry, Physics, and Statistics, who help identify good candidates. All GA applicants provide a resumé and transcript and are interviewed by the directors.

Tutor training programs

To provide a more intensive training program than just occasional workshops, the associate directors created and regularly teach a 1-unit interdisciplinary course "Methods of Q Tutoring." All tutors who complete this course receive a salary raise of $1/hour. Participants read and discuss literature about effective tutoring and engage in a variety of activities such as role-playing. In more recent years the tutors have made training videos on models of good and poor tutoring practices and explanations of particularly tricky Q topics that frequently arise.

In addition, a number of training sessions, open to all current tutors, are offered including workshops on specific content areas which need the most review such as infinite series, combinatorics and discrete probability, and Gaussian elimination. The instructors aim to give the tutors some perspective on the material, not just to explain it at the level they will tutor. For example, the workshop on probability covers multiple ways of solving problems, some based more on elementary counting, others on leveraging probabilistic insights.

4 Community Interactions

Although peer tutoring provides the most common form of interaction between the Q Center and the UConn community, there are other interactions as well.

Work with Faculty

While the Q Center's mission includes working with instructors in Q disciplines, the demands of coordinating the peer-tutoring program leaves the directors little time to work systematically in this direction. Also, there has been little overt demand from Q instructors for this service. Such interactions have so far been mostly one-on-one meetings with a few faculty regarding specific requests, along with occasional one-off lectures on topics such as how to use clickers. Occasionally opportunities arise for giving feedback to departments about particular courses such as when the Center experienced unexpected surges in student demand or tutoring staff reported some course-specific issue.

Relationship with other tutoring resources

Although the remarkable increase in volume of visits showed that there was a very large latent demand for tutoring in Q subjects when the Q Center was founded, there were already some tutoring resources on campus. Q departments tended to have small tutoring centers, staffed by some of their TAs, which handled a much smaller volume of traffic, less than 10% of the Q Center's steady-state. Possible reasons why the Q Center is so much more successful at attracting students and repeat customers include the following.

- The Q Center is spacious and conveniently located in the heart of campus, whereas departmental tutoring is often housed in small rooms which quickly become overcrowded.
- The Q Center's extensive number of open hours (including late into the evening) and drop-in availability have wide appeal unlike departmental tutoring which is largely limited to daytime hours.
- Departmental tutors are recruited from the same pool of TAs as those teaching recitations and are brought into their departments based on their research potential rather than teaching excellence. Students may already be unsatisfied with explanations provided by their TAs and appreciate the perspective of a fellow student, one whose struggles with particular topics are fresher and likely more substantial.
- The budget for departmental tutoring is usually a much lower priority than other needs of the departments such as research.

The key difference between departmental and Q Center tutoring is one of mission: at a public research institution, with high aspirations, research and grants are weighted most strongly in the evaluation of departments and faculty. This affects everything from budgetary priorities to hiring and graduate admissions. Consequently, teaching is not particularly important or rewarded beyond a certain minimum standard.

By contrast, funding for the Q Center comes from a separate budgetary stream, one designated to support student learning in undergraduate education. For the Q Center, the important metrics are the numbers of students served and the quality of those services, for which the IT infrastructure provides excellent data. As such, the Center has generally been able to get funding to meet the latent demand and to prioritize the quality of learning. This mission was essential to the Center's rapid growth.

Networking, Reputation, and Esprit de Corps

The intangible aspects of leadership should also be mentioned. In the early years, the director worked to develop strong ties with Q departments and with the wide variety of other units that provide support for teaching and learning throughout the University. The time and energy put into building those relationships often paid off when an informal conversation could lead to a significant synergy later on.

For example, upon the director's arrival, there was a small pre-existing tutoring program funded through Residence Life and administered through the Institute for Student Success. Through various conversations, it became clear that the Q Center was better equipped to hire and train quality tutors; so the Institute and Residential Life funded the Center to provide tutoring in one of the dining halls. This helped extend the Center's reach in the early years while building its reputation. This program continued until Residential Life was forced to cut funding for tutoring because of a large systemic budget shortfall.

The development of the Center's reputation over time was made clear by a conversation with an advisor from UConn's Academic Center for Exploratory Students, which handles advising for students who have yet to pick a major. In the first couple of years, when talking to students who came in with low Q placement, she would say, "I'm not sure, but you might want to try going to the Q Center for tutoring." A couple of years later she would say unequivocally, "Go to the Q Center!"

The Q Center endeavors to create a community of teachers and learners at every level. Listening to tutors and ensuring their efforts are acknowledged and appreciated is important to getting their best efforts. The importance of treating everyone with respect is a lesson regularly communicated to tutors and GAs. Occasional breaches of respect, sometimes detailed by student complaints, are highlighted in various tutor training meetings to good effect. The esprit de corps among Q tutors is clear from their interactions with the directors and each other. They are quick to point out issues that they see and suggest solutions for any problems that come up.

5 Assessment

Data Gathering & Analysis

In order to assess the Center's work, the directors gather significant amounts of data regarding students served and the quality of tutoring. Initially, visitors signed in on sheets of paper, but the data entry and analysis was time-consuming and only available for time periods significantly in the past. As introduced in Subsection 2, the IT infrastructure includes an online sign-in system which allows the tracking of individual students who visit the Center.

Tutees provide their demographic information only on the first visit each semester; on later visits they are asked different sets of questions regarding the quality of their previous visit to Q tutoring or their experience in their current Q course. Various tools allow the directors to run real-time reports on the data, disaggregating visits by course, department, gender, ethnicity, time period within the semester, etc. Side-by-side comparisons of the week-by-week visitor numbers for two different semesters (broken down by discipline) are particularly helpful.

These reports are of great use in evaluating the quality of peer tutoring. In the past, the average rating of tutees previous experiences dropped during periods when the Center's budget and hiring had not yet caught up to demand. Such data provided evidence to build a strong case for additional resources.

In addition to quantitative data, qualitative data, in the form of student comments solicited through the online system, proves quite informative. The qualitative data may be categorized broadly into positive comments (sometimes praising a particular tutor), negative comments, and the number of requests for more tutors, as in the table below.

Term	# visits	# tutors	#half GAs	#Pos	#Neg	#More tutors
Fall 08	9279	60–80	13	14	12	40
Spring 09	7895	80	13	23	4	25
Fall 09	11,334	80–90	12	24	12	123
Spring 10	8012	80–90	12	13	4	93

Evaluation & Management

Besides quantitative measures, the directors receive weekly feedback from the GA supervisors about how things are going in the Center. GAs meet regularly as a group with the directors and program assistant to discuss recent issues and possible solutions. Problems with individual tutors sometimes require a director's intervention. Issues with courses or homework sometimes call for a delicate discussion with a professor teaching a course. Data culled for specific courses using the online tools provides objective feedback for discussions with TAs or faculty from that discipline. A large increase in the volume of visitors for a particular course often points to a change in textbook, of professor, or a move to online homework.

Since the database can be queried in real time, directors can see the current state of tutoring such as the list of visitors (by their NetID's) and tutors who are signed in. When the Center unexpectedly receives higher tutee volume than it can handle, a director can send a general announcement to tutors asking for additional help from those willing to take on an extra shift.

Additional online tools include one that allows GAs to evaluate each of the tutors on their shifts and one that allows tutors to evaluate the GAs who supervise them. These evaluations take place at the end of each semester and are only seen by the directors who use them to make personnel decisions. In the aggregate, these can also indicate whether any training or intervention might be called for. Later on, the evaluations form the basis of letters of recommendation written by the directors when tutors apply to graduate schools or for jobs.

In addition to seeking qualitative comments, the evaluation requests that respondents provide numerical rankings on a five-point scale for each of the following categories: professionalism, attendance, knowledge, behavior, and attitude. Tutors and GAs receive email reminders (the first few are automatically generated) until they fill them out. The rankings for each tutor (or GA) are summarized in a table, along with all the specific comments made about that person.

Between the online database, analytic tools, and the constant communication between GAs and administrators, the directors are able to keep a good handle on day-to-day Q Center operations. Unfortunately, the directors did not solicit an outside evaluation during the 5th or 6th year of the Center's operation. In addition to aiding the directors' long-term strategic thinking, the external validation would likely have been helpful in convincing newly appointed upper administrators of the quality of the Center's work. While, the writing center did organize an outside evaluation to good effect, at that time, it was somewhat harder to organize an outside evaluation for the Q Center because there was little to no intercollegiate infrastructure and communication between Q directors in different locations. (Something this Handbook is taking the lead in addressing.)

6 Additional Topics

Research in Q education

The directors had hoped to eventually leverage the Center as a possible lab for research in Q learning, taking advantage of the confluence of GAs from different disciplines, the large number of students taking courses that are fairly standard across the country, and the robust systems for gathering and analyzing data. It seemed likely that the directors could obtain extramural funding to support such research after meeting the hurdles of the Institutional Review Board, etc. Unfortunately, just as it became possible to take on the project, there was a change in the upper administration and the Center was expressly forbidden to do such research (under the rationale that Undergraduate Education and Instruction should not be supporting research). A couple of years later, that decision was reversed after another change in the upper administration. But at the time of the initial mandate, directors were left to redirect all of their research efforts solely into disciplinary work within mathematics.

Split positions

The Q directors' positions are split 50-50 between an administrative role in the Center and a tenure-track faculty role in the department. This split has a number of advantages. Faculty can talk as peers with faculty from other departments, and they can exert some influence to improve the teaching structures within their departments based on their experience with the Center.

On the other hand, the split positions have led to serious evaluation issues for the faculty in these roles. In theory, the directors should be evaluated half on their administrative skills and the success of the Center and half on the usual mix of research, grants, teaching, and service (which at UConn is weighted much more heavily towards the first two). But, in practice, Q Center directors are held to the same standards for PTRM (Promotion, Tenure, Retention, & Merit) as faculty who serve full-time in the department. The associated course release (1 course per semester) directors receive is typically far less time and effort than the amount needed to keep the Center running smoothly.

7 Bibliography

[1] T. Deans and T. Roby, "Learning in the commons," *Inside Higher Ed*, November 2009.

8 Acknowledgments

The author is extremely grateful to all of the people who work so hard within the Q Center. In particular Álvaro Lozano-Robledo, who arrived as associate director in 2008 and continues now as director, has made a lasting impact on the center and the lives of many students. He was ably preceded by Kristen Sellke (2007–08) and Sarah Frey (2006–07), who started and maintained many initiatives during their too-brief tenures.

The author would have been lost at the beginning without the first permanent program assistant, Christy Hilliard. Her detailed knowledge of UConn's administration and ability to create new infrastructure as well as her dedication and reliability were outstanding. As she memorably put it: "We arrived to a kitchen with a small microwave and some paper plates, which we then had to transform into a functioning restaurant kitchen." She and her successor Diane Briody both have an excellent way of connecting with tutors and GAs that makes all the difference.

Of course the tutors and GAs, too numerous to mention (but still enshrined in the website's archives), are the ones in the trenches. Most of them do fantastic work, worthy of much higher pay than they receive. But there are other rewards.

Among the many administrators who made things happen, former vice-provost for Undergraduate Education & Instruction Veronica Makowsky deserves considerable kudos. Although her speciality in English was far from Q, she understood the importance of the work and how to support it. Her vision, caring, and advocacy on behalf of undergraduates leaves a continuing legacy. My immediate supervisor in ITL, Keith Barker, was also an excellent source of information, advice, and humor even in the face of the high bureaucratic viscosity typical of large public institutions.

Finally, the author is grateful to the editors of this handbook who made a number of fantastic suggestions for improvement. Their efforts to create a QMaSC community that effectively shares knowledge will have a profoundly positive effect on the lives of many students.

Case Study: Peterson Quantitative Resource Center at St. Lawrence University

© Michael E. Schuckers,
St. Lawrence University

1 Introduction

The Peterson Quantitative Resource Center (PQRC) at St. Lawrence University (SLU) was founded as the Quantitative Resource Center in August of 2007. SLU is a liberal arts college in Canton, New York with approximately 2400 students. The mission of the QRC has been to "*facilitate, encourage, and develop quantitative reasoning, skills, and applications for the St. Lawrence University Community.*" As a consequence of this mission, the PQRC supports a variety of quantitative courses and initiatives across campus, not just mathematical ones. This wide net of campus support is intentional and is one that is meant to make our services and our programs broadly appealing.

A plan for a QMaSC at SLU had been in existence for many years prior to the founding of the PQRC; however, in those times when funds were available for the center, no appropriate space was attainable. In 2007, a new science building, the Johnson Hall of Science, housing the Biology, Neuroscience, Chemistry and Biochemistry programs, was completed. As a result, space was available for a QMaSC center in Bewkes Hall, a neighboring building where the Biology Department had previously been located. In 2008, the then QRC moved into a renovated space that was made

Suggested Citation: Michael E. Schuckers (2016), "Case Study: Peterson Quantitative Resource Center at St. Lawrence University", `http://dx.doi.org/10.5038/9780977674435.ch31` in G. Coulombe, M. O'Neill, M. Schuckers (Eds.) *A Handbook for Directors of Quantitative and Mathematical Support Centers*, Neck Quill Press, `http://scholarcommons.usf.edu/qmasc_handbook`.

This material is based upon work supported, in part, by the National Science Foundation under Grant DUE-1255945. Any opinions, findings, and conclusions or recommendations expressed in this material are those of the author(s) and do not necessarily reflect the views of the National Science Foundation

possible by a generous donation from Martie and Gregg Peterson. The PQRC was renamed for them in 2012 on the occasion of Martie's 50th class reunion.

The PQRC is currently located in Valentine Hall, which is the primary classroom building for the sciences at St. Lawrence University. The exact location is at the corner of two highly trafficked hallways, outside of the Mathematics, Computer Science and Statistics Department office, and is a very visible location for that part of campus. The PQRC has approximately 2500 sq. ft of space, which is divided into several parts. The main space is an open room of approximately 1500 sq. ft that has movable furniture that can be arranged into tables or spread apart for a seminar. Just off the main space we have a small library where we try to keep reference materials as well as copies of textbooks for more popular courses. There is an office for the assistant director and two group study spaces off of the main room. Outside the entrance to the PQRC there is a lounge that has a couch, two love seats and several work tables. This lounge was intentionally designed to be a welcoming and casual entrance to the PQRC and has proved to be very popular. Finally, there is a seminar room that seats up to 12 that the PQRC shares with the Department of Mathematics, Computer Science, and Statistics. The department has scheduling priority during the day, and the PQRC has scheduling priority in the evenings.

St. Lawrence University is a private, liberal arts college in northern New York. We have a strong Department of Mathematics, Computer Science, and Statistics which regularly produces 40 to 50 majors per year with approximately the same number of minors. That said, we are part of a campus that has a long tradition of the liberal arts taken very broadly. Until recently, St. Lawrence did not have a quantitative graduation requirement, but instead had a mathematics or foreign language requirement. An impetus for the creation of the PQRC was a finding that graduating seniors at St. Lawrence felt less confident in their quantitative reasoning skills relative to when they started. In this environment, we began the work of the PQRC by taking a broad approach to quantitative matters. We emphasized and encouraged these topics in a variety of ways. One example of this early in the life of the PQRC was the distribution to SLU faculty of books with quantitative content, but with that content not being the main focus. Among these books were Andrew Gelman's *Red State, Blue State, Rich State, Poor State* and Edward Tufte's *Visual and Statistical Thinking: Displays of Evidence for Making Decisions*. Faculty who received these books were asked to come to a small group discussion of the books, led by another non-PQRC affiliated faculty member. We have also for several years now had a graphic feature called "On Q" in the weekly student newspaper that covers some topic of relevance from a quantitative viewpoint. For example, one of these was a bar chart showing the ten most popular candies at Halloween. The point of these features is to give examples of uses of quantitative material that is informative and appropriate.

In 2012, the Faculty of St. Lawrence passed a quantitative and logical reasoning requirement as part of a new set of general education requirements. This change will impact the role of the

PQRC on campus and we have begun planning for this new environment.

2 Center organization and services

The PQRC at St. Lawrence is a peer-led drop-in tutoring center. The staff is composed of between 20 and 25 undergraduate students, a director, and an assistant director. The Director has been a faculty member from the Mathematics, Computer Science, and Statistics Department who receives a course reduction each semester for their role at the center. The usual faculty teaching load is three courses per semester. The author of this chapter has been the only director since the founding of the PQRC. The work of the director is to oversee the staff, set the direction for the center, and interact with faculty and administration. The director reports to the Dean of Academic Affairs. The PQRC also has an advisory board of faculty and staff that meets at the end of the semester and gets a report from the director regarding accomplishments over the course of the semester. The original advisory board was quite large with representation from nearly every department or program with quantitative content. This was done to create awareness and support for the center. Recently, though, the board has shifted into a smaller group to assist the director in transitioning to the new general education requirements.

The Assistant Director is a graduate student in the Education department at SLU who reports to the director. The Education Department at SLU is the only department to offer graduate degrees and certifications. The assistant director is paid hourly and gets a tuition remission for six courses per academic year. The expectation is that this individual will work approximately 15 to 20 hours per week in the PQRC during the academic year. Summer duties vary depending upon need. The role of the assistant director is to help manage the center primarily through coordination of the day-to-day operations of the center, which includes tasks of staff scheduling and payroll. Both the director and the assistant director take active roles in the training and hiring of our staff.

The staff of undergraduate mentors has as their primary duty the one-on-one mentoring of drop-in students. Mentors report to both the assistant director and the director. The mentors have a variety of backgrounds, but they must have both quantitative and communication skills. Since the PQRC supports not just math, computer science, and statistics courses, but also economics, biology, and psychology courses, among others, we need a staff that has a range of backgrounds. In addition to their work as mentors, we also have our staff run many of the workshops that we lead. In the past, these workshops have covered topics such as trigonometry, logarithms, LaTeX, designing a poster, Excel, and the quantitative reasoning section of the GRE. Our general model for these workshops is to have the first couple of versions designed and run by the director and/or the assistant director. Beginning with the second or third iteration, we include mentors in the process and have them involved in developing materials and running the workshops. We then transition the workshops to be run as much as possible by the mentors. Some care and judgment are necessary in organizing mentor-led workshops, as some of our more formal faculty members prefer not to have students lead their classes; however, there are several benefits to having the mentors run these workshops. Foremost among these benefits is the experience the mentors gain from the planning and execution of the workshops.

3 Staffing, Hiring and Training

The staff for the PQRC, as mentioned in the previous section, consists of about twenty-five undergraduates, an assistant director, and a director. The assistant director's position is normally for two years, though in one exceptional case it was extended to three years. As noted above, there has been only one director for the PQRC since its founding. The PQRC mentors, who comprise the rest of the staff, are usually hired, and remain at the PQRC until they graduate. The PQRC is generally open from 10 am to 11 pm on Monday through Thursday, from 10 am to 3 pm on Friday and from 2 pm to 11 pm on Sundays. We ensure that there are never fewer than two mentors working at a given time. The number of mentors on staff varies with mentee attendance but generally there is a period in the evening when four or five mentors are on staff with a ramp up and ramp down period of fewer mentors working around those busiest hours. We have an electronic sign-in protocol, so that we are able to track when students come into the center. Initial staff scheduling decisions are done at the first staff meeting at the beginning of the semester, based upon sign-in traffic data from previous semesters. At that same meeting we set the weekly semester schedule, so that each mentor has the exact same hours every week during the semester. This has two benefits. First, mentors have a consistent schedule that is built around their classes and activities, and second, mentees who find a particular mentor to be helpful know when they can find these mentors again. When mentors have an exam or practice which conflicts with their schedules, we have a listserv that allows mentors to communicate the need for someone to cover a given shift. We also use this listserv for sharing of general center information. Mentors are expected to check their email regularly for updates.

Hiring for PQRC mentors is typically done every spring semester for the following year. Occasionally it is necessary to also hire in the fall semester for the spring semester. Before each hiring process begins, we do an internal assessment of the number and types of students that we will need. If necessary, we will try to recruit applications for specific areas, such as psychology, economics, or computer science, where we may need additional coverage. Our process for hiring is to ask applicants to submit a form (located in appendix of chapter), along with a copy of their unofficial transcript. After these applications have been submitted, we review each of them, solicit reference from relevant faculty and match qualifications with needs. We then interview a subset of these applicants. Our interview process is 30 minutes long and consists of two parts. In the first part, the candidate meets with the assistant director, the director, or both to discuss his or her background, interest in the PQRC, and any questions about the position. In the other half of the interview, candidates meet with two to four current mentors and are presented with a simple topic from a course they have previously taken, usually Calculus I or Statistics. They are asked to explain a problem as they would to someone taking that class. Also, in this half of the interview candidates have an opportunity to ask mentors about the work environment at the PQRC. Once all the candidates have been interviewed, all of the interviewers convene to discuss each candidate and to rank the candidates. This ranking is tentative, and further discussion usually takes place between the director and assistant director before final offers are made.

Hiring for the assistant director follows a roughly similar process. Once the approval to hire is given by our human resources department, applications are taken and reviewed by a committee. In this case, the committee is usually comprised of the director, the chair of the Mathematics, Computer Science and Statistics Department, two or three other quantitative faculty members, and two PQRC mentors. Once qualified candidates have been interviewed by each member of the committee, a decision is made, and the candidate is offered the position. This protocol much more closely matches a faculty search at SLU.

Initial training of mentors is done at the beginning of their first semester at the PQRC. Typically we have two training sessions. The first covers the basic rules and expectations of working at SLU and of working at the PQRC including PQRC disciplinary policies. We have created a 20-page manual of rules and procedures for our mentors. We give them a copy of this manual and cover the topics in that manual during the first session. The second session deals with how to have a good mentoring session with a student. We talk about creative ways to engage students and how to solve problems that may arise. One of the highlights of the second session is a panel that we create of current experienced mentors. The director and assistant director lead the first part of the panel discussion, and then there is a closed door session without administrators, where new mentors can ask experienced mentors about working at the PQRC. The second session also has a bit of role playing to practice some circumstances that arise commonly while working. Each of these sessions lasts approximately 2.5 hours.

Further training is done at our bi-weekly, with candy – Take 5™is the official candy of the PQRC — staff meeting which have traditionally been held on Sunday evenings. These all-hands-on meetings are nominally required and typically last about 40 minutes. Students are paid for their attendance at these meetings. During this time we cover incidents that have occurred since the last meeting and do additional training on professionalism in the workplace or ways to engage with different learning styles, for example. We also view these meetings as opportunities for team building. We have intentionally placed a good deal of emphasis on building our staff into a cohesive unit, since we feel that the more comfortable our staff is with each other, the more inviting the PQRC will feel. To that end we also have a social committee of mentors who schedule events for mentors such as a barbeque, a bowling night, or a formal.

4 Community Interactions

As our mission states, we consider our community to be the entire St. Lawrence Community. Certainly our primary focus has been and will continue to be the support of academic work at St. Lawrence University. To that end, we interact extensively with faculty who are teaching quantitative courses at St. Lawrence. This interaction takes several different forms. At the beginning of each year we interact with faculty in biology, chemistry, and physics as we run refreshers for some high school mathematics topics. Upon the request of some faculty, we set our schedule to have certain mentors available at certain times so that there is support for their courses at appropriate moments, such as within the evening hours before weekly quizzes. We have made classroom visits to discuss data visualization and to introduce Excel. Some faculty leave copies of solutions to homework or assignments with us, and we add them to our library. We also have longer-running interactions with courses that take place over the course of the semester which involve multiple visits to classes or specific workshops. As the director, I also consult with faculty on their courses. One of the benefits – and there are drawbacks, too — to having a faculty member as a director is that faculty consider you a peer and are willing, in most cases, to consider your opinion.

The director of the PQRC is also a member of a group at SLU called 'Critical Literacies' which includes the director of the writing center, the director of the arts technology center, and the head of the science library, among others. Much of the people in this group are involved, to one degree or another, in academic support; in some ways it is a support group for academic support. The bi-weekly meetings are chaired by the Dean of Academic Affairs. These meetings often focus on ways to ensure that students and faculty are getting the support to carry out their roles on campus.

At a couple of times over the course of the PQRC we have helped local community organizations with quantitative matters. We have helped a local bank analyze patterns of traffic at their different branches and we have assisted a local center for at-risk youth to analyze a survey of their clientele. We view our participation in these projects as being a good citizen in our community. There are probably more such projects that we could be doing, but we do not currently have the time or resources to make them a reality.

5 Assessment at the PQRC

Assessment at the PQRC takes a couple of forms. First, we measure the number of drop-in visits we receive each semester; we view this as one important measure of what we do. When students sign in, we acquire some information about them and about why they are visiting the PQRC on this particular occasion. The PQRC now has approximately 3300 student drop-in visits per year. Second, there is the assessment of how our mentors perform. After each visit, mentees are encouraged to complete an online evaluation of their experience. They are asked to rate the quality of the mentor, and the quality of the session, and they have the opportunity to give comments. The ratings are on a one to five scale with five being the highest rating. Our form for mentor evaluation can be found at: `http://web.stlawu.edu/qrcint/evaluation.php`. Over the past couple of years our average ratings have been about 4.8 for both mentors and session quality; however, we only get about 15% of students who sign in to complete an evaluation of their session. In many cases, this is due to mentees coming in only to use our resources, specifically the books in our small library, or because they were in a hurry to leave. There is some self-selection bias in our summaries, since those who do not complete the form are potentially not as satisfied with their experience at the PQRC as those that do. One area on which we need to improve is in evaluating this aspect of our operation. Third, we are assessed by our advisory board of faculty and staff. Twice a year, at the end of each semester, the director covers the goals for the upcoming semester, presents a summary of activities for the current semester and gets feedback on the performance of the PQRC from this board. This is the only form of external assessment that the PQRC has had.

It is worth noting here that with the implementation of the new general education requirement at St. Lawrence, we have convened a new advisory board of six faculty members to consider the future of the PQRC and its role on campus. The previous board had one member from each department that had some quantitative content as well as representatives from Information Technology and our Rhetoric and Communication Initiative. To facilitate the work of this new advisory board, we wrote a three page memo describing the current work of the PQRC and asked this group to consider the following questions:

1. What should be the mission of the PQRC for the next 5 to 10 years?
2. What should be the specific objectives of the PQRC over that time period?
3. What changes, if any, would you recommend to the structure of the PQRC in order to fulfill the mission and to achieve the objectives outlined in 1) and 2)?
4. What specific methods should the PQRC leadership undertake to fulfill the mission and to achieve the objectives outlined in 1) and 2)?

We hope that the answers to these questions will provide a basis and direction for leading the PQRC for the foreseeable future.

6 Acknowledgements

I would like to thank Martie and Gregg Peterson for their generosity which has allowed the PQRC to prosper. I also wish to acknowledge the tremendous Assistant Directors that the PQRC has had since 2007. These are (reverse chronologically): Kerrin Ehrensbeck, McKayla Nuffer, Alli Morrow, and Tim Morse. They have been immeasurably helpful in the success of the PQRC. I am lucky to have worked with such wonderful people and luckier to call them friends.

7 Appendix: PQRC Mentor Application Form

PQRC Mentor Job Description: PQRC mentors work with students to develop their quantitative skills (usually in the context of a particular quantitative course). Additionally, mentors will occasionally be involved in workshops, presentations, research, and community outreach. The characteristics that we are looking for in our mentors include a strong quantitative background (math major not required), good communication skills, patience, enthusiasm, and the ability to work well in a dynamic group situation. Mentors will be expected to work a minimum of 5 hours per week at the PQRC as well as attend bi-weekly staff meetings.

Name________________________ Class ____________ CMR# ____________
Email________________________ Cell Phone # ____________
Major(s)/Minor(s) ________________________________
Advisor(s) ________________________________

1) **Quantitative Course History** Please include all relevant quantitative courses (math, statistics, computer science, economics, psychology, physics, biology, chemistry, geology) **AND** attach an unofficial copy of your SLU transcript. Please include courses in which you are currently enrolled.

Dept.	**Course#**	**Professor**	**Semester**	**Year**	**Final Grade**

Continued on back

2) **Campus Work History**

Position	Dept	Supervisor	Semester	Year

3) **Do you have any formal or informal tutoring experience?** (If so, please elaborate)

____

____

____

____

4) **Have you used software in courses that you feel comfortable helping students with?** (If so, please elaborate)

____

____

____

5) **Additional Faculty/Staff References** (Please include contact information)

____

____

____

6) **What is the best way to contact you over summer break?**

____

____

If you have any questions about the PQRC or about this application, please refer to the PQRC website www.stlawu.edu/qrc OR contact us at qrc@stlawu.edu.

Contributor Biographies

Editor Biographies

Grace Coulombe is the founding director of the Mathematics and Statistics Workshop at Bates College in Lewiston, Maine. She is also a Lecturer in Mathematics teaching calculus, probability, statistics, introduction to abstract math, and several q-courses. Her interests include promoting quantitative reasoning across the curriculum and exploring the mathematical practices of other cultures. She has been at Bates since 2000.

Mary B. O'Neill was the director of the Quantitative and Symbolic Reasoning Center at Hamilton College from 1994 to 2013. She expanded these programs and helped to usher in a move to a new and enlarged center. Mary's main interests are quantitative literacy across the curriculum and tutor training. A founding member of the Northeast Consortium for Quantitative Literacy (NECQL), Mary has frequently presented papers and served as a panelist at meetings of NECQL and at Joint Mathematics Meetings. She is co-author, with Rob Kantrowitz, of "Quantitative Literacy at Hamilton College" in Current Practices in Quantitative Literacy Programs (MAA Notes Publication). She has been an invited speaker and an external evaluator at several colleges in the Northeast.

Michael Schuckers is currently Director of the Martha E. '62 and Gregg E. Peterson Quantitative Resource Center at St. Lawrence University. He is also a Professor of Statistics there. His research interests are in bioauthentication, statistical methods for sports, particularly ice hockey, and the application of statistical methods generally. He has been at St. Lawrence in Canton, NY since 2002.

Author Biographies

Darry Andrews is the director of the Mathematics and Statistics Learning Center at The Ohio State University in Columbus, Ohio. His interests include undergraduate mathematics education, tutor preparation, and outreach to STEM minority students.

Maria Belk is the Director of Quantitative Literacy and a Visiting Assistant Professor of Mathematics at Bard College. At Bard, she supervises the math and science tutors, which includes coordinating the Math Study Room, and she teaches mathematics courses. Her research interests include discrete geometry, computational geometry, and graph theory.

Kathy Bjorge is a higher education professional focused on student strengths and success. She received her Master of Education in Student Affairs Counseling at Bridgewater State University where she was a Graduate Assistant for Math Services. Her experience includes ongoing roles at both Bristol Community College and Roger Williams University instructing, advising and providing program administration.

Margot Black is the founding director of the Symbolic & Quantitative Resource Center at Lewis & Clark College in Portland, Oregon. She is also the primary and founding instructor of Foundations in Quantitative Reasoning, a new course at Lewis & Clark designed to replace Intermediate Algebra as the proficiency course for mathematically underprepared students. Her interests include peer education, and numeracy, and promoting quantitative reasoning across the curriculum.

Tiana Bosley is the program coordinator of the Mathematics Resource Center at Wayne State University in Detroit, Michigan. She is also a Ph.D. candidate at the university working under the direction of Shlomo S. Sawilowsky in the the area of Educational Evaluation and Research. When not working on her thesis, she enjoys spending time with her family and helping others become more comfortable with mathematics.

Patrice Conrath is an Associate Professor in the Math and Computer Science Department at Bethel University and founding director of the Math and Computer Science services (1989-present). She specializes in the teaching of Calculus and Operations Research which has provided efficiency solution modeling for organizations such as Feed My Starving Children and a clinic in Nigeria. She recently became Director of Advising for the undergraduate programs at Bethel.

Thomas Dick is a professor and former chair of Oregon State University's department of mathematics. He also serves as the Oregon State University's Coordinator of Collegiate Mathematics Education, Faculty Director of the OSU Math Learning Center and the OSU Math Excel (Treisman Emerging Scholars) program. His main interests in mathematics education research are in the use of technology to enhance mathematics teaching and learning.

Rachel Dunwell is the director of the Math Support Center at Rhodes College in Memphis, Tennessee.

Eric Gaze directs the Quantitative Reasoning (QR) program at Bowdoin College, is Chair of the Center for Learning and Teaching, and is a Lecturer in the Mathematics Department. He is the current President of the National Numeracy Network (NNN 2013 - 2017), and a past chair of SIGMAA-QL (2010–12). He writes a column, *Ratiocination*, for the NNN website: `http://serc.carleton.edu/nnn/columns.html`, and has a QR textbook published with Pearson, *Think-*

ing Quantitatively: Communicating with Numbers, with blog `https://thinkingquantitatively.wordpress.com/`. Eric has given talks and led workshops on the topics of QR Across the Curriculum, Creating a QR Entry Point Course, Writing with Numbers, QR Assessment, and Running a QR Program; and has served on review teams of QR programs. Eric is the Principal Investigator for a NSF TUES Type I grant (2012-14), *Quantitative Literacy and Reasoning Assessment* (QLRA) DUE 1140562. This collaborative project builds on Bowdoin College's QR instrument which is used for advising purposes and is available to interested schools. Prior to coming to Bowdoin, Eric led the development of a Masters in Numeracy program for K–12 teachers at Alfred University as an Associate Professor of Mathematics and Education.

Ja'Wanda Grant served as the founding director of the Vassar College Quantitative Reasoning Center from 2011–2015. While in her role as QRC director, she was a member of the Northeast Consortium for Quantitative Literacy and also hosted and participated in the annual Mid-Hudson Area Tutor Educators (MATE) Tutor Training Conference. Ja'Wanda was able to grow the QRC staff, resources, and space during her tenure and in turn secured a $250K endowment to support its continued growth. She is currently the Director of the Center for Undergraduate Research at Xavier University of Louisiana. Ja'Wanda still applies her experience in academic support in her role as advisor to student researchers pursuing graduate and professional school.

Cymra Haskell has been Director of the Math Center at the University of Southern California since its inception in the year 2000. She is also Associate Professor (Teaching) of Mathematics. Her research includes statistical properties of dynamical systems and modeling of biological processes. She is also interested in K-12 mathematics, in particular the preparation of teachers of mathematics.

Gail Y. Hilyard is the Academic Coordinator of the Math Center at Holyoke Community College, MA where she also teaches as an Adjunct in the Mathematics Department. She is interested in developed programming to help students to be successful in learning mathematics, particularly developmental mathematics. Gail is committed to fostering inclusivity for minority and underrepresented groups within STEM education. Gail's research interests are in examining strategies to reduce math anxiety and in exploring ways to build trust in teaching and learning.

Carolyn Johns is the assistant director of the Mathematics and Statistics Learning Center at The Ohio State University in Columbus, Ohio. Her interests include undergraduate mathematics education, tutoring best practices, and educational technology.

Dennis Lewandowski is currently Director of the Mathematics Learning Center at Michigan Technological University in Houghton, Michigan. He is also a Senior Lecturer there. His subject areas include Vector Calculus, Linear Algebra and Differential Equations.

Vera Mayes is the current Learning Commons Coordinator for the Math and Science area at Tallahassee Community College, which she helped design and implement. She is also a mathematics instructor at TCC. She has been involved with several transformative re-designs in the math program at TCC as well as two QEP projects. Her prior work included working with a Trio grant, including writing a successful re-application grant at Southeastern Louisiana University and a NSF grant at Illinois state university focusing on the Supplemental Instruction program. She has been involved with training tutors and providing quality learning resources for over thirty years.

Catherine McCune is the founding director of the Spinelli Center for Quantitative Learning at Smith College. Her interests include quantitative literacy, tutor training, math self-efficacy, differential geometry, and dancing tango. She has been at Smith since 2002.

Johnathan Nelson was the Director of the Mathematics Resource Center at the University of St. Thomas in Saint Paul, MN from 2011 to 2015. He has taught mathematics at every level from 6th grade to college calculus. He continues to work in mathematics education as a high school math teacher. His interests include college readiness in math, mathematics placement, and math remediation.

McKayla Nuffer is a Precalculus teacher at the Hyde School in Bath, ME. She received her bachelor's and master's degrees from St. Lawrence University in 2012 and 2014, respectively. During her master's degree, she served as the Assistant Director of the Martha E. '62 and Gregg E. Peterson Quantitative Resource Center at St. Lawrence University, focusing on tutor training and creating workshops and tutorials on various quantitative topics.

Laurie Wern Overmann is a mathematics instructor at Southeast Missouri State University where she is involved in all aspects of teaching and learning in lower division mathematics courses.

Patrick Perkins is a Senior Lecturer in Mathematics at the University of Washington. He was the Director of the Math Study Center at the UW from 2000 to 2013. His research interests are in Non-commutative Rings, Algebraic Combinatorics and Discrete Optimization. He has lived in Seattle, WA since 1994.

Eileen Phillips is Director of Developmental Mathematics Programs at Keene State College in Keene, NH. As director, she oversees the drop-in peer tutoring, Peer Course Assistant (PCA) program, and the math assessment testing for several majors on campus. She also teaches the mathematics courses for the early childhood and elementary education majors. She is co-author, with Joseph Witkowski, of *Applied Algebra and Trigonometry 2nd edition*, Kendall Hunt, 2012. Eileen has worked at Keene State College since 1999.

George Recck is the Director of the Math Resource Center and a Lecturer in the Math & Science division at Babson College in Wellesley, MA. He is also a double alumnus from Babson receiving is B.S. in 1982 and MBA in 1984. Georges interests include innovative teaching methods, assessment and data analytics in sports.

Thomas Roby is Associate Professor of Mathematics at the University of Connecticut, where he has been since 2005. His primary interests are in algebraic combinatorics and mathematics education. He was the Director of UConn's Quantitative Learning Center from 2005 to 2012, during its initial phase of rapid growth. He has spent many summers directing and working in math programs for both K–12 teachers and mathematically curious middle and high-school students. He has also been on the author team for several series of K–12 textbooks.

Ibrahim Rosic is currently the Director of the Learning Commons and Tutoring Services at Mohawk Valley Community College (MVCC) in Utica, NY. He is also an Adjunct Instructor in Mathematics at MVCC and State University of New York Polytechnic Institute in Utica, NY.

Matthew Salomone is associate professor of mathematics, Director of Math Services, and founding coordinator of Quantity Across the Curriculum at Bridgewater State University in Bridgewater, Massachusetts. His interests include applications of geometry and promoting quantitative literacy outside of mathematics and the sciences.

Rosalie Tepper is the director of the Math Learning Center at Shoreline Community College in Shoreline, Washington. Her interests include professional development for tutors with an emphasis on embedding practices to offset the effects of math anxiety and negative student beliefs about their abilities. She is also involved in developing and modifying curriculum, promoting depth of understanding, and quantitative literacy.

Christine Tinsley is currently the Director of the Mathematics Center for the University of Oklahoma Math Department. Before becoming the director of the Mathematics Center she helped develop and coordinate the Math Center for the University of Oklahoma Athletic Department from 1995 to 2010. She is also Lecturer in Mathematics.

Frederick Van Swearingen is the current director of the Quantitative Skills Center of Winston-Salem State University. He has been at WSSU in Winston-Salem, NC since 1997, when he started as an adjunct physics instructor. He was then recruited to lead the traditional tutoring program in University College in 2006. His interests and publications are in the areas of physics instruction, medical physics, and safety.

Index

Made in the USA
Monee, IL
08 November 2026

68020218R00240